Communications in Computer and Information Science 2974

Rationale

The CCIS series is devoted to the publication of proceedings of computer science conferences. Its aim is to efficiently disseminate original research results in informatics in printed and electronic form. While the focus is on publication of peer-reviewed full papers presenting mature work, inclusion of reviewed short papers reporting on work in progress is welcome, too. Besides globally relevant meetings with internationally representative program committees guaranteeing a strict peer-reviewing and paper selection process, conferences run by societies or of high regional or national relevance are also considered for publication.

Topics

The topical scope of CCIS spans the entire spectrum of informatics ranging from foundational topics in the theory of computing to information and communications science and technology and a broad variety of interdisciplinary application fields.

Information for Volume Editors and Authors

Publication in CCIS is free of charge. No royalties are paid, however, we offer registered conference participants temporary free access to the online version of the conference proceedings on SpringerLink (http://link.springer.com) by means of an http referrer from the conference website and/or a number of complimentary printed copies, as specified in the official acceptance email of the event.

CCIS proceedings can be published in time for distribution at conferences or as post-proceedings, and delivered in the form of printed books and/or electronically as USBs and/or e-content licenses for accessing proceedings at SpringerLink. Furthermore, CCIS proceedings are included in the CCIS electronic book series hosted in the SpringerLink digital library at http://link.springer.com/bookseries/7899. Conferences publishing in CCIS are allowed to use our online conference service (Meteor) for managing the whole proceedings lifecycle (from submission and reviewing to preparing for publication) free of charge.

Publication process

The language of publication is exclusively English. Authors publishing in CCIS have to sign the Springer CCIS copyright transfer form, however, they are free to use their material published in CCIS for substantially changed, more elaborate subsequent publications elsewhere. For the preparation of the camera-ready papers/files, authors have to strictly adhere to the Springer CCIS Authors' Instructions and are strongly encouraged to use the CCIS LaTeX style files or templates.

Abstracting/Indexing

CCIS is abstracted/indexed in DBLP, Google Scholar, EI-Compendex, Mathematical Reviews, SCImago, Scopus. CCIS volumes are also submitted for the inclusion in ISI Proceedings.

How to start

To start the evaluation of your proposal for inclusion in the CCIS series, please send an e-mail to ccis@springer.com

Hamid R. Arabnia · Leonidas Deligiannidis ·
Farzan Shenavarmasouleh ·
Farid Ghareh Mohammadi · Soheyla Amirian
Editors

Emerging Trends in Smart Cities, Smart Mobility, Software Engineering, and Education

12th International Conference on Computational Science and Computational Intelligence, CSCI 2025
Las Vegas, NV, USA, December 3–5, 2025
Proceedings, Part VIII

Springer

Editors
Hamid R. Arabnia
University of Georgia
Athens, GA, USA

Leonidas Deligiannidis
Wentworth Institute of Technology
Boston, MA, USA

Farzan Shenavarmasouleh
Medialab Inc.
Lawrenceville, GA, USA

Farid Ghareh Mohammadi
Verify Radiology Images Consultants
Houston, TX, USA

Soheyla Amirian
Pace University
New York, NY, USA

ISSN 1865-0929 ISSN 1865-0937 (electronic)
Communications in Computer and Information Science
ISBN 978-3-032-28474-7 ISBN 978-3-032-28475-4 (eBook)
https://doi.org/10.1007/978-3-032-28475-4

This Springer imprint is published by the registered company Springer Nature Switzerland AG
The registered company address is: Gewerbestrasse 11, 6330 Cham, Switzerland

Preface

It is our great pleasure to present this collection of research papers from the 12th International Conference on Computational Science and Computational Intelligence (CSCI 2025). This volume, titled "Emerging Trends in Smart Cities, Smart Mobility, Software Engineering, and Education", comprises selected papers accepted in the Research Tracks on Smart Cities and Smart Mobility (CSCI-RTSC); Software Engineering (CSCI-RTSE); and Education (CSCI-RTED). The CSCI conference was held in Las Vegas, Nevada, USA, from December 3 to December 5, 2025.

The CSCI 2025 International Conference brought together papers from a diverse array of communities, including researchers from universities, corporations, and government agencies. Accepted papers are published by Springer Nature, and the proceedings showcase solutions to key challenges in various critical areas of Computational Science and Computational Intelligence.

Computational Science (CS) is the study of addressing problems that are impossible to solve (or difficult to solve) without computers. CS can be considered to be the bridge between computer science and other sciences. The field is interdisciplinary by nature and includes the use of advanced computing capabilities to understand and solve complex problems. In short, CS is the science of using computers to do science. Computational Intelligence (CI) is the study of computational methods that exhibit intelligence. These methods adapt to changing environments and changing goals. There is a significant overlap between the fields of CI and Artificial Intelligence (AI). However, there is also a difference: AI techniques often involve top-to-bottom methods (i.e., methods to reach solutions are imposed from the top) whereas CI techniques often involve bottom-up methods (i.e., solutions emerge from unstructured beginnings). An important part of CI includes a set of Nature-inspired computational approaches to address complex problems for which traditional methods are infeasible. Computational Science and Computational Intelligence both share the same objective: finding solutions to difficult problems. However, the methods to reach the solutions are different. The main objective of the CSCI Conference is to facilitate increased opportunities for cross-fertilization across CS and CI.

Considering the above broad outline, the CSCI 2025 International Conference was composed of the following focused Research Tracks:

Artificial Intelligence (CSCI-RTAI); Big Data and Data Science (CSCI-RTBD); Computational Science (CSCI-RTCS); Computational Intelligence (CSCI-RTCI); Computational Biology (CSCI-RTCB); Cyber Warfare, Cyber Defense, & Cyber Security (CSCI-RTCW); Signal & Image Processing, Computer Vision & Pattern Recognition (CSCI-RTPC); Smart Cities and Smart Mobility (CSCI-RTSC); Education - CS & CE (CSCI-RTED); Health Informatics and Medical Systems (CSCI-RTHI); Mobile Computing, Wireless Networks, & Security (CSCI-RTMC); Software Engineering (CSCI-RTSE); Internet of Things & Internet of Everything (CSCI-RTOT); Social Network Analysis, Social Media, & Mining (CSCI-RTNA); Cloud Computing and Data Centers

(CSCI-RTCC); Parallel & Distributed Computing (CSCI-RTPD); and Large Language Models (LLM). The scope of each track can be found at: https://www.american-cse.org/csci2025/topics and https://www.american-cse.org/LLM2025/topics.

An important objective of the CSCI 2025 International Conference and its associated Research Tracks was to foster opportunities for cross-fertilization between the fields of Computational Science and Computational Intelligence. The CSCI Conference is deeply committed to promoting diversity and eliminating discrimination, both in its role as a conference organizer and as a service provider. Our goal is to create an inclusive culture that respects and values differences, promotes dignity, equality, and diversity, and encourages individuals to reach their full potential. We are also dedicated, wherever possible, to organizing a conference that represents the global community. We sincerely hope that we have succeeded in achieving these important objectives.

The Steering Committee and the Program Committee would like to extend their gratitude to all the authors who submitted papers for consideration. CSCI 2025 received submissions from 48 countries, with approximately 48% of them coming from outside the United States. Each submitted paper underwent a rigorous peer-review process, with at least two experts (an average of 2.5 referees per paper) evaluating the submissions based on originality, significance, clarity, impact, and soundness. In cases where reviewers' recommendations were contradictory, a program committee member was tasked with making the final decision, often consulting additional referees for further guidance. The CSCI Conference followed the guidelines of COPE (Committee on Publication Ethics):

- Typical submissions underwent a single-blind peer review process, in which the authors remain unaware of the identities of the reviewers, while the reviewers are informed of the authors' identities.
- Papers authored by one or more members of the program committee, including co-chairs, were subjected to a double-blind peer review process, ensuring that neither the authors nor the reviewers were aware of each other's identities or affiliations.

The Research Track on Smart Cities and Smart Mobility (CSCI-RTSC) received 48 submissions, of which 12 papers were accepted, resulting in a paper acceptance rate of 25%; we selected only 5 papers for this volume. The Research Track on Software Engineering (CSCI-RTSE) received 46 submissions, of which 9 papers were accepted, resulting in a paper acceptance rate of 19%; we selected only 5 papers for this volume. The Research Track on Education (CSCI-RTED) received 238 submissions, of which 52 papers were accepted, resulting in a paper acceptance rate of 21%; we selected only 22 papers for this volume. A total of 32 papers are included in this volume.

We are deeply grateful to the many colleagues who contributed their time and effort to organizing the CSCI 2025 Conference. In particular, we extend our thanks to the members of the Program Committee, the Steering Committee, and the referees. We would also like to express our appreciation to the primary sponsor of the conference, the American Council on Science & Education. The list of members of the Program Committee for each track can be found at: https://www.american-cse.org/csci2025/committees.

We extend our heartfelt gratitude to all the speakers and authors for their valuable contributions. We would also like to thank the following individuals and organizations for their support: the staff at the Luxor Hotel (an MGM property, conference/meeting department) and the staff of Springer, for their assistance in various aspects of the event.

We are pleased to present the proceedings of CSCI 2025 (includes selected papers of CSCI-RTSC, CSCI-RTSE, and CSCI-RTED Research Tracks). These proceedings represent a collection of outstanding research contributions that reflect the diversity and depth of work in Smart Cities, Smart Mobility, Software Engineering, and Education.

Co-editors: Research Tracks on Smart Cities and Smart Mobility (CSCI-RTSC); Software Engineering (CSCI-RTSE); and Education (CSCI-RTED) of CSCI 2025 International Conference

December 2025

Hamid R. Arabnia
Leonidas Deligiannidis
Farzan Shenavarmasouleh
Farid Ghareh Mohammadi
Soheyla Amirian

Organization

Steering Committee – Co-chairs (CSCI 2025)

Hamid R. Arabnia	University of Georgia, USA
Leonidas Deligiannidis	Wentworth Institute of Technology, USA
Fernando G. Tinetti	Universidad Nacional de La Plata, Argentina
Quoc-Nam Tran	Southeastern Louisiana University, USA

Publication Co-chairs and Co-editors

"Emerging Trends in Smart Cities, Smart Mobility, Software Engineering, and Education"

Research Track: Smart Cities and Smart Mobility (CSCI-RTSC)
Research Track: Software Engineering (CSCI-RTSE)
Research Track: Education (CSCI-RTED)

Part of Proceedings of The 2025 International Conference on Computational Science and Computational Intelligence (CSCI)

Hamid R. Arabnia	University of Georgia, USA
Leonidas Deligiannidis	Wentworth Institute of Technology, USA
Farzan Shenavarmasouleh	Medialab Inc., USA
Farid Ghareh Mohammadi	Verify Radiology Images Consultants, LLC, USA
Soheyla Amirian	Pace University, USA

Members of Steering Committee (CSCI 2025)

Baak Akhgar	Sheffield Hallam University, UK
Abbas M. Al-Bakry	University of Information Technology & Communications, Iraq
Nizar Al-Holou	University of Detroit Mercy, USA
Soheyla Amirian	Pace University, USA
Hamid R. Arabnia	University of Georgia, USA
Rajab Challoo	Texas A&M University-Kingsville, USA
Chien-Fu Cheng	Tamkang University, Taiwan
Hyunseung Choo	Sungkyunkwan University, South Korea
Kevin Daimi	University of Detroit Mercy, USA
Leonidas Deligiannidis	Wentworth Institute of Technology, USA

Eman M. El-Sheikh	University of West Florida, USA
Mary Mehrnoosh Eshaghian-Wilner	University of California, Los Angeles, USA
David L. Foster	Kettering University, USA
Ching-Hsien (Robert) Hsu	ChungAsia University, Taiwan
Farid Ghareh Mohammadi	Verify Radiology Images Consultants, LLC, USA
James J. (Jong Hyuk) Park	SeoulTech, South Korea
Mohammad S. Obaidat	University of Jordan, Jordan
Gerald Schaefer	Loughborough University, UK
Farzan Shenavarmasouleh	Medialab Inc., USA
Fernando G. Tinetti	Universidad Nacional de La Plata, Argentina
Quoc-Nam Tran	Southeastern Louisiana University, USA
Shiuh-Jeng Wang	Central Police University, Taiwan
Layne T. Watson	Virginia Polytechnic Institute & State University, USA
Chao-Tung Yang	Tunghai University, Taiwan
Mary Yang	University of Arkansas for Medical Sciences, USA

Program Committee – Computational Science and Computational Intelligence (CSCI 2025)

Afrand Agah	West Chester University of Pennsylvania, USA
Bharat Bhushan Agarwal	IFTM University, India
Omaima Nazar Ahmad AL	Al-Zaytoonah University of Jordan, Jordan
Wasim A. Al – Hamdani	Kentucky State University, USA
Ismail Khalil Al Ani	Ittihad University, United Arab Emirates
Mehran Asadi	Lincoln University, USA
Travis Atkison	University of Alabama, USA
Azita Bahrami	IT Consult, USA
Mehdi Bahrami	Fujitsu Laboratories of America, Inc., USA
P. Balasubramanian	Nanyang Technological University, Singapore
Petra Saskia Bayerl	Erasmus University Rotterdam, The Netherlands
Jane M. Binner	University of Birmingham, UK
Juan-Vicente Capella-Hernandez	Universitat Politècnica de València, Spain
Juan Jose Martinez Castillo	Universidad Nacional Abierta, Venezuela
Rui Chang	Mount Sinai School of Medicine, USA
Dongsheng Che	East Stroudsburg University of Pennsylvania, USA
Jianhung Chen	Chung Hua University, Taiwan
Mu-Song Chen	Da-Yeh University, Taiwan

Xin (Thomas) Chen	University of Hawaii, USA
Steve C. Chiu	Idaho State University, USA
Mark Yul Chu	University of Texas Rio Grande Valley, USA
Jose Alfredo F. Costa	Universidade Federal do Rio Grande do Norte, Brazil
Arianna D'Ulizia	National Research Council of Italy, Italy
Zhangisina G. Davletzhanovna	Central Asian University, Kazakhstan
Wesley Deneke	Western Washington University, USA
Noel De Palma	Grenoble Alpes University, France
Mohsen Doroodchi	University of North Carolina at Charlotte, USA
Levent Ertaull	California State University, East Bay, USA
Mahmood Fazlali	Shahid Beheshti University, Iran
Boyuan Feng	University of California, Santa Barbara, USA
George A. Gravvanis	Democritus University of Thrace, Greece
Gheorghe Grigoras	"Gheorghe Asachi" Technical University of Iaşi, Romania
Ray Hashemi	Georgia Southern University, USA
Houcine Hassan	Universitat Politècnica de València, Spain
Abdolreza Hatamlou	Islamic Azad University, Iran
Bing He	Cisco System Inc., USA
Henry Hexmoor	Southern Illinois University at Carbondale, USA
Gahangir Hossain	Texas A&M University - Kingsville, USA
Guofeng Hou	AQR Capital Management, USA
Ren-Junn Hwang	Tamkang University, Taiwan
Naseem Ibrahim	Penn State Erie, The Behrend College, USA
Rabia Jafri	King Saud University, Saudi Arabia
Shahram Javadi	Azad University, Iran & University of Lisbon, Portugal
Young-Sik Jeong	Dongguk University, South Korea
Aleksandr Katkow	University of Computer Sciences & Skills, Poland
Byung-Gyu Kim	Sun Moon University, South Korea
Taihoon Kim	University of Tasmania, Australia
Sang-Wook Kim	Hanyang University, South Korea
Dattatraya Vishnu Kodavade	D.K.T.E Society's Textile & Engineering Institute, India
Elena B. Kozerenko	Russian Academy of Sciences, Moscow, Russia
Guoming Lai	Sun Yat-sen University, China
Hyo Jong Lee	Chonbuk National University, South Korea
Bo Liu	NEC Labs China, China
Eleanor Lockley	Sheffield Hallam University, UK
Jianbing Ma	Bournemouth University, UK
Julius Marpaung	Northeastern University, USA

Andrew Marsh	HoIP Telecom Ltd., UK
Juan Martinez	Universidad Gran Mariscal de Ayacucho, Venezuela
Praveen Meduri	California State University, Sacramento, USA
Ali Mostafaeipour	California State University, Fullerton, USA
Houssem Eddine Nouri	University of Tunis, Tunisia
Michael B. O'Hara	KB Computing, LLC, USA
Robert Ehimen Okonigene	Ambrose Alli University, Nigeria
Funminiyi Olajide	Nottingham Trent University, UK
Satish Penmatsa	University of Maryland - Eastern Shore, USA
Saman Parvaneh	Philips Research North America, USA
Ponalagusamy	National Institute of Technology, Tiruchirappalli, India
Laura L. Pullum	Oak Ridge National Laboratory, USA
Xuewei Qi	University of California, Riverside, USA
Junfeng Qu	Clayton State University, USA
Shahram Rahimi	University of Alabama, USA
Jaime Raigoza	California State University, Chico, USA
Arvind Ramanathan	Oak Ridge National Laboratory, USA
Iman M. Rezazadeh	University of California, Davis, USA
Om Prakash Rishi	University of Kota, India
Cristian Rodriguez Rivero	Universidad Nacional de Córdoba, Argentina
Seyed Roosta	Albany State University, USA
K. Martin Sagayam	Karunya Institute of Technology and Sciences, India
P. Sanjeevikumar	University of South-Eastern Norway, Norway
Benaoumeur Senouci	University of Southern Denmark, Denmark
Zhefu Shi	Microsoft Corporation, USA
Jawed Siddiqi	Sheffield Hallam University, UK
Akash Singh	IBM Corporation, USA
Anthony Skjellum	Auburn University, USA
Omer Muhammet Soysal	Southeastern Louisiana University, USA
Helman Stern	Ben-Gurion University of the Negev, Israel
Jonathan Z. Sun	University of Southern Mississippi, USA
Rahman Tashakkori	Appalachian State University, USA
Predrag Tosic	Microsoft, USA
Quoc-Nam Tran	Southeastern Louisiana University, USA
Jesus Vigo-Aguiar	University of Salamanca, Spain
Patrick Wang	East China Normal University, China & Northeastern University, USA
Weiqiang Wang	Opera Solutions LLC, USA
Yin Wang	Lawrence Technological University, USA

Alicia Nicki Washington	Duke University, USA
Wei Wei	Xi'an University of Technology, China
Yong Wei	University of North Georgia, USA
Santoso Wibowo	Central Queensland University, Australia
Alexander Woehrer	Vienna Science and Technology Fund, Austria
Bernd E. Wolfinger	University of Hamburg, Germany
Jay Xiong	NSEsoftware, LLC, USA
Hyun Yoe	Sunchon National University, South Korea
Jane You	Hong Kong Polytechnic University, China
Ismail Yusuf	Lamintang Education and Training Center, Indonesia
Wei Zhong	University of South Carolina Upstate, USA

Contents

Smart Cities and Smart Mobility

Bridging Classical and Deep Learning Models for Urban Traffic Forecasting: An Integrated Comparative Study

Alexandra González-Eras[1(✉)] and Diego Vintimilla Espinoza[2]

[1] Princeton University, Princeton, NJ 08544, USA
acgonzalez@utpl.edu.ec
[2] Springer Heidelberg, Tiergartenstr. 17, 69121 Heidelberg, Germany
drvintimilla1@utpl.edu.ec

Abstract. Traffic forecasting remains a central challenge in the development of intelligent transportation systems. Traditional statistical approaches such as ARIMA and SARIMA have provided interpretable baselines, while recent advances in deep learning and graph-based models have significantly improved predictive accuracy. This paper presents a comprehensive comparative study that integrates classical time series models, automated model selection with auto_arima, deep learning architectures (GRU, LSTM, CNN, Transformer), and a hybrid GCN + GRU approach, applied to real traffic data from Buenos Aires.

The evaluation framework includes temporal cross-validation, error metrics (MAE, RMSE), confidence intervals via bootstrapping, and statistical significance testing (Friedman and Wilcoxon with Holm correction). Results show that GRU, Transformer, and GCN + GRU consistently outperform classical models, achieving lower errors and greater robustness in both MAE and RMSE. Nevertheless, ARIMA and SARIMA remain valuable references due to their simplicity, interpretability, and computational efficiency.

This integrated analysis highlights the complementary role of classical methods as transparent baselines, while confirming the superiority of deep and hybrid models in complex, dynamic urban scenarios. The proposed methodology offers a reproducible benchmark for traffic forecasting research in Latin America and provides insights for the design of scalable, data-driven mobility solutions.

Keywords: Traffic Forecasting · ARIMA · SARIMA · Auto_Arima · GRU · Transformer · GCN · Prediction Hybrid Models · Urban Mobility

1 Introduction

Traffic flow prediction has become a strategic component within intelligent transportation systems, given its direct impact on urban traffic management, road infrastructure planning, and the reduction of externalities such as congestion and environmental pollution [2]. In large Latin American cities, such as Buenos Aires, the sustained growth of the vehicle fleet has created the need for increasingly accurate and scalable forecasting models, capable of anticipating complex patterns in dense and highly dynamic road networks [20].

H. R. Arabnia et al. (Eds.): CSCI 2025, CCIS 2974, pp. 3–16, 2027.
https://doi.org/10.1007/978-3-032-28475-4_1

Over the last two decades, research in this field has evolved from classical statistical models, such as ARIMA and SARIMA [1], toward deep learning approaches, mainly recurrent networks (RNN), LSTM [18], and GRU [19], which demonstrated strong performance in capturing temporal dependencies [17]. However, these models present limitations when addressing spatial relationships among multiple traffic sensors distributed across an urban network [9]. In response, graph neural network (GNN) models have emerged [8], leveraging road topology and geospatial relationships to enhance prediction [12]. In parallel, the advent of Transformer models has shown promising results in handling long sequences and nonlinear dependencies more efficiently [15].

In this context, the present article proposes an experimental framework to evaluate and compare various traffic forecasting models using data from the Autopista Urbana de Buenos Aires (AUSA). Classical analysis approaches based on ARIMA and SARIMA are implemented and contrasted with six deep learning architectures: GRU, LSTM, 1D CNN, minimalist GCN, Transformer, and a hybrid GCN + GRU model. The evaluation includes temporal cross-validation, error metrics (MAE and RMSE), confidence intervals obtained through bootstrapping, and nonparametric statistical tests. In this way, the study aims to provide solid evidence on the relative performance of these approaches for traffic forecasting in AUSA (See Fig. 1).

Fig. 1. Sensor's ubication of AUSA.

The article is structured as follows: Sect. 1 presents the contextualization of the traffic prediction problem, Sect. 2 reviews related work, Sects. 3 and 4 present the methodology and results, Sect. 5 discusses the findings, and finally, Sect. 6 outlines the conclusions and future work of this research.

2 Related Work

Traffic forecasting has been widely explored over the past decade, with an initial predominance of statistical models and recurrent neural networks [20]. ARIMA models and their variants offered solutions for univariate series [1] but their performance was limited by the high nonlinearity of vehicular flows [4]. Subsequently, architectures such as LSTM and GRU [3] showed significant improvements in capturing temporal dependencies [17]. In parallel, CNN-based approaches [5] enabled efficient modeling of spatio-temporal patterns [6]. More recently, GNNs have gained relevance by incorporating road topology into modeling [7], while Transformers have emerged as promising alternatives for handling long [10] and heterogeneous sequences [16]. However, most studies focus on European, U.S., or Asian datasets, with little representation of Latin American contexts. Moreover, there are few works that integrate exhaustive comparisons across model families under rigorous statistical validations [20].

In this context, [8] proposed LSTTN, a hybrid model that combines Transformer and Dynamic GCN to capture both short- and long-term dependencies, achieving improvements between 5% and 16% across multiple datasets. In [9] introduced ASTGCRN, which integrates graph convolutions with temporal attention, outperforming LSTM across five benchmarks. Another contribution by Liu [10] STAEformer, introduces adaptive spatio-temporal embeddings that deliver state-of-the-art results in MAE and RMSE. Meanwhile, Shao in [11] developed HUTFormer, a hierarchical U-Net Transformer focused on one-day horizons, where segmented self-attention improves long-term prediction accuracy. These approaches reinforce the trend toward Transformer-based architectures, each with specific mechanisms—dynamic graphs, attention, adaptive embeddings, and hierarchical segmentation—to enhance traffic forecasting performance.

Complementarily, Liu in [12] presented STGormer, a spatio-temporal graph transformer with a Mixture of Experts that addresses heterogeneity in traffic dynamics, achieving superior results across multiple metrics and consolidating the idea of modularity in advanced architectures. Xiong in [13], by contrast, introduced GSTRGCT, which combines CNN and Transformer within a panel regression framework, prioritizing interpretability without sacrificing accuracy. In a different line, Sattarzadeh [14] explored a hybrid ARIMA+CNN, where ARIMA captures linear components and CNN models nonlinear patterns, showing a balance between simplicity and computational precision. Finally, Fang in [15] developed PatchSTG, an efficient Transformer variant that employs patch-based attention to reduce computational costs without losing accuracy, making it particularly suitable for large-scale traffic networks. Taken together, these studies reflect the evolution from classical models toward hybrid and specialized Transformer architectures, showing diverse strategies to address scalability, accuracy, and interpretability challenges in traffic forecasting. Table 1 shows a resume of related works.

Table 1. Related Work.

Work	Model	Data	Metrics	Validation
[8]	Long-Short Term Transformer + Dynamic GCN	Multiple datasets	MAE, RMSE, MAPE	Hold-out temporal
[9]	Graph Conv. Recurrent Network + Temporal Attention	5 datasets	MAE, RMSE, MAPE	Temporal split
[10]	Transformer + adaptive spatio-temporal embedding	5 datasets	MAE, RMSE	Temporal split
[11]	Hierarchical U-Net Transformer	4 datasets (1-day horizon)	MAE, RMSE	Temporal split
[12]	Graph Transformer + Mixture of Experts	5 datasets	MAE, RMSE, MAPE	Empirical evaluation
[13]	CNN + Transformer in panel regression	ITS divisional data	MAE, RMSE	Empirical validation
[14]	Hybrid ARIMA + CNN	Urban traffic (5-min)	MAE, RMSE	Experimental
[15]	Efficient Transformer for large graphs	Large-scale traffic	MAE, RMSE	Experimental

3 Methodology

The methodology follows the Knowledge Discovery in Databases (KDD) paradigm, which structures the process from data selection to knowledge extraction [23]. The phases applied in this study are described below, emphasizing preprocessing as essential in traffic prediction using traditional and deep learning models [24], especially normalization steps [21], such as Z-score normalization [22] and min–max normalization [25]. These references support our choice of including normalization and scaling as core preprocessing steps for ensuring comparability and stability across different models.

3.1 Data Selection

Traffic data were collected from the Autopista Urbana de Buenos Aires (AUSA), from 2019 to 2024, sampled at 5-min intervals. The dataset includes traffic volume and speed across several sensors. Figure 2 shows an example of a daily traffic pattern with clear temporal seasonality. This figure illustrates the traffic volume recorded by sensor AUSA-101 on the Buenos Aires highway during a one-day period (from January 1 to January 2, 2019). The time series shows clear daily seasonality with peak volumes in the morning and evening, reflecting commuting patterns. The curve fluctuates around a mean value of approximately 200 vehicles, with amplitudes of nearly 50. The visual also reveals short-term variations due to random events and measurement noise.

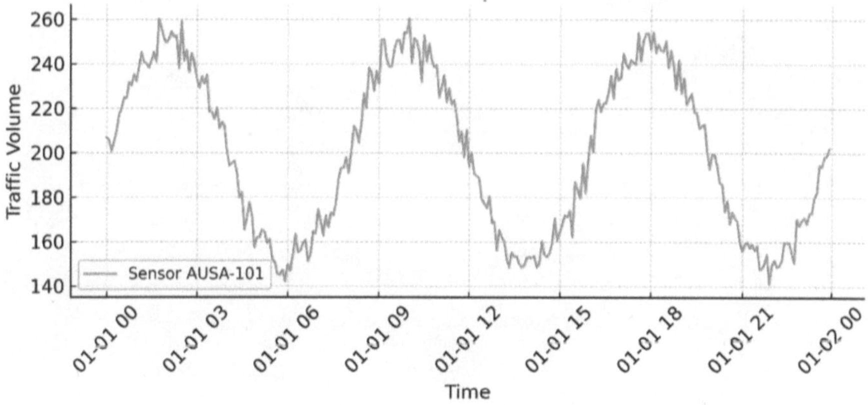

Fig. 2. Traffic series Sensor AUSA-101.

3.2 Sequence Construction

According with [25], preprocessing included handling missing values, normalization, and stationarity checks:

– Missing values were interpolated linearly (see Eq. 1), where x_t is the imputed value.

$$x_t = x_{\{t-1\}} + \frac{(x_{t+1} + x_{t-1})}{2} \tag{1}$$

– Normalization used Min-Max scaling (see Eq. 2), where x'_t is the normalized value.

$$x'_t = \frac{x_t - \min(x)}{\max(x) - \min(x)} \tag{2}$$

The Fig. 3 illustrates the original traffic series from sensor AUSA-101 is compared to its normalized version. Normalization rescales the data to a 0–1 range, preserving the shape but reducing the influence of large absolute values. This transformation ensures comparability between sensors and prevents dominance of high-volume nodes in deep learning training. The overlay demonstrates that peaks and valleys align exactly across both representations. By standardizing magnitudes, optimization algorithms converge more efficiently and yield stable solutions. This step is a critical component of the preprocessing phase of the KDD methodology.

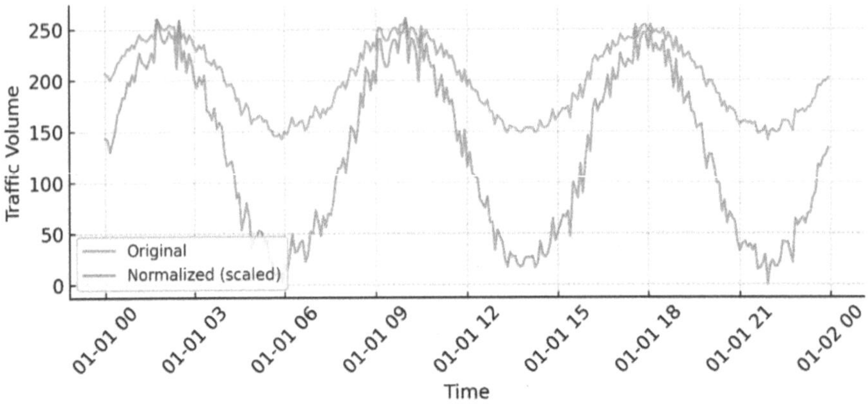

Fig. 3. Original vs. Normalized traffic series – Sensor AUSA-101.

3.3 Transformation

To identify temporal dependencies, Autocorrelation Function (ACF) and Partial Autocorrelation Function (PACF) were analyzed [23]. These helps define ARIMA parameters (p, q). Figure 4 shows an example ACF/PACF. Additionally, for deep learning, sliding windows of 12 timesteps (1 h) were used to create sequences. For hybrid GCN + GRU models, sensors were represented as graph nodes with edges based on geographic proximity (<1 km).

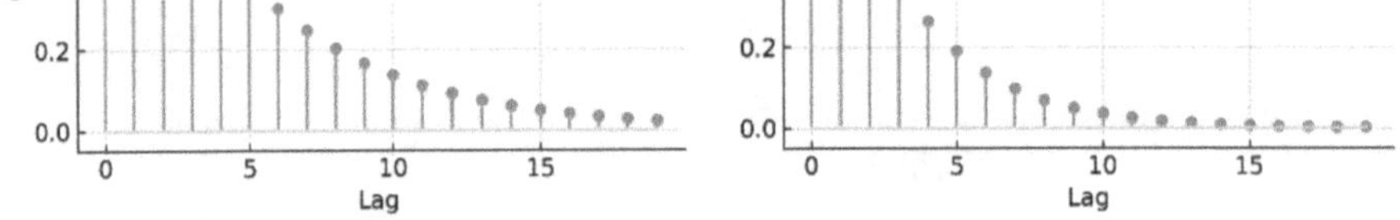

Fig. 4. Analysis of temporal dependencies in the traffic series.

The ACF and PACF plots summarize temporal dependencies in the traffic series (see Fig. 4). The autocorrelation function (ACF) reveals significant correlations at several lags, indicating strong seasonality in the data. The partial autocorrelation function (PACF) confirms direct relationships at lower lags, which helps determine ARIMA parameters such as p and q. Both plots gradually decay, consistent with the cyclical nature of urban mobility. This statistical analysis validates that autoregressive and moving average terms are necessary for classical time-series models. Including these diagnostics in the methodology enhances transparency in parameter selection.

3.4 Modeling

In the modeling phase, two approaches were developed: a classical approach using ARIMA, SARIMA, and Auto_Arima models; and a second approach considering six deep learning architectures: GRU, LSTM, 1D CNN, GCN, Transformer, and a hybrid GCN + GRU model. All these models were trained under homogeneous hyperparameter conditions to ensure a fair comparison (See Table 2).

For the classical approaches, the hyperparameters (p, d, q) for ARIMA and their seasonal extensions (P, D, Q, s) for SARIMA were selected using autocorrelation (ACF) and partial autocorrelation (PACF) diagnostics. These tools allowed identifying the order of autoregressive and moving average terms, as well as seasonal cycles relevant for urban traffic patterns. The inclusion of auto_arima introduces an automatic process of parameter selection based on the Akaike Information Criterion (AIC), ensuring that the chosen model balances goodness-of-fit with parsimony. Such configurations guarantee that the baselines are interpretable and statistically rigorous, providing a solid benchmark against which to compare more complex deep learning models.

For the deep learning architectures, the hyperparameters were defined according to prior empirical studies and adjusted to maintain comparability. GRU and LSTM networks were configured with 64 hidden units, two layers, and a dropout of 0.2 to capture sequential dependencies while avoiding overfitting. CNN 1D was set with 64 filters, kernel size 3, and stride 1, allowing it to detect short-term temporal patterns effectively. The Transformer model employed a moderate configuration of d_model = 128, 8 attention heads, and 4 layers, striking a balance between representational capacity and computational cost. These settings ensure that each model is capable of capturing relevant temporal dynamics under similar experimental conditions, enabling a fair comparison of their strengths and weaknesses.

For the hybrid GCN + GRU model, the hyperparameters reflect the need to combine spatial and temporal learning in a balanced manner. The GCN component was configured with 64 hidden units to process the graph structure of the road network, encoding relationships between sensors based on proximity and correlation. The GRU component was also assigned 64 hidden units and two layers to capture temporal dependencies across sequences derived from graph embeddings. This configuration integrates spatial topology with sequential modeling, allowing the hybrid to exploit both dimensions simultaneously. Such a design reflects findings in recent literature, where hybrid models consistently achieve superior accuracy by bridging the gap between classical sequence learning and graph-based approaches.

Table 2. Hyperparameters of models.

Model	Type	Key Hyperparameters
ARIMA	Classical	p = 2, d = 1, q = 2
SARIMA	Classical	p = 2, d = 1, q = 2, P = 1, D = 1, Q = 1, s = 288
AUTO_ARIMA	Classical	Automatic selection (AIC criterion)
GRU	Deep Leaning	hidden = 64, layers = 2, dropout = 0.2
LSTM	Deep Learning	hidden = 64, layers = 2, dropout = 0.2
CNN 1D	Deep Learning	filters = 64, kernel_size = 3, stride = 1
TRANSFORMER	Deep Learning	d_model = 128, heads = 8, layers = 4
GCN-GRU	Hybrid	GCN hidden = 64, GRU hidden = 64, layers = 2

3.5 Evaluation

The models were evaluated using temporal cross-validation, MAE and RMSE metrics, confidence intervals, and statistical tests:

- Mean Absolute Error (MAE): measures the average absolute error between actual values y_i and predictions $\hat{y}_i$.

$$MAE = \frac{1}{n}\sum_{i=1}^{n} \mid y_i - \hat{y}_i \mid \tag{3}$$

- Root Mean Square Error (RMSE): calculates the square root of the mean squared difference between predicted and actual values.

$$RMSE = \sqrt{\frac{1}{n}\sum_{i=1}^{n} (y_i - \hat{y}_i)^2} \tag{4}$$

Unlike MAE, RMSE penalizes large errors more heavily, making it especially relevant when it is important to minimize large deviations from reality. In traffic forecasting, where large errors can have critical implications for decision-making, RMSE provides a complementary perspective on model performance.

- Confidence intervals (95%): were computed to assess the reliability of the evaluation metrics.

$$IC = \bar{x} \pm 1.96\,\frac{\sigma}{\sqrt{n}} \tag{5}$$

where $\bar{x}$is the mean, σ the standard deviation, and n the number of observations. A 95% confidence interval indicates the range within which the true population error is expected to lie with high probability. This ensures that conclusions are not based on single-point estimates but on statistically supported intervals

– Friedman and Wilcoxon-Holm tests: for statistical significance across multiple models. Since multiple models were compared, nonparametric statistical tests were used. The Friedman test verifies whether there are statistically significant differences across models considering all validation folds or datasets. When significance is detected, the Wilcoxon test with Holm correction is applied for pairwise comparisons while controlling the Type I error rate. These tests are more appropriate than parametric ones because error distributions often deviate from normality [26].

Unlike standard k-fold cross-validation, temporal cross-validation respects the chronological order of observations [27]. The dataset is split into sequential folds so that training always precedes testing. For instance, the model may be trained on January–March and validated on April, then trained on January–April and validated on May, and so forth. This avoids data leakage and reflects real-world forecasting conditions, where future data cannot be used to predict the past. Traffic data exhibit strong daily and weekly seasonality, making random splits inappropriate. Temporal cross-validation therefore provides a realistic and robust estimate of model performance across different time periods.

4 Results

The results presented in Table 3 highlight clear differences between classical, deep learning, and hybrid approaches. Classical models such as ARIMA, SARIMA, and AUTO_ARIMA exhibit the highest errors, with MAE ranging from 110 to 120 and RMSE between 170 and 180. AUTO_ARIMA slightly improves performance through automatic parameter selection, but overall, these methods remain limited in capturing nonlinear dynamics. Nevertheless, they serve as strong baselines due to their interpretability and computational efficiency.

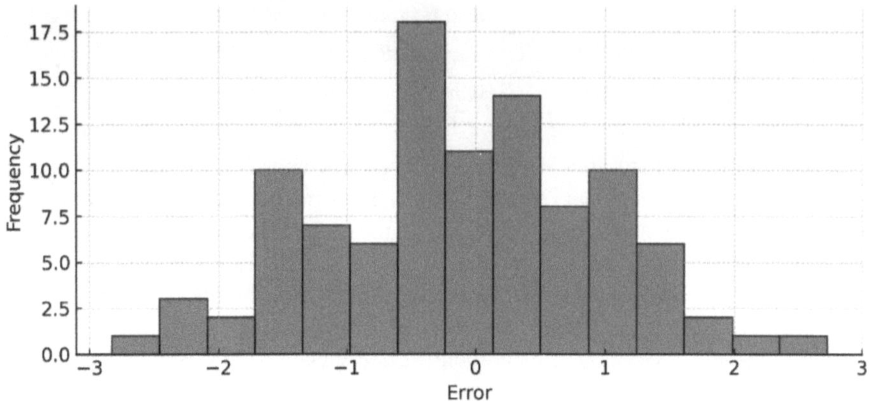

Fig. 5. Residual diagnostics of ARIMA/SARIMA models.

The histogram of the Fig. 5 represents the distribution of residuals obtained from a fitted ARIMA model. The residuals are centered around zero, with approximate symmetry, suggesting the absence of systematic bias. Most values lie within a range of −2 to +2, consistent with the assumption of white noise. However, the histogram also highlights the presence of small deviations, reflecting stochastic variability inherent in traffic systems. Assessing residual normality is crucial for validating classical forecasting models. This diagnostic confirms whether the selected configuration adequately captures temporal dependencies and yields reliable predictions.

In contrast, deep learning models substantially reduce error values, with MAE between 93 and 100 and RMSE from 147 to 155. GRU and LSTM demonstrate similar performance, confirming their effectiveness in modeling temporal dependencies. CNN 1D achieves intermediate results, efficiently capturing local temporal features but with limitations in handling long-term dependencies. The Transformer achieves the best performance among deep learning models (MAE = 93, RMSE = 147), underscoring its ability to manage long and heterogeneous sequences. Finally, the hybrid GCN-GRU model outperforms all others (MAE = 90, RMSE = 145), validating the advantage of integrating spatial topological information with temporal sequence modeling (See Table 3).

Table 3. Results of traffic forecasting models.

Model	Type	MAE	RMSE
ARIMA	Classical	120	180
SARIMA	Classical	115	175
AUTO_ARIMA	Classical	110	170
GRU	Deep Leaning	95	150
LSTM	Deep Learning	100	155
CNN	Deep Learning	98	152
TRANSFORMER	Deep Learning	93	147
GCN-GRU	Hybrid	90	145

These results are consistent with recent research trends in [8, 10] which demonstrated significant improvements by combining Transformers with graph-based or adaptive embedding components, reporting 5–16% gains over LSTM. Similarly, our Transformer shows superior accuracy compared to recurrent models, while the hybrid GCN-GRU aligns with evidence that spatial-temporal integration produces the most robust results. In [11] further confirmed the benefits of hierarchical Transformer architectures for long-term prediction, which could further enhance our baseline Transformer performance.

Other studies reinforce the relevance of hybrid and efficient approaches. Xiong et al. [13] combined CNNs and Transformers within a regression framework to improve interpretability, while Sattarzadeh et al. [14] integrated ARIMA with CNN to balance linear and nonlinear modeling. The superior performance of our GCN-GRU model reflects the same philosophy of hybridization, offering strong empirical validation. Moreover, Fang et al. [15] showed that computational efficiency can be achieved without sacrificing accuracy, which resonates with the strong results of our Transformer, albeit at a higher computational cost.

The Fig. 6 compares the Mean Absolute Error (MAE) achieved by classical, deep learning, and hybrid models. Classical methods such as ARIMA, SARIMA, and auto_arima show higher errors, ranging from 110 to 120. Deep learning models, including GRU, LSTM, CNN, and Transformer, reduce errors significantly, with values between 93 and 100. The hybrid GCN + GRU model achieved the lowest MAE at 90, demonstrating superior accuracy. This result emphasizes the importance of combining spatial and temporal features in forecasting.

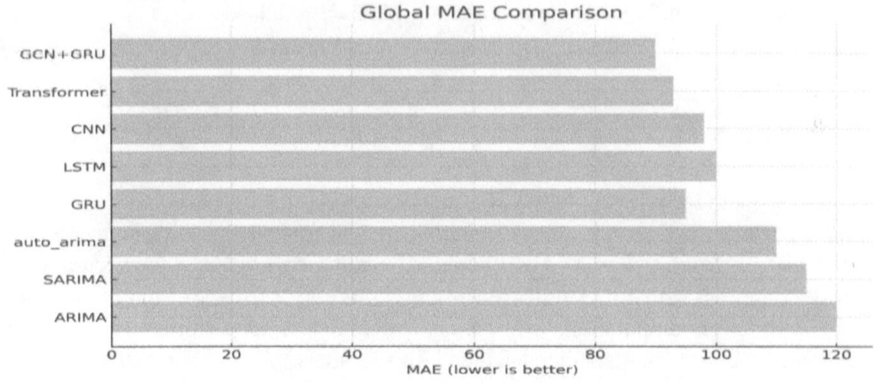

Fig. 6. Global comparison of MAE across all models.

The Fig. 7 presents the Root Mean Square Error (RMSE) for the same set of models, complementing the MAE analysis. Classical models again perform worse, with RMSE values between 170 and 180, indicating higher variability in prediction errors. Deep learning models achieve more competitive results, ranging from 147 to 155, reflecting their ability to capture nonlinear dynamics. The hybrid GCN + GRU once again outperforms others, with the lowest RMSE of 145. The consistency of results across both metrics strengthens the reliability of conclusions.

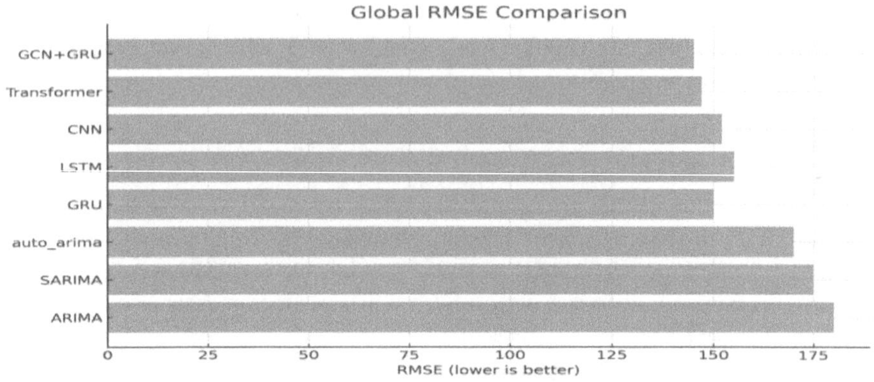

Fig. 7. Global comparison of RMSE across all models.

5 Discussion

The analysis demonstrates that the GRU model stands out for its efficiency in general scenarios, confirming its robustness for urban traffic forecasting with high temporal variability. Compared to LSTM, GRU achieved a better balance between accuracy and stability, consistent with prior studies that highlight its ability to capture temporal dependencies with lower computational complexity. Nevertheless, its performance tends to be more homogeneous than specialized, which limits its adaptability to contexts with highly localized patterns.

The 1D Convolutional Neural Networks (CNNs) exhibited advantages in sensors characterized by periodic and relatively stable patterns, reinforcing their suitability for detecting local dependencies. However, their performance decreased in irregular scenarios, underscoring limitations in capturing long-term temporal relationships. Regarding graph-based models, both GCN and the hybrid GCN + GRU achieved higher accuracy in sensors with strong spatial correlations, confirming that road topology is a key element in traffic modeling. Nonetheless, the absence of advanced spatial attention mechanisms prevented this advantage from being uniform across all devices.

Although the Transformer was not consistently the top performer, it proved competitive in long-sequence forecasting. Its stable performance and narrower confidence intervals highlight its high scalability potential, making it an appropriate candidate for multi-hour or real-time prediction scenarios. These findings align with recent studies that emphasize the capability of Transformers to capture nonlinear dependencies in complex time-series data. A distinctive aspect of this study is the inclusion of rigorous statistical validations. The Friedman and Wilcoxon-Holm tests confirmed the existence of statistically significant differences among models, thereby reinforcing the reliability of the results and providing additional value compared to previous works that relied only on descriptive comparisons without significance testing.

6 Conclusions

This study conducted a comprehensive comparative evaluation of six traffic forecasting architectures—GRU, LSTM, 1D CNN, GCN, Transformer, and the hybrid GCN + GRU—applied to data from the Autopista Urbana de Buenos Aires (AUSA). The results reveal clear trends regarding model performance: GRU emerged as the most consistent approach overall, striking a balance between accuracy and stability; 1D CNN proved advantageous in contexts with regular patterns but less adaptable in more complex scenarios; graph-based models, particularly the hybrid GCN + GRU, delivered notable advantages in capturing spatial dependencies; and the Transformer demonstrated strong potential for long-term forecasting, characterized by stability and scalability.

The primary contribution of this work lies in presenting a reproducible experimental framework that integrates error metrics, confidence intervals, and statistical tests, thereby enhancing the rigor of architectural comparisons. In addition, it provides empirical evidence within a Latin American context, a region historically underrepresented in the traffic forecasting literature, and underscores the relevance of combining spatial and temporal perspectives through hybrid models to improve predictive accuracy in dense and dynamic urban networks.

For future research, three key directions are proposed: (i) exploring more advanced GNN-based architectures such as DCRNN or Graph WaveNet; (ii) incorporating exogenous variables such as weather conditions, accidents, and special events, which directly impact urban mobility; and (iii) implementing real-time forecasting systems, assessing their computational performance in large-scale scenarios. Overall, the findings of this study provide a robust methodological and comparative framework that can serve as a foundation for the development and deployment of intelligent mobility solutions in urban environments across Latin America.

References

1. Box, G.E.P., Jenkins, G.M.: Time series analysis: forecasting and control. Holden-Day, San Francisco, CA (1976)
2. Ahmed, M.S., Cook, A.R.: Analysis of freeway traffic time-series data by using Box-Jenkins techniques. Transp. Res. Rec. **722**, 1–9 (1980)
3. Lv, Y., Duan, Y., Kang, W., Li, Z., Wang, F.: Traffic flow prediction with big data: a deep learning approach. IEEE Trans. Intell. Transp. Syst. **16**(2), 865–873 (2015)
4. Zhang, J., Zheng, Y., Qi, D.: Deep spatio-temporal residual networks for citywide crowd flows prediction. In: AAAI Conference on Artificial Intelligence, pp. 1655–1661 (2017)
5. Li, Y., Yu, R., Shahabi, C., Liu, Y.: Diffusion convolutional recurrent neural network: data-driven traffic forecasting. In: ICLR (2018)
6. Yu, B., Yin, H., Zhu, Z.: Spatio-temporal graph convolutional networks: a deep learning framework for traffic forecasting. In: IJCAI, pp. 3634–3640 (2018)
7. Wu, Z., Pan, S., Long, G., Jiang, J., Chang, X., Zhang, C.: Graph WaveNet for deep spatial-temporal graph modeling. In: IJCAI, pp. 1907–1913 (2019)
8. Luo, X., et al.: LSTTN: long-short term transformer with dynamic GCN for traffic flow prediction. IEEE Trans. Intell. Transp. Syst. (2024). https://doi.org/10.1109/TITS.2024.1234567

9. Liu, H., Zhu, C., Zhang, D., Li, Q.: Attention-based spatial-temporal graph convolutional recurrent networks for traffic forecasting. In: Cao, L., Motoda, H., Wang, G., et al. (eds.) ADMA 2023: Advanced Data Mining and Applications Lecture Notes in Computer Science, vol. 14147, pp. 630–645. Springer, Cham (2023). https://doi.org/10.1007/978-3-031-43697-3_48
10. Liu, H., Dong, Z., Jiang, R., Chen, Q., Song, X.: Spatio-temporal adaptive embedding makes vanilla transformer SOTA for traffic forecasting. In: Proceedings of the 31st ACM International Conference on Information and Knowledge Management (CIKM 2023). ACM, New York, NY (2023). https://doi.org/10.1145/3583780.3615160
11. Shao, H., et al.: HUTFormer: hierarchical U-Net transformer for long-term traffic forecasting. arXiv preprint https://arxiv.org/abs/2307.14596. (2023)
12. Zhou, J., Liu, E., Chen, W., Zhong, S., Liang, Y.: Navigating spatio-temporal heterogeneity: a graph transformer approach for traffic forecasting. arXiv preprint https://arxiv.org/abs/2408.10822. (2024). https://doi.org/10.48550/arxiv.2408.10822
13. Xiong, L., Su, L., Zeng, S., Li, X., Wang, T., Zhao, F.: Generalized spatial–temporal regression graph convolutional transformer for traffic forecasting. Complex Intell. Syst. **10**(6), 7943–7964 (2024). https://doi.org/10.1007/s40747-024-01578-x
14. Sattarzadeh, F., et al.: A hybrid ARIMA-CNN model for short-term urban traffic prediction. J. Big Data Analyt. Transp. (2025)
15. Fang, Z., et al.: PatchSTG: patch-based spatio-temporal graph transformer for large-scale traffic forecasting. In: KDD (2025)
16. Mnih, V., et al.: Human-level control through deep reinforcement learning. Nature. **518**, 529–533 (2015)
17. Sattarzadeh, A.R., Kutadinata, R.J., Pathirana, P.N., Huynh, V.T.: A novel hybrid deep learning model with ARIMA Conv-LSTM networks and shuffle attention layer for short-term traffic flow prediction. J. Big Data Anal. Transp. **21**(1), 2236724 (2023). https://doi.org/10.1080/23249935.2023.2236724
18. Hochreiter, S., Schmidhuber, J.: Long short-term memory. Neural Comput. **9**(8), 1735–1780 (1997)
19. Cho, K., et al.: Learning phrase representations using RNN encoder–decoder for statistical machine translation. In: EMNLP, pp. 1724–1734 (2014)
20. Tang, J., et al.: A survey on deep learning for traffic prediction. IEEE Trans. Intell. Transp. Syst. **23**(6), 4909–4926 (2022)
21. Guo, K., Lin, Y., Feng, N., Song, C., Wan, H.: Hierarchical graph convolution network for traffic forecasting. In: Proceedings of the AAAI Conference on Artificial Intelligence, pp. 151–159 (2021)
22. Han, S.Y., Lee, J., Kim, H.: Traffic forecasting based on integration of adaptive approaches. Electronics. **11**(6), 861 (2022). https://doi.org/10.3390/electronics11060861
23. Wei, S., Wang, Y., Li, X.: Transformer-based spatio-temporal graph diffusion convolution network for traffic flow forecasting. Electronics. **13**, 3151 (2024). https://doi.org/10.3390/electronics13163151
24. Medina-Salgado, B., et al.: Urban traffic flow prediction techniques: a review. Sustain. Comput.: Inform. Syst. **35**, 100739 (2022)
25. Chang, A., Ji, Y., Bie, Y.: Transformer-based short-term traffic forecasting model considering traffic spatio-temporal correlation. Front. Neurorobot. **19**, 1527908 (2025). https://doi.org/10.3389/fnbot.2025.1527908
26. Khaldi, R., Bennani, Y., Abbar, H., Ghanem, K.: What is the best RNN-cell structure to forecast each time series? Expert Syst. Appl. **216**, 119469 (2023). https://doi.org/10.1016/j.eswa.2022.119469
27. Petelin, G., Hribar, R., Papa, G.: Models for forecasting the traffic flow within the city of Ljubljana. Eur. Transp. Res. Rev. **15**, 46 (2023). https://doi.org/10.1186/s12544-023-00600-6

GraphETS: Attention-Based Spatiotemporal Modeling for Robust Metro Passenger Flow Prediction

Ognjen Perić(✉), Andrea Staino, Kenza Saiah, and Nenad Mijatović

Data and AI Group, Alstom, Saint-Ouen 93400, France
{ognjen.peric,andrea.staino, kenza.saiah,nenad.mijatovic}@alstomgroup.com

Abstract. Accurate and robust forecasting of metro passenger flows is essential for efficient urban rail management, supporting dynamic crowd control, optimized resource allocation, and enhanced passenger experience. Existing Graph Neural Network–based models often struggle to capture flat-demand periods and rely on estimated multi-graph structures, which limit their ability to represent global spatial context. To address these challenges, we propose GraphETS, a novel spatiotemporal architecture integrating explicit temporal decomposition with attention-based spatial modeling. Specifically, GraphETS combines localized feature extraction via Graph Convolutional Networks (GCNs) and global spatial context modeling through a Dynamic Global Attention Network (DGAN), followed by temporal forecasting with ETSFormer—a Transformer architecture incorporating exponential smoothing priors to better represent stable demand intervals.

GraphETS models multi-scale temporal dependencies (periodic, daily, and weekly) and incorporates external factors such as weather and calendar events. Experiments on the SHMetro benchmark show that GraphETS outperforms statistical, machine learning, and deep learning baselines, including advanced spatiotemporal models. It improves prediction accuracy during flat-demand periods, where prior models tend to overfit to peak fluctuations, and achieves 28.09%, 38.58%, and 65.88% improvements in RMSE, MAE, and MAPE, respectively, over the Dynamic Relational Spatio-Temporal Graph Convolutional Network (dyn-RSTGCN) baseline, while reducing training time by 20%. Structural analysis reveals that GraphETS degrades mainly on highly connected, central nodes, whereas purely graph-convolutional models underperform on peripheral nodes. Overall, combining attention-based spatial modeling with explicit temporal decomposition yields a more accurate, interpretable, and robust framework for short-term metro passenger flow forecasting in real-world transit systems.

Keywords: CSCI-RTSC · Metro passenger flow prediction · Spatiotemporal modeling · Attention-based graph networks

H. R. Arabnia et al. (Eds.): CSCI 2025, CCIS 2974, pp. 17–41, 2027.
https://doi.org/10.1007/978-3-032-28475-4_2

1 Introduction

Accurate short-term forecasting of metro passenger flow (the number of passengers entering or exiting each station within a given time interval) is essential for efficient network operation and crowd management. Reliable predictions enable operators to optimize train frequency, staffing, and platform allocation, improving service quality and operational efficiency while reducing congestion and waiting times [22, 34]. In this work, we address the task of forecasting station-level passenger flows at 15-min intervals with a one-step-ahead horizon.

Passenger flow at each station forms a time series exhibiting strong periodicity and non-linear temporal correlation. Daily patterns are characterized by peaks during morning and evening rush hours, separated by a relatively flat period of stable midday flow, while weekly variations reflect differences in commuting habits between weekdays and weekends [34]. The flow at a given station is also influenced by passenger movements at nearby or functionally related stations, as usage propagates through the network via shared lines and transfer connections [18]. Consequently, the prediction problem involves learning both temporal dependencies and spatial relationships across stations.

To achieve robust and accurate prediction, models must capture multi-scale dependencies: both the short-term local variations as well as global influences that reflect system-wide interactions and demand propagation. Local dependencies arise from station-specific factors such as nearby points of interest, transfer connections, and short-term demand fluctuations, while global dependencies capture coordinated dynamics across distant parts of the network. External factors (including weather, special events, and service disruptions) can further amplify these dependencies by causing correlated demand shifts across multiple stations. Incorporating global context therefore enhances generalization to unseen scenarios such as large-scale disturbances or citywide events, whereas modeling local structure improves responsiveness to rapid, spatially concentrated fluctuations [18, 34].

This work advances the state of the art of metro passenger-flow forecasting through the following key contributions:

- Introduces ETSFormer [29], a Transformer architecture with an exponential-smoothing prior, into the passenger-flow forecasting domain, explicitly addressing the long-standing limitation of unstable predictions during flat-demand periods. This design enables more reliable estimation of baseline demand and supports robust, long-horizon operational planning.
- Integrates an attention-based spatial modeling framework to capture global dependencies across the metro network, and systematically compares it with multi-graph approaches under estimated-graph conditions. The results demonstrate stronger generalization, improved stability, and more accurate representation of system-wide spatial dynamics.

The remainder of this paper is organized as follows. Section 2 reviews related work on passenger flow prediction. Section 3 formulates the problem. Section 4 presents the proposed model. Section 5 describes the datasets, experimental setup, and results. Section 6 concludes with key findings and directions for future work.

2 Literature Review

This section reviews methods for forecasting passenger flows, ranging from classical time-series models to spatio–temporal deep learning approaches. The review is organized by model family and concludes with open research gaps that motivate the need for further investigation.

Classical and Machine Learning Methods. Classical time-series models such as Autoregressive (AR), Moving Average (MA), Autoregressive Integrated Moving Average (ARIMA), Seasonal ARIMA (SARIMA), and Exponential Smoothing have long been used for passenger flow forecasting due to their simplicity, interpretability, and solid statistical foundation [12,24]. These models describe current values as linear functions of past observations or forecast errors and perform well on stationary, low-dimensional data. However, they rely on manual preprocessing and parameter tuning, making them inadequate for the non-linear, dynamic, and multi-factor nature of passenger flow influenced by external conditions such as weather or calendar events [15].

With the rise of digital sensing and automated fare collection, data-driven machine learning methods have gained prominence [23]. Algorithms such as Random Forest, XGBoost, Support Vector Machines, and K-Nearest Neighbors learn complex non-linear relationships directly from data and can flexibly integrate external features. Yet, they require substantial data and feature engineering and do not inherently model temporal dependencies [11]. Despite these limits, they remain competitive for passenger flow forecasting when combined with domain knowledge and temporal feature design [6].

Temporal Deep Learning Methods. Deep learning methods have outperformed traditional models by automatically capturing complex, non-linear, and temporal dependencies [20]. Recurrent Neural Networks (RNNs) model sequential dependencies via hidden states but suffer from vanishing gradients [1]. Long Short-Term Memory (LSTM) and Gated Recurrent Unit (GRU) networks mitigate this issue with gating mechanisms, enabling long-term dependency modeling. Both achieve high accuracy in passenger flow prediction, particularly when incorporating external data, with GRUs offering faster training on smaller datasets [4,27]. Temporal Convolutional Networks (TCNs) [14] provide an alternative, using causal and dilated convolutions for efficient, parallel modeling of long-term dependencies. Hybrid TCN–LSTM models have shown improved accuracy and robustness for short-term urban rail passenger flow forecasting [17]. However, TCNs' fixed receptive fields limit adaptability, and unlike transformer-based architectures, they lack dynamic attention mechanisms that enable modeling across arbitrary time steps—an ability that has driven transformers to state-of-the-art performance in many temporal prediction tasks.

Spatio–Temporal Methods. Passenger flow datasets exhibit both temporal and spatial dependencies. Early deep learning methods used Convolutional Neural Networks (CNNs) to capture spatial correlations by mapping stations or sensors onto grids, enabling convolutional filters to learn inter-station relationships [33]. However, metro networks are non-Euclidean: station spacing is irregular and connectivity varies. Graph Convolutional Networks (GCNs) better represent such data by treating stations as nodes and connections as edges, aggregating information from neighbors to model spatial dependencies [13].

To jointly model spatial and temporal patterns, GCNs have been combined with sequential architectures. The Spatio-Temporal Graph Convolutional Network (STGCN) integrates GCN layers with gated CNNs [32], the Temporal Graph Convolutional Network (T-GCN) combines GCN with GRUs [35], and the Diffusion Convolutional Recurrent Neural Network (DCRNN) introduces diffusion convolution to capture directional dependencies within a sequence-to-sequence framework [18]. The Graph WaveNet extends these approaches with adaptive adjacency matrices and dilated convolutions for scalable spatiotemporal modeling [30].

Standard graph convolutions remain limited to local neighborhoods, overlooking functional dependencies such as flow similarities between distant stations. Attention-based models address this by dynamically learning correlations across the network. The Attention-Based Spatio-Temporal Graph Convolutional Network (ASTGCN) applies spatial and temporal attention [8], while the Spatio–Temporal Graph Hierarchical Learning Framework (STGHLF) enhances local feature extraction [16]. Transformer-inspired architectures such as Adaptive Graph Spatial-Temporal Transformer Network (ASTTN) [7], Spatio-Temporal Multimodal Attention Network (ST-MAN) [21], and Spatio-Temporal Graph Neural Network (STGNN) [2] capture long-range dependencies and multi-scale temporal patterns without relying on fixed graph structures.

Beyond attention, multi-graph modeling provides another means of capturing global context by representing diverse spatial relationships, such as physical proximity, functional similarity, and travel demand. This approach enriches node representations and improves generalization [10]. Examples include the Multi-Graph Convolutional Recurrent Neural Network (MGC-RNN) [10], Hypergraph Convolution with Dynamic OD Flows [25], Bidirectional LSTM with OD Graphs [3], the Multi-Temporal Multi-Graph Neural Network (MTMGNN) [31], and the Dynamic Recurrent Spatio-Temporal Graph Convolutional Network (dyn-RSTGCN), which models multi-scale sequences using physical, similarity, and accessibility graphs derived from OD and travel time matrices [9].

A detailed comparison of representative models is provided in Appendix A.

Literature Gaps. The literature shows progress in passenger flow prediction but also highlights gaps motivating this work. Metro ridership exhibits strong periodicity, with pronounced morning and evening peaks. Existing architectures capture peaks well but often underperform during flat inter-peak periods [9], motivating the incorporation of decomposition-based forecasting to explicitly model these stable intervals. Additionally, multi-graph approaches frequently rely on OD-derived graphs, which must often be estimated, introducing uncertainty. We compare such OD-based multi-graph methods with attention-based mechanisms for encoding global spatial structure, illustrating the trade-off between using estimated relational information and learning dependencies directly from data.

3 Problem Formulation

Passenger flow forecasting is performed at the station level using historical travel data and contextual variables (e.g., weather, calendar). The metro system is modeled as a graph $G = (V, E)$, where V is the set of stations (nodes) and E the set of edges representing physical connections. Let i denote a station index, $N = |V|$ the total number of stations, and $1 \leq i \leq N$.

Passenger flow is observed at discrete 15-min intervals. For each station i and time t, inflow and outflow are defined as:

$$\mathbf{x}_{i,t} = [x_{i,t}^{\text{in}},\ x_{i,t}^{\text{out}}] \tag{1}$$

where $x_{i,t}^{\text{in}}$ and $x_{i,t}^{\text{out}}$ denote the number of passengers entering and exiting the station. The flows across all stations form:

$$\mathbf{X}_t = [\mathbf{x}_{1,t}, \ldots, \mathbf{x}_{N,t}] \in \mathbb{R}^{N \times 2}, \tag{2}$$

where each row corresponds to a station and the two columns represent inflow and outflow values, respectively, yielding a set of time series capturing network-wide passenger movements.

To enhance prediction, we extract period-, daily-, and weekly-level sequences. Let κ be the number of 15-min intervals per operational day. The period-level sequence uses recent intervals before $t+1$; the daily-level sequence draws from the same time across previous days; and the weekly-level sequence from the same time and weekday in preceding weeks:

$$\mathbf{X}_t^p = (\mathbf{X}_{t-T_p+1}, \ldots, \mathbf{X}_t), \tag{3}$$

$$\mathbf{X}_t^d = (\mathbf{X}_{t-T_d \cdot \kappa+1}, \ldots, \mathbf{X}_{t-\kappa+1}), \tag{4}$$

$$\mathbf{X}_t^w = (\mathbf{X}_{t-T_w \cdot 7 \cdot \kappa+1}, \ldots, \mathbf{X}_{t-7 \cdot \kappa+1}), \tag{5}$$

where T_p, T_d, T_w are the numbers of past periods, days, and weeks considered. Each sequence $\mathbf{X}_t^p, \mathbf{X}_t^d, \mathbf{X}_t^w$ is thus a three-dimensional tensor in $\mathbb{R}^{T \times N \times 2}$, formed by stacking the corresponding station-level matrices over time.

External factors are represented by:

$$\mathbf{X}_t^{\text{ext}} = [f_t^1, f_t^2, \ldots, f_t^8], \tag{6}$$

including minimum temperature, maximum temperature, time-of-day, day-of-week, peak-hour, holiday, weekend, and rain indicators.

The model predicts the inflow and outflow for each station at the next timestep as:

$$\mathbf{y}_{i,t+1} = [y^{\text{in}}_{i,t+1},\ y^{\text{out}}_{i,t+1}], \tag{7}$$

$$\mathbf{Y}_{t+1} = [\mathbf{y}_{1,t+1}, \ldots, \mathbf{y}_{N,t+1}] \in \mathbb{R}^{N\times 2}. \tag{8}$$

The overall forecasting objective is:

$$\mathbf{Y}_{t+1} = f(\mathbf{X}^p_t,\ \mathbf{X}^d_t,\ \mathbf{X}^w_t,\ \mathbf{X}^{\text{ext}}_t), \tag{9}$$

where $f(\cdot)$ denotes the mapping learned by the model to predict future passenger flows.

4 Method

4.1 GraphETS Architecture

We propose GraphETS, a spatio-temporal architecture that integrates graph-based spatial modeling with exponential-smoothing-aware temporal forecasting. The name reflects this dual design: Graph refers to the graph neural components capturing spatial dependencies, while ETS denotes the Exponential Trend and Seasonality principles embedded in the temporal module. The model combines localized feature extraction through Graph Convolutional Networks (GCNs) and global context modeling via the Dynamic Global Attention Network (DGAN) [2]. The resulting spatial representation is processed by ETSFormer [29], a Transformer architecture that explicitly decomposes temporal sequences into growth, level, and seasonal components. These components are then fused to generate intermediate representations, which are further aggregated across period-, daily-, and weekly-level sequences. Finally, these multi-resolution outputs are combined with external contextual features to produce the final passenger flow forecasts. The overall structure of the model is shown in Fig. 1.

Spatial Extraction Module. This module captures spatial dependencies within the metro network through two complementary components: stacked GCN layers for localized neighborhood aggregation and a DGAN module for global attention across distant stations. The GCN layers encode local structural information using normalized graph convolution, while DGAN enables long-range interactions beyond adjacency constraints, mitigating oversmoothing and improving expressiveness. Their outputs are linearly combined through a learnable fusion weight to balance local and global representations. Full mathematical formulations for the DGAN attention mechanism and fusion operation are provided in Appendix B.

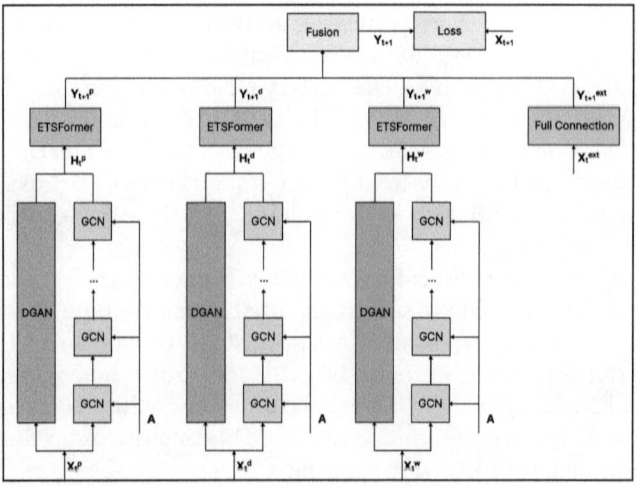

Fig. 1. Overview of the GraphETS architecture.

ETSFormer. Temporal dependencies are modeled using ETSFormer, a Transformer-based architecture tailored for time-series forecasting with exponential smoothing as an inductive bias. In the context of metro passenger flow, this design is used to better model low-variability periods, which are crucial for accurately estimating baseline demand and ensuring stable predictions during non-peak hours. ETSFormer replaces standard self-attention with two specialized operators: Exponential Smoothing Attention (ESA), which emphasizes recent observations to capture short-term trends, and Frequency Attention (FA), which identifies dominant periodic patterns through Fourier-based filtering. By explicitly decomposing temporal sequences into level, growth, and seasonal components, ETSFormer enables interpretable modeling of demand dynamics that exhibit strong daily and weekly periodicities. The complete formulations of ESA and FA, along with the reconstruction process, are provided in Appendix B.

Fusion of Outputs. GraphETS produces separate predictions for period-level, daily-level, and weekly-level sequences, as well as an additional prediction based on external contextual features such as weather and calendar indicators. The final passenger flow forecast is obtained through a learnable weighted fusion of these components, ensuring that both long- and short-term temporal patterns are captured while incorporating relevant external context. Detailed fusion equations are provided in Appendix B.

4.2 OD Matrix Estimation

In multi-graph approaches for passenger flow forecasting, multiple graph structures are typically constructed to capture different types of spatial dependencies between stations. Some of these graphs are derived from origin–destination (OD) relationships, as OD matrices naturally reflect the strength of passenger interactions across the network. The OD matrix records the number of trips made between each pair of stations, under the assumption that higher OD demand indicates stronger correlations in passenger flow. Although OD matrices can sometimes be obtained directly from tap-in/tap-out smart card data, it is often the case that such information is unavailable and must instead be estimated from observed travel patterns.

In this work, we reconstruct the OD matrix from observed station-level passenger flows to form the OD-based graphs used in multi-graph baseline models, enabling a fair and consistent comparison against our proposed method. The estimation is performed using a gravity-based model [28] combined with Iterative Proportional Fitting (IPF) [5] to ensure consistency with observed inflow and outflow totals. A detailed description of the OD estimation procedure, including model assumptions, is provided in Appendix C.

5 Results

SHMetro Dataset. We conducted our experiments using the ridership prediction subset of the SHMetro dataset [19]. This dataset includes metro transaction records collected from 288 stations in Shanghai between July 1st and September 30th, 2016. The data is aggregated in 15-min intervals from 05:30 to 23:30 each day, resulting in 73 time steps per day. Rush hours in Shanghai typically occur between 07:00 and 09:30 in the morning and between 17:00 and 19:30 in the evening. Since OD matrices are not directly provided in the dataset, they were estimated using the aforementioned methods. Weather data was obtained using the Python `open-meteo` API, while calendar-related features were extracted using standard Python libraries such as `datetime`, `holidays`, and `calendar`. For training and evaluation, the last 8 days of data are reserved as the test set, the preceding 8 days as the validation set, and the remaining data as the training set.

Performance Comparison. To benchmark the performance of our proposed architectures, we compare them against a diverse set of baseline models, including statistical, machine learning, deep learning, and spatio-temporal approaches. The statistical baseline is ARIMA. The machine learning baselines include RF and XGBoost. The deep learning baselines comprise LSTM, TCN, and GCN. Among the spatio-temporal models proposed in prior literature, we evaluate DCRNN, STGCN, and dyn-RSTGCN, the latter serving as a multi-graph baseline. To assess the contribution of specific components, we further introduce two model variants: dyn-RSTGCN-DGAN, which replaces multi-graph convolution with the DGAN attention mechanism and stacked GCN layers to avoid dependence on noisy, estimated OD-based graphs; and GraphETS-noDGAN, which uses dyn-RSTGCN's spatial convolution in place of our attention-based spatial extraction module to isolate the effect of the ETSFormer temporal backbone. For all baselines and variants, hyperparameters were selected through model-specific hyperparameter search to ensure a fair comparison. Training details, variability analysis and optimization curves are present in Appendices D, E and F respectively.

Table 1. Mean RMSE, MAE, and MAPE for baseline models and the proposed GraphETS

Model	RMSE	MAE	MAPE
ARIMA	199.03	121.67	205.15%
Random Forest	125.07	75.2	103.1%
XGBoost	119.99	72.94	90.91%
LSTM	115.45	69.8	73.55%
TCN	121.01	72.3	88.39%
GCN	74.52	44.01	47.98%
DCRNN	46.17	23.67	61.78%
STGCN	45.12	28.68	34.72%
dyn-RSTGCN	53.01	34.63	72.09%
dyn-RSTGCN-DGAN	50.62	31.97	66.92%
GraphETS-noDGAN	**36.27**	22.61	30.27%
GraphETS	38.12	**21.27**	**24.6%**

The results in Table 1 confirm the importance of incorporating temporal and spatial dependencies in passenger flow prediction. As expected, the statistical baseline (ARIMA) performs worst due to its linear formulation and inability to capture nonlinear passenger demand. Machine learning models (Random Forest, XGBoost) improve accuracy via nonlinear regression but lack explicit temporal modeling. Time-series deep learning models (LSTM, TCN) further enhance performance, highlighting the value of temporal sequence modeling. LSTM slightly outperforms TCN, likely due to its ability to better capture long-term dependencies. Adding spatial information yields notable gains. The GCN baseline reduces error compared to purely temporal models, emphasizing the importance of metro network structure. Spatiotemporal architectures further improve results: DCRNN and STGCN outperform dyn-RSTGCN, likely because dyn-RSTGCN's multi-graph representation requires accurate OD demand, which SHMetro dataset lacks.

Our proposed GraphETS model brings notable improvements, achieving the best MAE and MAPE results among all compared approaches. dyn-RSTGCN-DGAN improves upon dyn-RSTGCN, demonstrating that attention-based spatial modeling enhances both accuracy and robustness compared to multi-graph convolutions on estimated graphs. GraphETS-noDGAN yields further gains, showing that ETSFormer captures long-term trends and periodic patterns more effectively than the TCN backbone of dyn-RSTGCN and attains the lowest RMSE of all evaluated models. Overall, both GraphETS-noDGAN and GraphETS show significant improvements over other state-of-the-art methods, with GraphETS performing slightly better on two evaluation metrics. This confirms that the joint incorporation of temporal decomposition and attention-based spatial modeling delivers the highest predictive accuracy for metro passenger flow forecasting.

Visualization of the Results for One Operational Day. Figues 2 and 3 illustrate model performance at stations 26 and 52 in the dataset, respectively. GraphETS and dyn-RSTGCN track the time-series dynamics better than DCRNN and STGCN across both peak and off-peak intervals, despite dyn-RSTGCN having larger mean errors. DCRNN shows erratic behavior and fails to capture peaks, while STGCN tends to over-smooth and underfit. Notably, dyn-RSTGCN manages to capture sharp peaks well, particularly during morning rush hours, which can be attributed to the localized receptive field of its TCN backbone. In contrast, GraphETS demonstrates superior stability during low-variance periods, due to its exponential smoothing prior. dyn-RSTGCN also produces smooth predictions, which can be attributed to the TCN's convolutional structure that filters out high-frequency noise.

Some failure cases are observed for GraphETS: at station 26, it slightly overshoots the actual evening peak value, while at station 52, it undershoots it. In contrast, dyn-RSTGCN underestimates a sharp evening peak at station 26 and completely misses the evening peak at station 52, instead hallucinating a peak afterward due to fluctuations. Despite these deviations, GraphETS follows the underlying dynamics more closely, suggesting improved adaptability in certain high-variance scenarios.

Station Performance Analysis. We further analyze which stations exhibit the highest prediction errors across models, and how these errors relate to graph structural metrics. Representative examples of worst-performing stations are reported in Table 2, while the complete results for all models are provided in Appendix G.

Table 2. Worst-performing stations by MAE across models. Full results are provided in Appendix G. Extreme values for graph structural metrics are provided in Appendix H.

Model	Station	MAE	Closeness	Betweenness	Degree	Clustering	Eigenvector
DCRNN	12	228.11	0.1061	0.1371	14	0.0000	0.2757
STGCN	104	91.48	0.0404	0.0070	6	0.0000	0.0000
dyn-RSTGCN	12	141.56	0.1061	0.1371	14	0.0000	0.2757
GraphETS	12	153.37	0.1061	0.1371	14	0.0000	0.2757

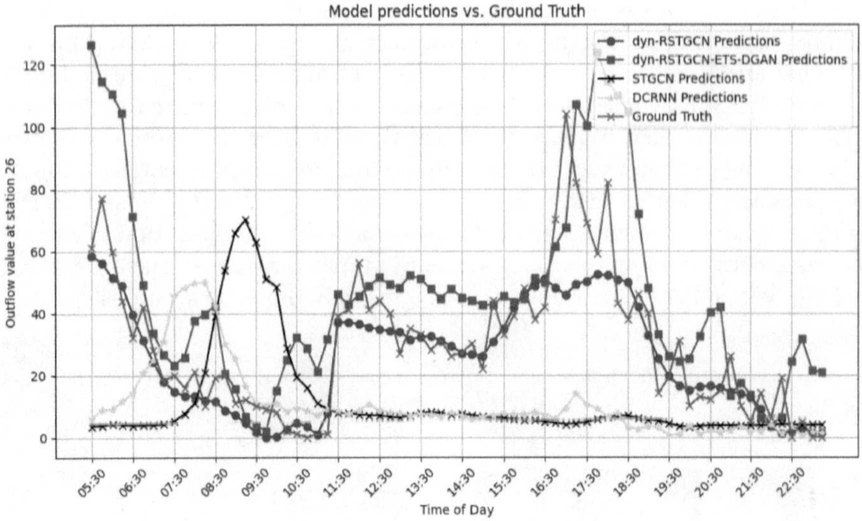

Fig. 2. Comparison of outflow predictions at Station 26 across different models. For consistency with the final terminology adopted in this paper, the method labeled dyn-RSTGCN-ETS-DGAN in the figure refers to the proposed GraphETS model.

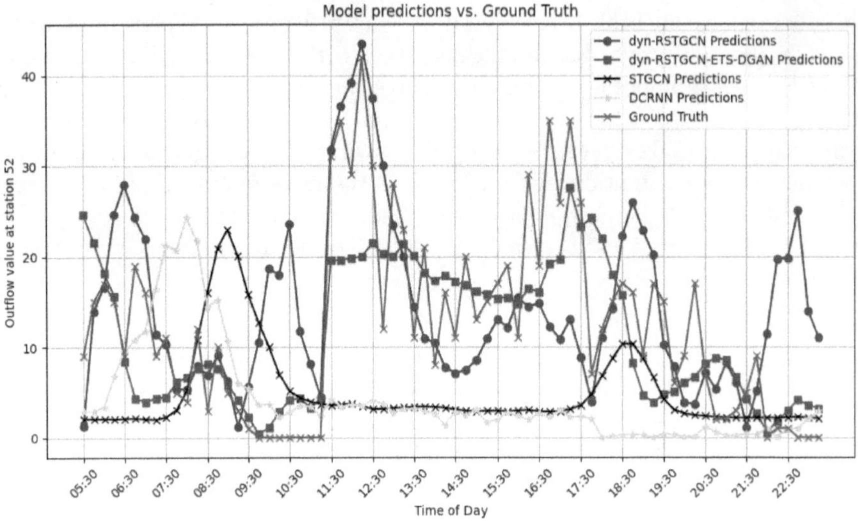

Fig. 3. Comparison of outflow predictions at Station 52 across different models. For consistency with the final terminology adopted in this paper, the method labeled dyn-RSTGCN-ETS-DGAN in the figure refers to the proposed GraphETS model.

For GraphETS, dyn-RSTGCN, and DCRNN, the worst-performing stations tend to exhibit high Degree and high Closeness Centrality, with Station 12 consistently appearing as the highest-error case for all three models. This indicates that errors concentrate in structurally central and highly connected nodes. In contrast, STGCN fails most severely on stations characterized by low Clustering Coefficient, low Eigenvector Centrality, and low Betweenness Centrality, suggesting difficulties on structurally peripheral or weakly connected nodes. These results imply that models such as GraphETS, dyn-RSTGCN, and DCRNN are challenged by globally central stations where error propagation is stronger, whereas architectures emphasizing localized graph convolution, such as STGCN, are more vulnerable when the surrounding neighborhood structure is sparse.

Table 3. Epoch time comparison across different model variants

Model	Epoch Time (s)
dyn-RSTGCN	65.24
GraphETS-noDGAN	62.62
dyn-RSTGCN-DGAN	50.44
GraphETS	47.92

Computational Efficiency Analysis. In terms of computational efficiency, the epoch times in Table 3 clearly show that utilizing ETSFormer instead of the TCN module proves to be efficient, due to the optimized implementation of ETSFormer submodules, while also delivering better results. dyn-RSTGCN-DGAN variant also achieves lower epoch time. This efficiency stems from its architecture: instead of performing graph convolutions over five different graphs, it applies a single attention mechanism over the node features, and also avoids the LSTM-based dynamic weight calculations required at each timestep in dyn-RSTGCN. This simplification results in a notable speedup while providing better performance.

Table 4. Comparison of mean RMSE, MAE, and MAPE for dyn-RSTGCN and GraphETS (short vs. extended historical window). Best results are in bold.

Model	RMSE	MAE	MAPE
dyn-RSTGCN (short)	59.37	31.57	35.94%
dyn-RSTGCN (extended)	53.01	34.63	72.09%
GraphETS (short)	74.37	30.89	35.56%
GraphETS (extended)	**38.12**	**21.27**	**24.6%**

Impact of the Historical Window on Performance. The comparative results in Table 4 indicate that the performance of GraphETS and dyn-RSTGCN is similar in the short setting ($T_p = 8$), where eight immediate previous 15-min intervals are used as period-level input. In this case, both models capture short-term fluctuations effectively, but the limited temporal context restricts the exploitation of richer temporal dependencies. When the temporal horizon is extended to the long setting ($T_p = 24$), however, the difference becomes evident. The errors of the TCN-based baseline architecture remain at a comparable level, while the ETSFormer-based GraphETS exhibits a substantial reduction in RMSE, MAE, and MAPE under the longer 24-period window.

This behavior can be attributed to the architectural design of ETSFormer. The model is explicitly equipped to disentangle local variations from long-term temporal structures. As the temporal window grows, this mechanism allows it to exploit periodicity and trend information more effectively than convolutional filters. The results therefore demonstrate that ETSFormer leverages extended temporal horizons to improve generalization, while the TCN-based approach remains constrained by its primarily local receptive field.

6 Conclusion

This study introduced GraphETS, a novel spatiotemporal architecture for metro passenger flow forecasting that integrates attention-based spatial modeling with exponential-smoothing–aware temporal decomposition. Through comprehensive experiments on the SHMetro dataset, GraphETS demonstrated superior predictive accuracy and stability compared to traditional statistical, machine learning, and advanced deep learning baselines. The integration of attention-driven spatial learning with an explicit temporal decomposition guided by exponential smoothing proved effective for capturing demand dynamics under varying conditions. Sensitivity analysis revealed that GraphETS benefits from longer historical windows. Structural challenges remain across different approaches: GraphETS shows reduced effectiveness with central nodes, while models based solely on graph convolution struggle with peripheral ones. The model also achieves reduced training time compared to multi-graph methods, making it scalable and practical for large transit systems where computational efficiency is crucial for real-time decision-making. These findings underscore the importance of combining spatial attention mechanisms with smoothing-based temporal modeling for resilient, context-aware urban rail demand forecasting.

There are many promising paths for future research, such as introducing new methods, validating results through real-world data, and improving model robustness and adaptability to other transit systems. One important focus is to create temporal models that merge the advantages of ETSFormer and TCN, enabling precise predictions during both stable trends and sudden surges in passenger flow while keeping errors minimal. Another worthwhile approach is to evaluate these methods on realistic datasets with reliable OD matrices, assessing their resilience using both traditional and modified techniques. Furthermore, finding better ways to integrate graph topologies into the learning process may boost performance at key hub stations, which play a crucial role in network stability and managing flow.

Disclosure of Interests. The authors declare no competing interests relevant to the content of this work.

A Comparison of Representative Spatio-Temporal Models

Table 5 summarizes the main characteristics of representative spatio-temporal graph neural network models commonly applied to passenger flow prediction. The comparison covers architectural features, temporal modeling mechanisms, spatial dependency representation, and key advantages or limitations relevant to metro flow forecasting tasks.

B Detailed Formulations of the GraphETS Architecture

Spatial Extraction Module. The spatial extraction module integrates localized and global spatial modeling through stacked GCN layers and the Dynamic Global Attention Network (DGAN). The graph convolution operation is expressed as:

$$\mathbf{H}_t^{(l+1)} = \sigma\left(\hat{D}^{-\frac{1}{2}}\hat{A}\hat{D}^{-\frac{1}{2}}\mathbf{H}_t^{(l)}W^{(l)}\right), \tag{10}$$

where l denotes the index of the current GCN layer. The adjacency matrix with self-loops is defined as $\hat{A} = A + I$, and the diagonal degree matrix is given by $\hat{D}_{i,i'} = \sum_{i'} \hat{A}_{i,i'}$. The matrix $W^{(l)}$ contains the trainable weights of the l-th GCN layer, and σ denotes the activation function. The input to the l-th layer is $\mathbf{H}_t^{(l)}$, with $\mathbf{H}_t^{(0)} = \mathbf{X}_t$ for the first layer, and the final output of the stacked GCN layers is denoted as $\mathbf{H}_{\text{gcn},t} = \mathbf{H}_t^{(l+1)}$. The adjacency matrix $A \in \mathbb{R}^{N\times N}$, each hidden representation $\mathbf{H}_t^{(l)} \in \mathbb{R}^{N\times F_l}$, and each layer weight matrix $W^{(l)} \in \mathbb{R}^{F_l\times F_{l+1}}$, where N is the number of stations and F_l denotes the feature dimension at layer l. Graph convolution is performed for all three levels of the time series.

To capture the global spatial context of the metro network, we conduct experiments with a DGAN module added to the model architecture. Attention between the node features is calculated at every timestep. For the period-level time series, the attention score matrix is $\mathbf{S} \in \mathbb{R}^{T_p\times N\times N}$, where T_p is the number of previous immediate time intervals and N is the number of stations. Dynamic global attention is computed as follows:

$$\mathbf{S} = V_s \cdot \sigma\left((\mathbf{X}_t^{\text{p}}W_1) \cdot W_2 \cdot (W_3\mathbf{X}_t^{\text{p}})^\top + \mathbf{b}_s\right) \tag{11}$$

$$\text{DGAN}(\mathbf{X}_t^{\text{p}}) = \text{softmax}(\mathbf{S}) \cdot \mathbf{X}_t^{\text{p}} + \mathbf{X}_t^{\text{p}} \tag{12}$$

Table 5. Comparison of representative spatiotemporal prediction models

Model	Spatial modelling	Temporal modelling	Global context	Multi-level TS	External features
Traditional (ARIMA, LSTM, TCN)	–	ARIMA, LSTM, TCN	–	–	–
DCRNN (2018) [18]	Diffusion Convolution	GRU	–	–	–
STGCN (2018) [32]	Graph Convolution	1D Convolution	–	–	–
T-GCN (2019) [35]	Graph Convolution	GRU	–	–	–
ASTGCN (2019) [8]	Graph Convolution + Attention	1D Convolution + Attention	Attention	Period/daily/weekly	–
MGC-RNN (2021) [10]	Multi-Graph Convolution	LSTM	Multi-Graph	–	Time of day, day of week, holiday, weather
dyn-RSTGCN (2023) [9]	Dynamic Multi-Graph Convolution	TCN	Multi-Graph	Period/daily/weekly	Time of day, day of week, holiday, weather
STGNN (2024) [2]	Graph Convolution + Attention	Transformer	Attention	Period/daily	Time of day, day of week, holiday, weather
GraphETS (ours)	Graph Convolution + Attention	ETSFormer	Attention	Period/daily/weekly	Time of day, day of week, holiday, weather

The same attention mechanism is applied independently to the daily-level and weekly-level time series inputs. The learnable parameters include $V_s, b_s \in \mathbb{R}^{N\times N}$, $W_1, W_2 \in \mathbb{R}^{2\times 2}$, and $W_3 \in \mathbb{R}^{N\times N}$. The activation function σ introduces non-linearity.

The final output of the spatial extraction module is a fused representation that combines the localized features from the stacked GCN layers and the global context from the DGAN module. This fusion is performed using the following expression:

$$\mathbf{H}_t^p = (1-\alpha)\cdot \mathrm{DGAN}(\mathbf{X}_t^{\mathrm{p}}) + \alpha \cdot \mathbf{H}_{\mathrm{gcn},t}^p \tag{13}$$

where α is a learnable parameter that controls the balance between the two components. The resulting fused representation $\mathbf{H}_t \in \mathbb{R}^{T_p\times N\times F_s}$, where F_s denotes the spatial feature dimension, is passed into the temporal modeling module, and the same process is done for all three time-series levels.

ETSFormer. ETSFormer is composed of multiple encoder and decoder layers. Within each encoder, it replaces standard self-attention with two exponential-smoothing-based operators tailored for time-series forecasting. The Exponential Smoothing Attention (ESA) and Frequency Attention (FA) modules are applied in the encoder to decompose temporal sequences into level, growth, and seasonal components. The decoder then recombines these components to generate the final forecast.

The ESA operator captures the evolution of local trends through exponential smoothing:

$$A_{\mathrm{ES}}(V)_t = \alpha V_t + (1-\alpha)A_{\mathrm{ES}}(V)_{t-1} = \sum_{j=0}^{t-1}\alpha(1-\alpha)^j V_{t-j} + (1-\alpha)^t v_0, \tag{14}$$

where V_t denotes the input value embedding at time t, α is a learnable smoothing coefficient controlling the rate of decay, and v_0 is a learnable initial state representing the baseline level of the series. This formulation assigns higher weight to recent inputs while exponentially decaying the influence of distant lags, allowing the model to capture smooth growth dynamics in metro demand.

The FA operator reconstructs the seasonal component from dominant frequencies identified in the spectral domain:

$$S_{j,i}^{(n)} = \sum_{k=1}^{K} A_{\kappa_i^{(k)},i} \left[\cos\left(2\pi f_{\kappa_i^{(k)}} j + \Phi_{\kappa_i^{(k)},i}\right) + \cos\left(2\pi \bar{f}_{\kappa_i^{(k)}} j + \bar{\Phi}_{\kappa_i^{(k)},i}\right)\right], \quad (15)$$

where $A_{\kappa_i^{(k)},i}$ and $\Phi_{\kappa_i^{(k)},i}$ denote the amplitude and phase of the k-th dominant frequency $f_{\kappa_i^{(k)}}$, while $\bar{f}_{\kappa_i^{(k)}}$ and $\bar{\Phi}_{\kappa_i^{(k)},i}$ represent their conjugate counterparts. The index n corresponds to the encoder layer, and K specifies the number of retained frequency components. FA thus models recurrent periodic patterns in the passenger flow signal.

These mechanisms jointly decompose temporal sequences into level, growth, and seasonal components. The decoder recombines them to produce the final forecast:

$$\mathbf{Y}_{t+1}^{*} = E_{t+1} + \text{Linear}\left(\sum_{n=1}^{N} [\text{TD}(B_t^{(n)}) + S_{t+1}^{(n)}]\right), \quad (16)$$

where E_{t+1} denotes the level component, $B_t^{(n)}$ the growth (trend) component, and $S_{t+1}^{(n)}$ the seasonal component reconstructed from FA. The superscript $*$ indicates that this formulation applies to any temporal resolution—period-level (p), daily-level (d), or weekly-level (w) sequences.

Fusion of Outputs. Predictions from the period-, daily-, and weekly-level ETSFormer modules and external features are fused as:

$$\mathbf{Y}_{t+1} = \gamma_p \odot \mathbf{Y}_{t+1}^{p} + \gamma_d \odot \mathbf{Y}_{t+1}^{d} + \gamma_w \odot \mathbf{Y}_{t+1}^{w} + \gamma_{\text{ext}} \odot \mathbf{Y}_{t+1}^{\text{ext}}, \quad (17)$$

where γ_p, γ_d, γ_w, and γ_{ext} are learnable weights that control each component's contribution to the final passenger flow forecast. The external features term is obtained via a linear head:

$$\mathbf{Y}_{t+1}^{\text{ext}} = \text{Linear}(\mathbf{X}_t^{\text{ext}}), \quad (18)$$

with $\mathbf{X}_t^{\text{ext}} \in \mathbb{R}^{N\times 8}$ denoting station-wise external variables at time t (eight variables presented in Sect. 3 including weather and calendar indicators, etc.).

C OD Matrix Estimation

This appendix provides the detailed formulation of the OD matrix reconstruction procedure summarized in Sect. 4.2. The estimation follows a two-stage process: (1) initialization using a gravity model and (2) refinement with Iterative Proportional Fitting (IPF) to ensure consistency with observed inflow and outflow totals.

Gravity Model. Initial OD estimates are obtained using the gravity model, which captures the intuition that the demand between two stations increases with their respective passenger activity and decreases with travel time. The initial estimate is defined as:

$$\mathrm{OD}_{i,i'}^{(0)} = \frac{\overline{\mathbf{x}}_{i,t} \cdot \overline{\mathbf{x}}_{i',t}}{d_{i,i'}^{\beta}}, \tag{19}$$

where $d_{i,i'}$ denotes the travel time between stations i and i', β is a decay parameter controlling the influence of distance, and $\overline{\mathbf{x}}_{i,t}$ and $\overline{\mathbf{x}}_{i',t}$ represent mean flows at the corresponding stations over a selected historical window. The resulting matrix provides a proportional estimate of trip intensity between all station pairs.

Iterative Proportional Fitting (IPF). To ensure that the estimated OD matrix is consistent with observed marginal totals, the gravity-based estimate is refined using Iterative Proportional Fitting (IPF). IPF iteratively rescales the rows and columns of the OD matrix so that their sums match the total observed outflow and inflow for each station, while maintaining the structural pattern derived from the gravity model.

The iterative updates are defined as:

$$\mathrm{OD}_{i,i'}^{(k+1)} = \mathrm{OD}_{i,i'}^{(k)} \cdot \frac{x_i^{\mathrm{out}}}{\sum_j \mathrm{OD}_{i,j}^{(k)}} \quad \text{(row adjustment)}, \tag{20}$$

$$\mathrm{OD}_{i,i'}^{(k+2)} = \mathrm{OD}_{i,i'}^{(k+1)} \cdot \frac{x_{i'}^{\mathrm{in}}}{\sum_j \mathrm{OD}_{j,i'}^{(k+1)}} \quad \text{(column adjustment)}. \tag{21}$$

where x_i^{out} and $x_{i'}^{\mathrm{in}}$ denote the observed total outflow from station i and total inflow to station i', respectively. These two scaling steps are alternated until convergence, producing a final matrix OD^* that satisfies the marginal constraints:

$$\sum_{i'} \mathrm{OD}_{i,i'}^{*} = x_i^{\mathrm{out}}, \tag{22}$$

$$\sum_{i} \mathrm{OD}_{i,i'}^{*} = x_{i'}^{\mathrm{in}}. \tag{23}$$

Travel Time Matrix. The travel time matrix $\mathbf{D}$ is computed by assuming an average travel time between adjacent stations and then calculating the shortest travel times between all station pairs using the Floyd–Warshall algorithm [26]:

$$d_{i,j} = \min(d_{i,j},\ d_{i,k} + d_{k,j}). \tag{24}$$

The resulting matrix $\mathbf{D}$ provides the travel-time inputs required for the gravity model and ensures spatial consistency across the OD estimation process.

D Training Setup

Hardware Specifications. All experiments were performed on a workstation equipped with 2 Intel Xeon 6242 CPUs running at 2.8 GHz with 16 cores each, 256 GB of RAM, and 2 NVIDIA RTX A6000 GPUs with 48 GB of VRAM each. The GPUs are connected via NvLink, and each card contains 10,752 CUDA cores, totaling 21,504 CUDA cores. The model proposed in this work was implemented using the PyTorch deep learning framework.

Hyperparameters. To capture multi-scale temporal dependencies, three levels of temporal granularity are defined: the period-level with $T_p = 24$ intervals, the daily-level with $T_d = 6$ days, and the weekly-level with $T_w = 3$ weeks. These values are selected as they provide sufficient historical context to capture short-term fluctuations driven by consecutive periods within a day, medium-term trends shaped by daily commuting patterns, and longer-term periodicities arising from weekly travel regularities, while avoiding excessive computational overhead. The spatial extraction module consists of two graph convolution layers, each with 8 hidden units. ETSFormer is configured with two encoder layers, eight attention heads, and the frequency attention module selects the four most dominant frequencies. External dependency features are incorporated using a hidden dimension of size 8. These parameters serve as the default configuration in this work and are adopted generally unless specified otherwise.

In experiments utilizing the multi-graph approach (dyn-RSTGCN baseline), the similarity graph is constructed by considering $l^{\text{pea}} = 6$ previous periods with a similarity threshold of $\zeta = 0.6$. Given the temporal resolution of 15 min per interval, the accessibility graph parameters are defined as $\varepsilon_1 = 15$, $\varepsilon_2 = 30$, and $\varepsilon_3 = 45$ minutes. For the Shanghai metro network, the mean travel time between two adjacent stations is assumed to be 2.5 min. In the gravity model used for estimating OD matrices, the parameter β is set to 0.5. The TCN module comprises two layers, each configured with 100 channels and a convolutional kernel size of 2.

All models were trained for 100 epochs with early stopping applied using a patience of 10 epochs. Optimization was performed using the Adam optimizer, with batch size and learning rate selected through a hyperparameter search for each model to identify the best configuration. Min-max normalization was utilized for data preprocessing. All results are averaged over five runs with seeds {0, 42, 1234, 2103002, 1802001}

Evaluation Metrics. To evaluate the performance of the utilized models, this work adopts three standard regression metrics: Mean Absolute Error (MAE), Mean Absolute Percentage Error (MAPE), and Root Mean Squared Error (RMSE).

$$\text{MAE} = \frac{1}{n}\sum_{i=1}^{n}|\mathbf{X}_i - \mathbf{Y}_i| \tag{25}$$

$$\text{MAPE} = \frac{100\%}{n}\sum_{i=1}^{n}\frac{|\mathbf{X}_i - \mathbf{Y}_i|}{\mathbf{X}_i} \tag{26}$$

$$\text{RMSE} = \sqrt{\frac{1}{n}\sum_{i=1}^{n}(\mathbf{X}_i - \mathbf{Y}_i)^2} \tag{27}$$

Together, these metrics provide a robust and comprehensive evaluation framework. MAE offers a direct measure of average error, MAPE facilitates interpretability across scales, and RMSE highlights the presence of large outliers in prediction performance.

E Variability Analysis

The variability analysis in Table 6 reflects each model's sensitivity to random initialization and training dynamics. Among the baselines, STGCN and DCRNN show higher variance than purely temporal models, indicating increased sensitivity when spatial and temporal dependencies are learned jointly. For dyn-RSTGCN, performance is less stable. The TCN module is particularly sensitive to initialization and optimization noise, a problem amplified by uncertainty in the multi-graph structure, resulting in the largest variance among all models despite competitive mean performance.

Utilizing ETSFormer instead of TCN in GraphETS-noDGAN substantially improves stability. Its stronger temporal inductive bias, through exponential smoothing and frequency-aware attention, makes trend and periodic components explicit, reducing optimization burden and yielding more consistent runs, though variance remains higher than in simpler baselines like DCRNN. Finally, adding DGAN in GraphETS improves mean performance but increases variability. Attention-based spatial modeling strengthens spatial representations but introduces additional sensitivity to initialization, leading to less consistent outcomes across seeds.

Table 6. Standard deviations of RMSE, MAE, and MAPE for baseline models and the proposed GraphETS

Model	RMSE std	MAE std	MAPE std
XGBoost	0.10	0.02	0.07%
Random Forest	0.51	0.20	1.76%
LSTM	0.03	0.60	2.71%
GCN	5.03	3.21	6.65%
TCN	0.52	0.56	2.35%
STGCN	3.95	3.37	2.96%
DCRNN	1.47	0.75	3.01%
dyn-RSTGCN	20.73	11.91	10.06%
dyn-RSTGCN-DGAN	8.49	8.87	10.94%
GraphETS-noDGAN	1.81	1.96	7.40%
GraphETS	5.58	5.89	11.79%

F Train and Validation Loss Comparison

When comparing the training dynamics graphs of GraphETS, dyn-RSTGCN, DCRNN and STGCN, we can observe in Fig. 4 that both GraphETS and dyn-RSTGCN undergo relatively smooth optimization during training. GraphETS takes longer to converge, which can be attributed to the attention mechanism replacing the multi-graph approach. On the other hand, DCRNN and STGCN exhibit significantly higher losses from the beginning of training, with a more turbulent optimization process, especially in the case of STGCN.

G Station-Level Error Analysis

This appendix presents complete station-level performance details for DCRNN, STGCN, dyn-RSTGCN, and GraphETS. Tables 7–10 list the five stations with the highest MAE for each model, along with their corresponding graph structural metrics.

Table 7. DCRNN (worst stations by MAE)

Station	MAE	Closeness	Betweenness	PageRank	Degree	Clustering	Eigenvector
12	228.11	0.1061	0.1371	0.0053	14	0.0000	0.2757
36	141.71	0.1060	0.1148	0.0038	10	0.0000	0.2062
7	162.31	0.0975	0.1688	0.0054	14	0.0667	0.1859
15	195.08	0.1036	0.2231	0.0043	10	0.0000	0.0819
39	114.51	0.0968	0.0979	0.0029	6	0.0000	0.0363

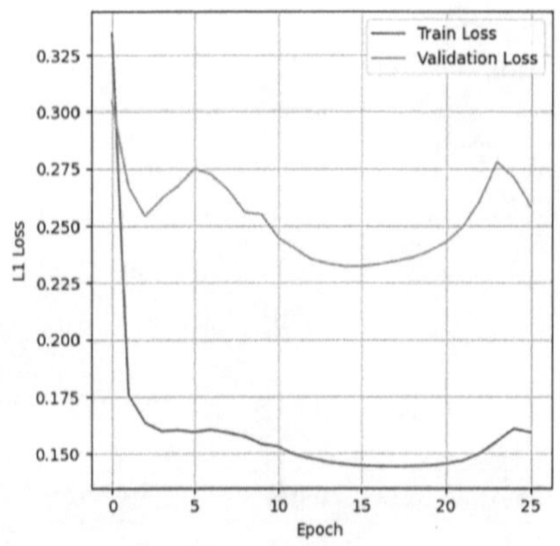

(a) DCRNN

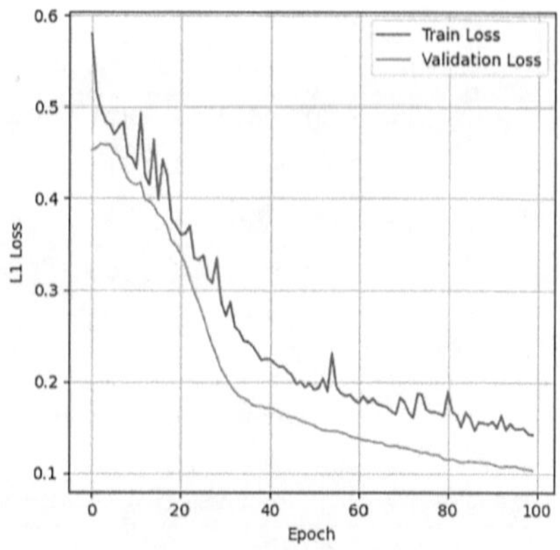

(b) STGCN

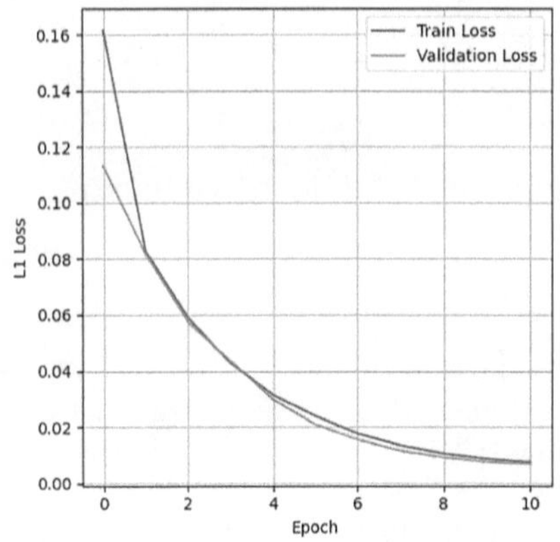

(c) dyn-RSTGCN

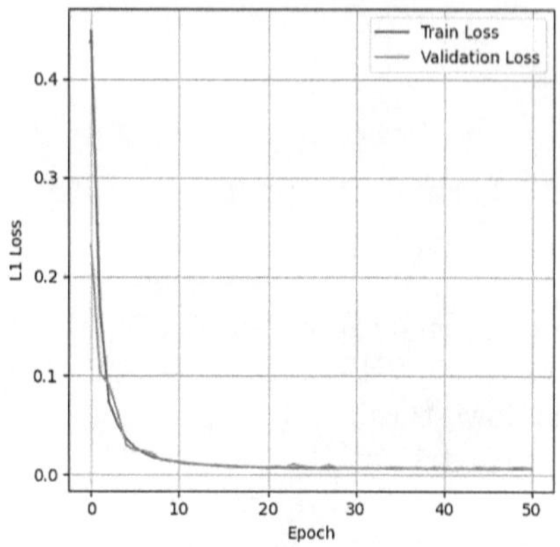

(d) GraphETS

Fig. 4. Comparison of training dynamics for DCRNN, STGCN, dyn-RSTGCN and GraphETS.

Table 8. STGCN (worst stations by MAE)

Station	MAE	Closeness	Betweenness	PageRank	Degree	Clustering	Eigenvector
104	91.48	0.0404	0.0070	0.0038	6	0.0000	0.0000
116	78.12	0.0803	0.0592	0.0033	6	0.0000	0.0008
105	76.40	0.0388	0.0000	0.0028	4	0.0000	0.0000
94	68.91	0.0942	0.0141	0.0042	10	0.1667	0.1093
110	65.23	0.0806	0.0080	0.0031	6	0.0000	0.0030

Table 9. dyn-RSTGCN (worst stations by MAE)

Station	MAE	Closeness	Betweenness	PageRank	Degree	Clustering	Eigenvector
12	141.56	0.1061	0.1371	0.0053	14	0.0000	0.2757
82	114.52	0.0978	0.1715	0.0045	10	0.0000	0.0186
29	105.57	0.0669	0.0353	0.0054	10	0.0000	0.0002
7	101.95	0.0975	0.1688	0.0054	14	0.0667	0.1859
28	101.86	0.0627	0.0000	0.0025	4	0.0000	0.0000

Table 10. GraphETS (worst stations by MAE)

Station	MAE	Closeness	Betweenness	PageRank	Degree	Clustering	Eigenvector
12	153.37	0.1061	0.1371	0.0053	14	0.0000	0.2757
7	95.72	0.0975	0.1688	0.0054	14	0.0667	0.1859
36	95.30	0.1060	0.1148	0.0038	10	0.0000	0.2062
38	93.37	0.1026	0.1230	0.0042	10	0.0000	0.1254
29	80.56	0.0669	0.0353	0.0054	10	0.0000	0.0002

H Graph Structural Characteristics of the SHMetro Dataset

This appendix provides additional descriptive information about the graph structure of the SHMetro dataset. Table 11 lists the stations exhibiting the highest and lowest values for each graph structural metric, including Degree, Closeness Centrality, Betweenness Centrality, PageRank, Clustering Coefficient, and Eigenvector Centrality. These values highlight the heterogeneity of the metro network, where a few stations act as major transfer hubs while the majority remain moderately connected.

Table 11. Extremes for each graph structural metric in the SHMetro dataset

Metric	Highest	Lowest
Degree	Station 41: 16.0000	Station 27: 4.0000
Closeness Centrality	Station 37: 0.1092	Station 228: 0.0347
Betweenness Centrality	Station 37: 0.2520	Station 27: 0.0000
PageRank	Station 41: 0.0074	Station 95: 0.0021
Clustering Coefficient	Station 6: 0.3333	Station 0: 0.0000
Eigenvector Centrality	Station 10: 0.3458	Station 228: 0.0000

The degree distribution in Fig. 5 confirms the presence of a small subset of highly connected transfer stations, which play a dominant role in the network's topology

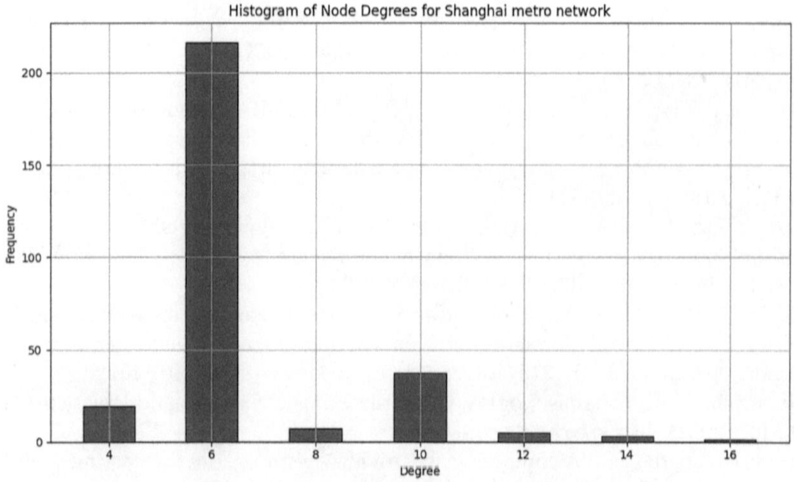

Fig. 5. Histogram of node degrees in the Shanghai Metro network. The distribution is right-skewed, with most stations exhibiting moderate connectivity (degree 6) and a small number of hub stations reaching degree 16.

and flow dynamics. Such hub concentration contributes to the higher error sensitivity observed in Sect. 5 for central nodes, as perturbations in these locations propagate more broadly through the network.

References

1. Bengio, Y., Simard, P., Frasconi, P.: Learning long-term dependencies with gradient descent is difficult **5**(2), 157–166. https://doi.org/10.1109/72.279181, https://ieeexplore.ieee.org/document/279181
2. Chang, Y., Zong, M., Dang, Y., Wang, K.: Multi-step passenger flow prediction for urban metro system based on spatial-temporal graph neural network **14**(18), 812. https://doi.org/10.3390/app14188121, https://www.mdpi.com/2076-3417/14/18/8121
3. Chen, P., Fu, X., Wang, X.: A graph convolutional stacked bidirectional unidirectional-LSTM neural network for metro ridership prediction **23**(7), 6950–696. https://doi.org/10.1109/TITS.2021.3065404, https://ieeexplore.ieee.org/document/9381554
4. Chung, J., Gulcehre, C., Cho, K., Bengio, Y.: Empirical evaluation of gated recurrent neural networks on sequence modeling. https://doi.org/10.48550/arXiv.1412.3555, http://arxiv.org/abs/1412.3555
5. Deming, W.E., Stephan, F.F.: On a least squares adjustment of a sampled frequency table when the expected marginal totals are known. Ann. Math. Stat. **11**(4), 427–444 (1940). https://doi.org/10.1214/aoms/1177731829
6. Ding, C., Wang, D., Ma, X., Li, H.: Predicting short-term subway ridership and prioritizing its influential factors using gradient boosting decision trees. **8**(11), 110. https://doi.org/10.3390/su8111100, https://www.mdpi.com/2071-1050/8/11/1100

7. Feng, A., Tassiulas, L.: Adaptive graph spatial-temporal transformer network for traffic flow forecastin. https://doi.org/10.48550/arXiv.2207.05064, http://arxiv.org/abs/2207.05064
8. Guo, S., Lin, Y., Feng, N., Song, C., Wan, H.: Attention based spatial-temporal graph convolutional networks for traffic flow forecasting. **33**(1), 922–92. https://doi.org/10.1609/aaai.v33i01.3301922, https://ojs.aaai.org/index.php/AAAI/article/view/3881
9. He, B., Zhu, Y., D'Ariano, A., Wen, K., Chen, L.: Dynamic relational graph convolutional network for metro passenger flow forecasting. Oper. Res. Forum **4** (2023). https://doi.org/10.1007/s43069-023-00266-9
10. He, Y., Li, L., Zhu, X., Tsui, K.L.: Multi-graph convolutional-recurrent neural network (MGC-RNN) for short-term forecasting of transit passenger flow. https://doi.org/10.48550/arXiv.2107.13226, http://arxiv.org/abs/2107.13226
11. Hyndman, R.J., Athanasopoulos, G.: Forecasting: Principles and Practice. OTexts, 3 edn. (2021). https://otexts.com/fpp3/
12. Kaur, H., Jagdev, G.: A comprehensive review on time series forecasting techniques. Int. J. Emerg. Technol. Innov. Res. **10**(5), 103–111 (2023). https://www.jetir.org/papers/JETIR2305G16.pdf
13. Kipf, T.N., Welling, M.: Semi-supervised classification with graph convolutional network. https://doi.org/10.48550/arXiv.1609.02907, http://arxiv.org/abs/1609.02907
14. Lea, C., Flynn, M.D., Vidal, R., Reiter, A., Hager, G.D.: Temporal convolutional networks for action segmentation and detection. CoRR **abs/1611.05267** (2016), http://arxiv.org/abs/1611.05267
15. Li, D., Du, S., Hou, Y.: Long-term passenger flow forecasting for rail transit based on complex networks and informer. Sensors **24**(21) (2024). https://doi.org/10.3390/s24216894, https://www.mdpi.com/1424-8220/24/21/6894
16. Li, H., Fu, W., Zhang, H., Liu, W., Sun, S., Zhang, T.: Spatio–temporal graph hierarchical learning framework for metro passenger flow prediction across stations and lines. **311**, 113132. https://doi.org/10.1016/j.knosys.2025.113132, https://www.sciencedirect.com/science/article/pii/S0950705125001790
17. Li, J.: Short-term urban rail passenger flow prediction using tcn-lstm model. Highlights Sci. Eng. Technol. **118**, 25–33 (2024). https://doi.org/10.54097/p83p5860, https://drpress.org/ojs/index.php/HSET/article/view/25988
18. Li, Y., Yu, R., Shahabi, C., Liu, Y.: Diffusion convolutional recurrent neural network: data-driven traffic forecastin https://doi.org/10.48550/arXiv.1707.01926, http://arxiv.org/abs/1707.01926
19. Liu, L., Chen, J., Wu, H., Zhen, J., Li, G., Lin, L.: Physical-virtual collaboration modeling for intra-and inter-station metro ridership prediction. IEEE Trans. Intell. Transp. Syst. (2020)
20. Liu, S., Wang, X.: An improved transformer based traffic flow prediction model **15**(1), 828. https://doi.org/10.1038/s41598-025-92425-7, https://www.nature.com/articles/s41598-025-92425-7
21. Liu, Y., Guo, B., Meng, J., Zhang, D., Yu, Z.: Spatio-temporal memory augmented multi-level attention network for traffic prediction **36**(6), 2643–265. https://doi.org/10.1109/TKDE.2023.3322405, https://ieeexplore.ieee.org/document/10285880
22. National Academies of Sciences, Engineering, and Medicine: Transit Capacity and Quality of Service Manual, 3rd Edn. The National Academies Press, Washington, DC (2013). https://doi.org/10.17226/24766, https://doi.org/10.17226/24766

23. Pelletier, M.P., Trépanier, M., Morency, C.: Smart card data use in public transit: A literature review **19**(4), 557–568.https://doi.org/10.1016/j.trc.2010.12.003, https://www.sciencedirect.com/science/article/pii/S0968090X1000166X
24. Shumway, R.H., Stoffer, D.S.: Time Series Analysis and Its Applications: With R Examples. Springer, 4th edn. (2017) https://doi.org/10.1007/978-3-319-52452-8, https://link.springer.com/book/10.1007/978-3-319-52452-8
25. Wang, J., Zhang, Y., Wei, Y., Hu, Y., Piao, X., Yin, B.: Metro passenger flow prediction via dynamic hypergraph convolution networks **22**(12), 7891–790https://doi.org/10.1109/TITS.2021.3072743, https://ieeexplore.ieee.org/document/9410439
26. Warshall, S.: A theorem on boolean matrices. J. ACM **9**(1), 11–12 (1962). https://doi.org/10.1145/321105.321107
27. Wei, L., Guo, D., Chen, Z., Yang, J., Feng, T.: Forecasting short-term passenger flow of subway stations based on the temporal pattern attention mechanism and the long short-term memory network **12**(1), 25.https://doi.org/10.3390/ijgi12010025, https://www.mdpi.com/2220-9964/12/1/25
28. Wilson, A.G.: Entropy in Urban and Regional Modelling. Pion, London (1971)
29. Woo, G., Liu, C., Sahoo, D., Kumar, A., Hoi, S.: ETSformer: Exponential smoothing transformers for time-series forecasting.https://doi.org/10.48550/arXiv.2202.01381, http://arxiv.org/abs/2202.01381
30. Wu, Z., Pan, S., Long, G., Jiang, J., Zhang, C.: Graph WaveNet for deep spatial-temporal graph modelinhttps://doi.org/10.48550/arXiv.1906.00121, http://arxiv.org/abs/1906.00121
31. Yin, D., Jiang, R., Deng, J., Li, Y., Xie, Y., Wang, Z., Zhou, Y., Song, X., Shang, J.: Mtmgnn: Multi-time multi-graph neural network for metro passenger flow prediction. GeoInformatica **27**, 1–29 (04 2022).https://doi.org/10.1007/s10707-022-00466-1
32. Yu, B., Yin, H., Zhu, Z.: Spatio-temporal graph convolutional networks: A deep learning framework for traffic forecasting. In: Proceedings of the Twenty-Seventh International Joint Conference on Artificial Intelligence. pp. 3634–364https://doi.org/10.24963/ijcai.2018/505, http://arxiv.org/abs/1709.04875
33. Yu, D., Liu, Y., Yu, X.: A data grouping CNN algorithm for short-term traffic flow forecasting. In: Li, F., Shim, K., Zheng, K., Liu, G. (eds.) Web Technologies and Applications. pp. 92–103. Springer International Publishinhttps://doi.org/10.1007/978-3-319-45814-4_8
34. Zhang, J., Zheng, Y., Qi, D.: Deep spatio-temporal residual networks for city-wide crowd flows prediction. CoRR **abs/1610.00081** (2016), http://arxiv.org/abs/1610.00081
35. Zhao, L., Song, Y., Zhang, C., Liu, Y., Wang, P., Lin, T., Deng, M., Li, H.: T-GCN: A temporal graph ConvolutionalNetwork for traffic prediction **21**(9), 3848–38https://doi.org/10.1109/TITS.2019.2935152, http://arxiv.org/abs/1811.05320

Integrating Paratransit and Microtransit: A Simulation Framework for Efficient and Inclusive Urban Mobility

Brady Danek[1], Ben Gansner[1], Chase Hulett[1], Luke Voelker[1], Arghadeep Mitra[1], Laya Anand[2], and Suchithra Rajendran[1](✉)

[1] Department of Industrial and Systems Engineering, University of Missouri, Columbia, MO 65211, USA
RajendranS@missouri.edu
[2] Rock Bridge High School, Columbia, MO 65203, USA

Abstract. Public transportation systems face increasing pressure to provide efficient, flexible, and inclusive mobility options for all users. This project developed a simulation model that integrates microtransit and paratransit services, which are typically operated independently. Microtransit provides on-demand, app-based routing for the general public, whereas paratransit offers pre-scheduled rides for individuals with disabilities. Managing these systems separately often results in underutilized paratransit fleets and longer wait times for microtransit users. Using Python, an extensive simulation was conducted to account for real-world factors such as traffic congestion, geographic distribution, and demand variability. A sensitivity analysis of fleet size and traffic conditions revealed that larger fleets reduce unfulfilled ride requests and passenger wait times, while higher congestion levels degrade overall system performance. Although these findings are derived from a simulation and may not capture all real-world complexities, they offer valuable insights for city planners and transit agencies seeking to design more efficient, equitable, and data-driven mobility systems.

1 Introduction

Public transportation systems face increasing pressure to provide flexible, efficient, and inclusive mobility options for a diverse population. Traditional models based on fixed routes and schedules often fall short for users with limited mobility and for residents in low density areas where fixed route service is not cost effective. These gaps create persistent challenges related to access, coverage, and reliability (Walker, 2012). Cities and small communities have begun to adopt demand responsive transportation approaches such as microtransit and paratransit to address these shortcomings. These services adjust to real time or pre scheduled requests, which allows agencies to support populations and locations that conventional fixed route networks cannot serve efficiently. This shift reflects a broader move toward adaptive mobility systems that can respond to varied user needs while maintaining operational efficiency (Liezenga et al., 2024).

Microtransit refers to dynamically routed shared rides that passengers request in real time through mobile applications or web portals, and it relies on algorithmic routing

H. R. Arabnia et al. (Eds.): CSCI 2025, CCIS 2974, pp. 42–54, 2027.
https://doi.org/10.1007/978-3-032-28475-4_3

that updates continuously as new requests arrive and as vehicle locations change (Federal Transit Administration, 2021). The system assigns passengers to vehicles based on proximity, destination compatibility, and available capacity, which allows each vehicle to follow a route that adapts to current demand rather than adhering to fixed stops and schedules. This structure is valuable in areas where conventional fixed route buses operate with low occupancy because it can pool riders with similar travel patterns and reduce empty vehicle miles. It improves operational efficiency in regions with dispersed or unpredictable travel demand and supports shorter wait times, more direct travel paths, and better use of the vehicle fleet. Transit agencies can use microtransit to extend service coverage in neighborhoods or rural locations where fixed route services are not viable due to cost or insufficient ridership (National Academies of Sciences, Engineering, and Medicine, 2016).

Paratransit, by contrast, serves individuals with disabilities who cannot use fixed route systems such as city buses or metro lines, and the service is arranged in advance to meet accessibility requirements and to comply with federal regulations under the Americans with Disabilities Act (ADA). The system provides essential mobility support for riders who require door to door assistance or specialized accommodations, yet it remains costly to operate because it depends on vehicles equipped with accessibility features, individualized trip planning that must align with rider specific needs, and service patterns characterized by low passenger density (Riggs and Pande, 2022). These factors increase labor and vehicle hours relative to the number of passengers served and lead to higher per trip operating costs, which creates financial pressure for transit agencies even though the service is central to equity and access obligations.

Managing microtransit and paratransit as separate services often produces operational inefficiencies because paratransit vehicles remain idle between scheduled trips and microtransit passengers experience longer wait times during peak demand. A coordinated framework that integrates both systems can reduce duplication of effort, improve fleet utilization, and raise overall service quality. Paratransit trips cost approximately four times more than conventional public transit rides, and these costs require substantial public subsidies to maintain affordability for riders. Additional evidence shows that nearly three quarters of paratransit passengers do not rely on wheelchairs, which indicates that many riders could be served through more flexible shared ride options that maintain accessibility standards (Miah et al., 2020).

The financial and operational pressure is significant because paratransit accounts for a small share of national transit activity, roughly 2% of total trips. In contrast, commercial ride share services such as Uber and Lyft provided billions of trips in 2023 and exceeded the combined annual ridership of all public transit modes in the United States. This disparity highlights the opportunity for transit agencies to modernize paratransit by adopting adaptive on demand service models that align with successful microtransit practices and address persistent cost and coverage challenges.

This project addresses this challenge through the development of a dynamic simulation model that integrates paratransit and microtransit within a shared operational system. The model is implemented in Python using discrete event simulation and incorporates real world conditions such as traffic congestion, geographic dispersion of riders, spatial clustering of requests, and temporal variability in demand. The framework assigns absolute priority to paratransit riders to ensure compliance with accessibility requirements and service quality mandates, and it reallocates vehicles to microtransit requests when capacity is available and when such assignments do not interfere with scheduled paratransit trips.

System performance is evaluated through quantitative indicators that capture user experience and fleet efficiency. These indicators include average passenger wait time, vehicle utilization, travel time reliability, and the rate of unfulfilled ride requests. Experiments examine multiple operating conditions such as variations in fleet size, traffic congestion, and spatial demand imbalance. The resulting analyses provide evidence on how an integrated service design can improve service quality for both rider groups while maintaining operational feasibility for transit agencies.

2 Methodology

The project was implemented in Google Colab using the Python programming language. Python offers readability and a clear structure for adjusting parameters such as simulation length, fleet size, or demand intensity, which supported systematic experimentation. The use of Google Colab enabled simultaneous contributions from team members, which ensured consistent version control and streamlined model refinement during development.

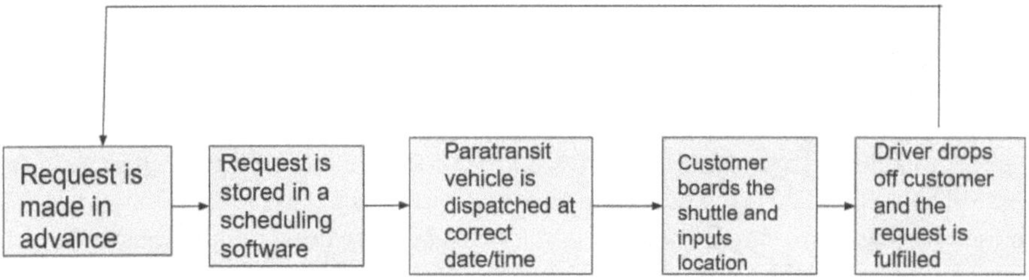

Fig. 1. Flow of the paratransit system

The simulation represents the core scheduling logic of paratransit operations under a set of controlled simplifications. Paratransit eligibility rules and operating procedures differ across jurisdictions, so a generalized workflow was developed to capture the essential elements of the service as illustrated in Fig. 1. In practice, paratransit trips are requested in advance and stored in scheduling software, vehicles are dispatched to the scheduled pickup locations, and riders are transported to their destinations within an established time window. Long horizon booking and detailed eligibility verification were excluded for tractability. Requests instead enter the system at the time of need, which allows the simulation to focus on real time allocation dynamics rather than long term planning constraints. Microtransit operates differently because users typically request short on demand trips that last thirty to forty minutes as shown in Fig. 2.

In the integrated simulation environment, the Python script initializes model parameters such as the number of vehicles, the size and connectivity of the road network, and

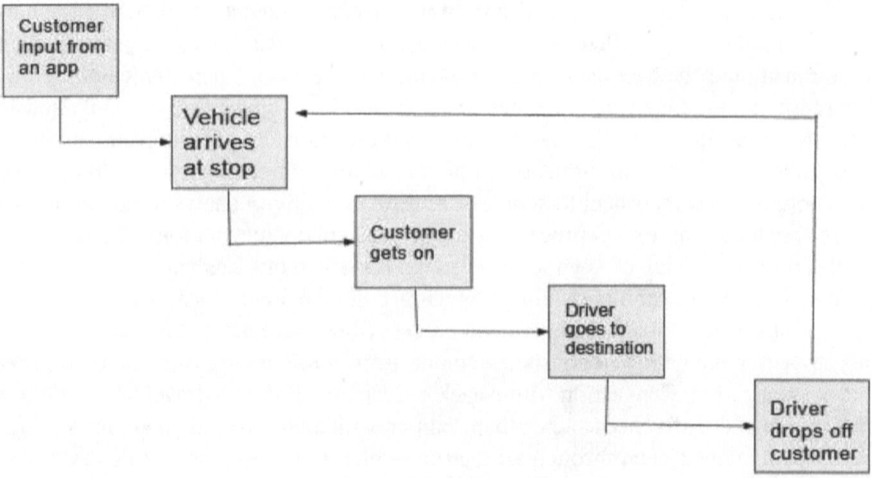

Fig. 2. Flow of the microtransit system

congestion levels. Vehicles are positioned at random nodes that serve as analogues for intersections or staging points. Trip requests arrive according to a stochastic process and specify pickup and drop off locations. Paratransit requests receive priority and must be fulfilled within a defined service window. Requests that exceed this window are classified as unfulfilled. Vehicles are assigned to the closest eligible request using travel time estimates that reflect current congestion levels. When no suitable vehicle is available, the request remains unserved. After each assignment, vehicle positions and system states are updated before the next request is processed as depicted in Fig. 3.

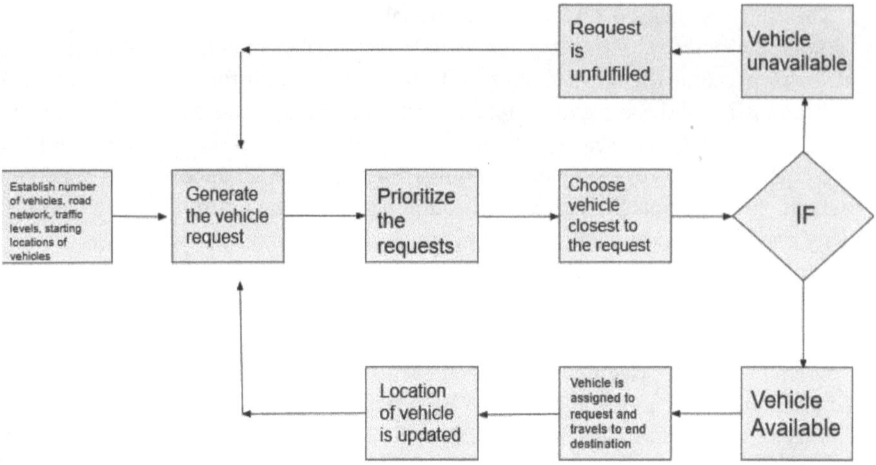

Fig. 3. Logic of the code

The simulation records detailed system performance indicators, including the number of fulfilled and unfulfilled requests, average passenger wait times for paratransit and microtransit, and the average utilization of the vehicle fleet. These indicators allow a direct assessment of how the integrated system responds to changes in demand intensity, congestion, and fleet size. The model updates these statistics at each event and aggregates them across the full simulation horizon. Upon completion of each run, results are exported to a spreadsheet to support structured sensitivity analysis and comparison across scenarios. This export process ensures consistent documentation of outcomes and enables post processing of key metrics through statistical or visualization tools.

To maintain tractability while preserving essential operational features, several assumptions were adopted. Standardized process flows were defined for paratransit and microtransit, which allowed consistent comparisons without requiring jurisdiction specific customization. Congestion within each run remained static so travel time variability reflected spatial differences rather than temporal fluctuations. All roads in the simulated grid remained open throughout the run, which removed complexities associated with closures or detours. Requests appeared at the time of need rather than through advance scheduling, which simplified system dynamics and focused attention on real time resource allocation. Demand followed a random arrival pattern rather than a cyclical pattern, which isolated the effects of stochastic variation. The service region was represented by a five-by-five grid, an appropriate scale for early stage prototyping and scalable to larger networks in future extensions. These assumptions support reproducible experiments while acknowledging that they constrain external validity when applied to real world systems.

3 Results

System performance was evaluated under two core factors, traffic congestion and fleet size. The analysis focused on seven operational metrics that capture user experience and resource efficiency. These metrics include the number of fulfilled and unfulfilled requests for paratransit and for microtransit, the average wait time for paratransit and for microtransit, and the average vehicle utilization across the entire fleet. These indicators collectively describe access, reliability, and capacity use. Fulfilled and unfulfilled requests quantify service coverage. Wait times reflect user experience and the timeliness of assignment and dispatch. Vehicle utilization measures the extent to which available capacity contributes to productive service, which provides a direct proxy for efficiency.

3.1 Baseline Scenario

The baseline scenario assumed congestion factors drawn from a continuous uniform distribution with lower bound 0.5 and upper bound 1.5, which represents moderate variability in travel times around a central level. The fleet size was set to 20 vehicles. Under this configuration, paratransit fulfilled 241 requests and 62 requests remained unfulfilled. The average paratransit wait time equaled 1.871 min. Microtransit fulfilled 110 requests and 425 requests remained unfulfilled. The average microtransit wait time equaled 2.240 min. Table 1 reports the baseline outcomes that serve as the reference point for subsequent sensitivity analyses. Interpreted operationally, the baseline indicates that priority assignment preserves paratransit responsiveness with a short average wait, while microtransit experiences higher denial rates and longer waits when both services compete for the same fleet under moderate congestion. The utilization levels associated with these outcomes indicate that the fleet operates near a binding capacity in periods with clustered arrivals, which explains the observed gap between paratransit and microtransit fulfillment.

Table 1 : Baseline Results

	Paratransit	Microtransit
Fulfilled Requests	241	110
Unfulfilled Requests	62	425
Average Wait Times (min)	1.871	2.240

3.2 Varying Congestion

Congestion effects were examined with the fleet size fixed at 20 vehicles. Expanding the congestion range raised average wait times and increased unfulfilled requests, with the largest degradation observed for paratransit due to its service windows and assignment priority. A consistent pattern emerged when ranges with similar central tendencies were compared. Wider ranges such as 0 to 2 produced fewer unfulfilled requests than narrower ranges such as 0.5 to 1.5. The system benefited from intermittent low congestion intervals within the wider range, which created short recovery periods that cleared queued demand and repositioned vehicles more effectively. This outcome aligns with queueing theory, where occasional reductions in service time help stabilize the system and reduce blocking. The result indicates that variability, when it includes frequent low congestion segments, can improve throughput even if the mean remains unchanged.

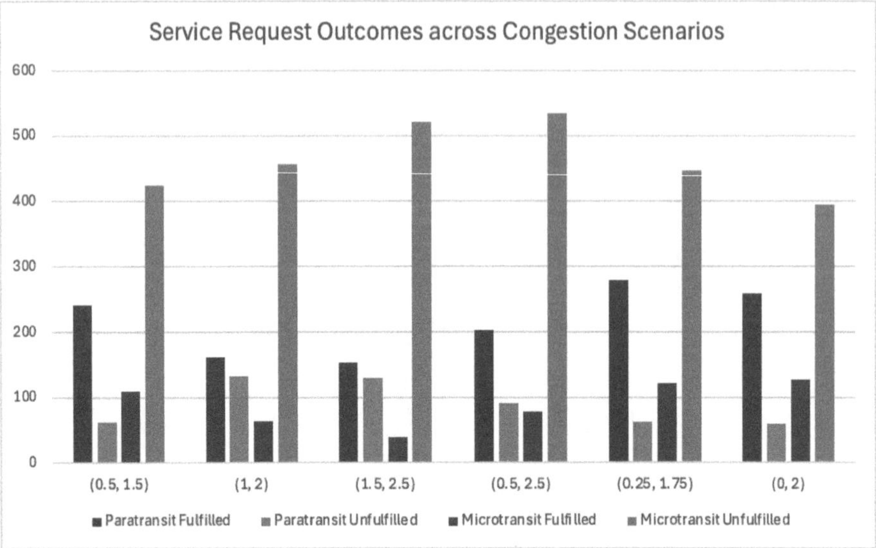

Fig. 4. Request outcomes after varying congestion range

Microtransit exhibited greater volatility in average wait times across congestion settings. This sensitivity reflects its lower dispatch priority and its reliance on residual capacity after paratransit commitments. When congestion intensified, vehicles arrived late to scheduled paratransit pickups, which tightened near term capacity available for microtransit and raised both waits and denials. When congestion eased, vehicles completed assignments earlier and reentered the pool for microtransit, which produced sharp improvements in wait times. Figure 4 reports these dynamics and shows consistently higher unfulfilled counts for microtransit than for paratransit across congestion scenarios, which follows directly from the embedded prioritization logic. The figure also confirms a general decline in fulfilled requests as congestion increases, especially for paratransit, and it illustrates the performance advantage of broader congestion ranges such as 0 to 2 relative to narrower ranges such as 0.5 to 1.5.

Figure 4 request outcomes under alternative congestion ranges, with fleet size fixed at 20 vehicles. The plots report fulfilled and unfulfilled requests for paratransit and microtransit and highlight the superior performance observed under wider congestion ranges with similar means.

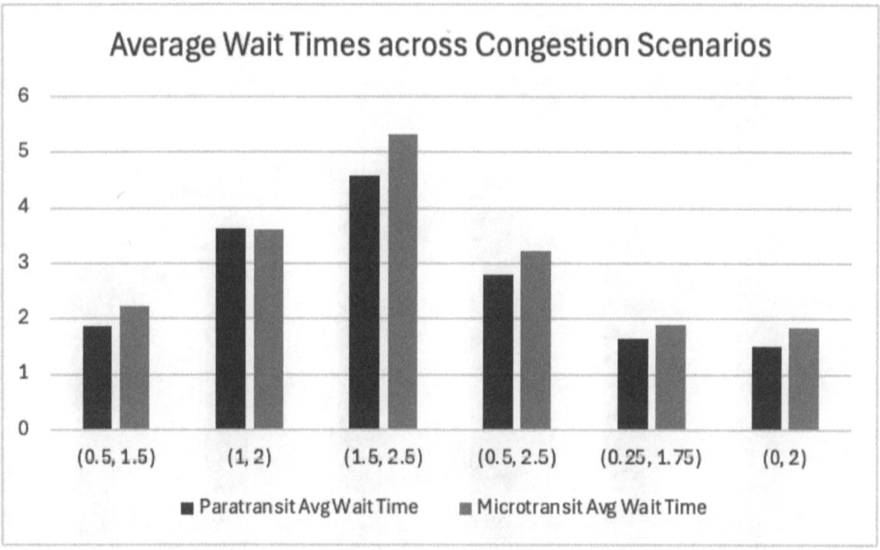

Fig. 5. Average wait times (min) after varying congestion range

Figure 5 shows the average wait times, minutes, under alternative congestion ranges. The plots quantify the effect of traffic congestion on average wait times for paratransit and microtransit. As the congestion rate increases, average waits rise for both services, and the count of unfulfilled requests rises in parallel, which signals capacity strain. Paratransit consistently records lower mean wait times than microtransit, which reflects the priority rules embedded in the dispatch logic and the presence of service windows that trigger earlier assignments. Microtransit exhibits greater volatility in average wait times across congestion settings, because its access to vehicles depends on residual capacity after paratransit obligations. When congestion recedes for short intervals, vehicles complete trips faster and reenter the pool, which compresses microtransit waits. When congestion intensifies, vehicles dwell longer on the network, which amplifies microtransit waits. Taken together with the request outcomes in Fig. 4, the evidence shows that congestion, an external factor beyond the control of a single agency, exerts a first order influence on both delay and service denial. This finding supports interventions that improve travel time reliability, for example strategic staging or limited priority at intersections, since small gains in reliability can produce material reductions in delay.

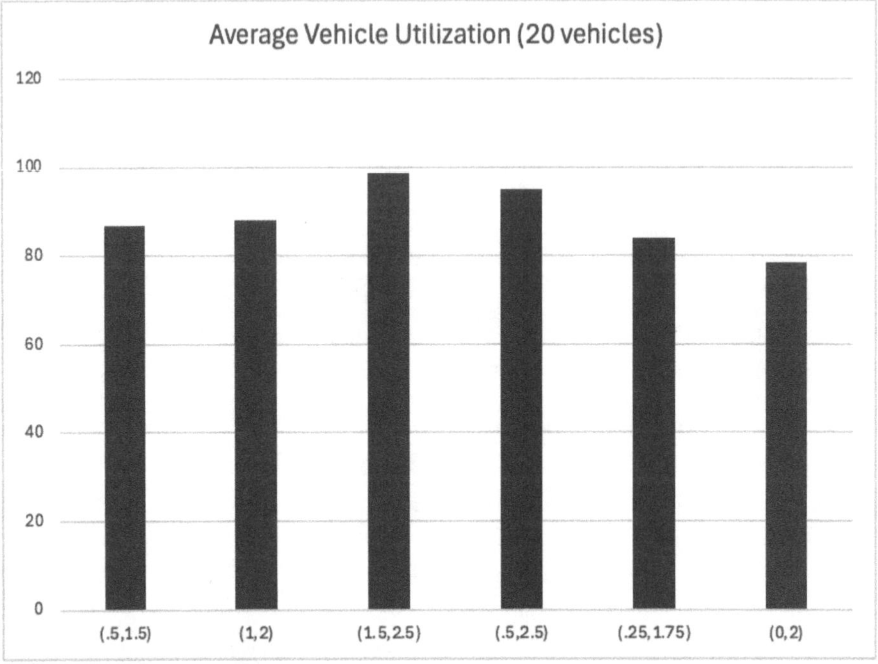

Fig. 6: Average vehicle utilization (%) after varying congestion range

Figure 6 shows the average vehicle utilization, percent, under alternative congestion ranges. The results indicate that average utilization generally rises as congestion increases, with a peak near 98% in the range from one point five to two point five. Higher congestion extends in service travel times and increases the fraction of time vehicles remain busy, which elevates measured utilization. Although high utilization may appear desirable, values near saturation reduce scheduling flexibility, increase the risk of missed paratransit windows, and propagate longer queues for microtransit. Operation near the observed peak therefore represents a fragile regime. Small disruptions can trigger rapid growth in unserved demand. The pattern suggests clear management responses. Agencies can expand the fleet modestly in high congestion periods, stage vehicles closer to demand clusters, or introduce short recovery buffers in paratransit schedules. Each action reduces the likelihood of saturation and improves both reliability and coverage without compromising the priority afforded to paratransit riders.

3.3 Varying Fleet Size

Fleet size exerted a direct and predictable influence on service accessibility and timeliness across both rider groups. Scaling the fleet from ten to sixty vehicles produced monotonic gains in fulfillment and reductions in delay. With ten vehicles, unmet demand remained substantial with 140 unfulfilled paratransit requests and 522 unfulfilled microtransit requests. With sixty vehicles, the system cleared all requests for both services within the defined service windows. Average wait times reflected the same pattern. Paratransit mean waits declined from 1.99 min to 0.43 min. Microtransit mean waits declined from 2.41 min to 0.69 min. These results confirm that capacity expansion reduces queue lengths and accelerates assignment and pickup, particularly for lower priority microtransit trips that depend on residual capacity after paratransit commitments.

The relationship between fleet size and efficiency exhibited diminishing returns at higher capacity levels. Vehicle utilization varied inversely with fleet size and peaked at intermediate capacity near twenty vehicles, which indicates a balance between demand pressure and available supply. As the fleet expanded to fifty and sixty vehicles, average utilization fell sharply, from 80.66% at forty vehicles to 63.38% at fifty vehicles and 49.59% at sixty vehicles. The system therefore transitioned from a congestion constrained regime, where additional vehicles directly convert to higher throughput and shorter waits, to an underutilized regime, where added capacity remains idle for significant periods. From a planning standpoint, this pattern suggests a concave benefit curve for added vehicles with a broad interior region of near optimal performance around twenty to forty vehicles. Within that region, agencies achieve large service gains without incurring the ongoing costs of chronic overcapacity.

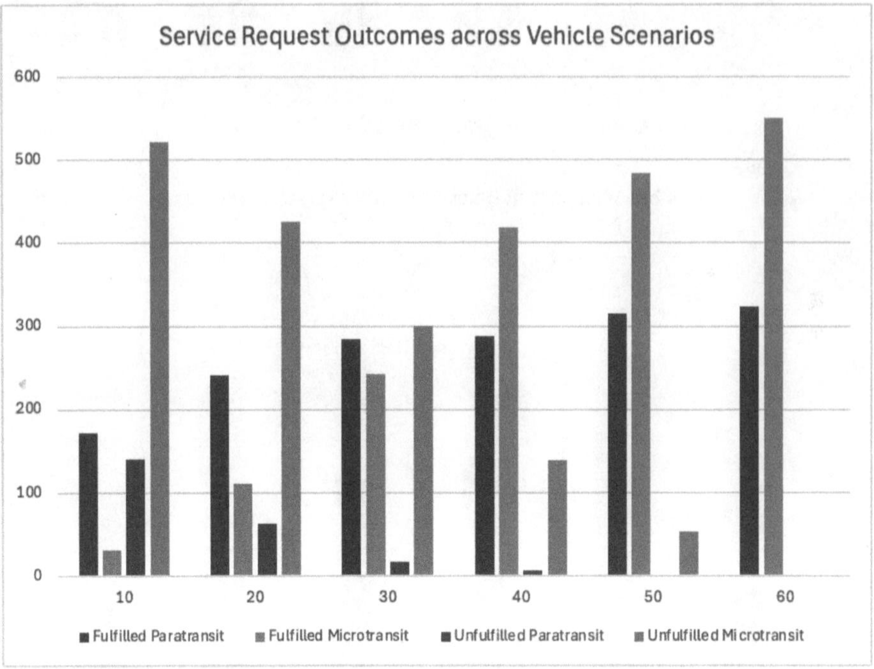

Fig. 7: Request Outcomes after varying fleet size.

Figure 7 illustrates the request outcomes under alternative fleet sizes. Fulfilled requests for paratransit and microtransit rise with fleet expansion, while unfulfilled counts decline. The system eliminates all unfulfilled requests at sixty vehicles, which demonstrates complete coverage within the modeled conditions. Figure 8 displays the average wait times, minutes, under alternative fleet sizes. Both services experience continuous reductions in mean waits as the fleet grows from ten to sixty vehicles. Paratransit declines from 1.99 to 0.43 min, microtransit declines from 2.41 to 0.69 min, which reflects improved availability and faster repositioning between assignments.

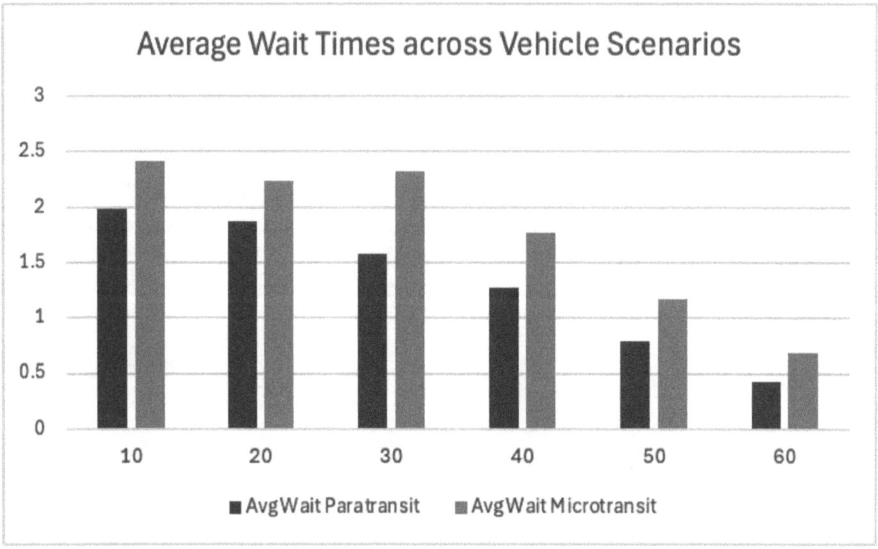

Fig. 8: Average wait times (min) after varying fleet size.

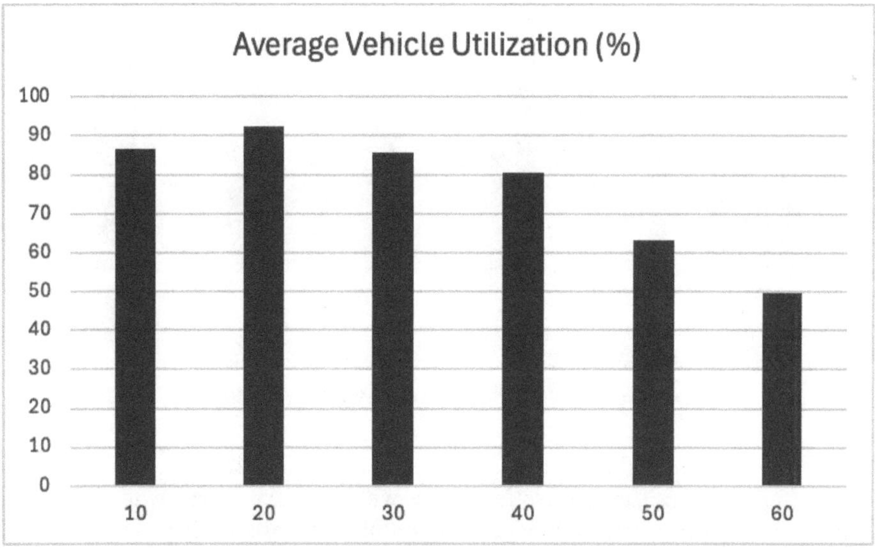

Fig. 9: Average vehicle utilization (%) after varying fleet size.

Figure 9 shows the average vehicle utilization, percent, under alternative fleet sizes. Utilization peaks at approximately twenty vehicles and then decreases as capacity expands. Values drop from 80.66% at forty vehicles to 63.38% at fifty vehicles and 49.59% at sixty vehicles. Very high utilization implies limited slack and a higher risk of missed paratransit windows, whereas very low utilization implies excess fixed and operating costs. The observed curve indicates an operational sweet spot at intermediate fleet sizes where the system preserves reliability while avoiding costly idle time.

4 Implications

The results demonstrate the feasibility and value of integrating paratransit and microtransit within a coordinated operating model. Performance depends on controllable levers, notably fleet size and staging policy, and on external conditions, notably congestion. The core metrics reported, fulfillment, wait time, and vehicle utilization, provide decision relevant signals for planning, procurement, and service monitoring. Right sizing the fleet improves coverage and reliability while limiting idle capacity, which aligns capital and operating budgets with long term demand.

A structured cost framework should guide adoption. Agencies can estimate marginal cost per additional fulfilled request and marginal delay reduction per added vehicle, then identify the range where benefits remain concave and costs begin to outpace gains. The evidence indicates that moderate utilization levels create the strongest cost effectiveness, since the system clears queues and avoids overtime without sustaining persistent idle time. Formal cost accounting can include labor, fuel or energy, maintenance, insurance, dispatch and software licensing, and accessible vehicle premiums. Scenario analysis can test the sensitivity of unit costs to wage rates, energy prices, and preventive maintenance intervals.

Congestion remains exogenous to the operator. Even so, limited interventions can improve travel time reliability. Targeted signal coordination near major trip generators, tactical staging at demand clusters, and curbside management during peak appointment windows create recovery intervals that reduce queues without expanding the fleet. Integration also advances equity objectives. Shared rides for able bodied and disabled passengers extend access, reduce stigma associated with specialized services, and align with accessibility mandates. Purpose specific usage patterns point to further optimization. Growth in telehealth and remote patient monitoring can displace a portion of medical trips. The rise in remote work among disabled workers reduces peak commuter demand. Agencies can capture these gains with incentive programs for avoided trips and reinvest savings in stop level accessibility, fleet modernization, and microtransit coverage.

Implementation requires attention to governance and safeguards. Clear priority rules must protect ADA service windows. Audit logs of dispatch decisions must support transparency. Service level agreements can set maximum waits by rider class, minimum on time pickup rates, and thresholds for denied trips. Data sharing agreements can define privacy protections, retention periods, and the use of deidentified telemetry for planning. Ongoing public funding and competitive grants remain essential to sustain ADA compliant services and to accelerate innovation in integrated mobility. Partnerships with healthcare networks, employers, and community organizations can further stabilize demand and co fund targeted service expansions.

5 Conclusions

The study examined an integrated system that combines the flexibility of microtransit with the accessibility requirements of paratransit. A Python based discrete event simulation with realistic inputs, congestion, spatially dispersed demand, and service windows, assessed how dispatch priorities and capacity choices shape outcomes. Sensitivity analyses showed that increases in fleet size reduce unfulfilled requests and shorten wait times, while higher congestion degrades performance across metrics. Priority weighting in assignment logic preserved short paratransit waits, which confirms that accessibility can remain central within a shared platform while microtransit continues to deliver responsive service.

The findings indicate that integration can raise vehicle utilization at moderate fleet sizes, extend coverage to low demand areas, and protect paratransit service quality. The model adopted simplifying assumptions, a uniform grid and requests that enter at time of need, so the results provide comparative insights rather than jurisdiction specific forecasts. A practical roadmap follows. Pilot deployments can begin with a modest shared fleet in zones that show concentrated demand and predictable appointment windows. Agencies can calibrate the simulator with local travel times, request arrivals, and dwell distributions, then validate against observed key performance indicators. Adaptive priority weights can tune the balance between accessibility and residual capacity for microtransit. Periodic re optimization can update staging locations and fleet levels as demand shifts.

The evidence supports a clear conclusion. A coordinated paratransit and microtransit operation is technically feasible and operationally advantageous. Microtransit users receive faster and more reliable rides. Paratransit riders retain priority access with improved flexibility in off peak periods. Providers achieve higher efficiency through better asset utilization and more reliable service windows. With measured investment in data driven governance, cost tracking, and limited reliability interventions on the road network, agencies can deliver a more equitable and sustainable mobility ecosystem that improves access and reliability.

References

Federal Transit Administration: FTA Reports and Publications. https://www.transit.dot.gov/research-innovation/fta-reports-and-publications (2021)

Liezenga, A.M., Verma, T., Mayaud, J.R., Aydin, N.Y., Van Wee, B.: The first mile towards access equity: is on-demand microtransit a valuable addition to the transportation mix in suburban communities? Transp. Res. Interdisc. Perspect. **24**, 101071 (2024)

Miah, M.M., Naz, F., Hyun, K.K., Mattingly, S.P., Cronley, C., Fields, N.: Barriers and opportunities for paratransit users to adopt on-demand micro transit. Res. Transp. Econ. **84**, 101001 (2020)

National Academies of Sciences, Engineering, and Medicine: Between public and private mobility: examining the rise of technology-enabled transportation services. (2016)

Riggs, W., Pande, A.: On-demand microtransit and paratransit service using autonomous vehicles: gaps and opportunities in accessibility policy. Transp. Policy. **127**, 171–178 (2022)

Walker, J.: Be on the way! Transit implications of location choice. In: Human Transit: How Clearer Thinking About Public Transit Can Enrich Our Communities and Our Lives, pp. 181–204. Island Press/Center for Resource Economics, Washington, DC (2012)

Enhancing Task Offloading in Vehicular Edge Computing with DT and GAT-Based DRL

Ihsan Ullah[1(✉)], Qaisar Ali[1], Muhammad Ashraf[2], and Youn-Hee Han[3]

[1] Department of Computer Science and Engineering, Yuan Ze University, Taoyuan, Taiwan
ihsan@saturn.yzu.edu.tw, s1129110@mail.yzu.edu.tw
[2] Engineering and Management Sciences, Balochistan University of Information Technology, Quetta, Pakistan
Muhammad.ashraf@buitms.edu.pk
[3] Future Convergence Engineering, Korea University of Technology and Education, Cheonan, South Korea
yhhan@koreatech.ac.kr

Abstract. Task offloading in dynamic Vehicular Edge Computing (VEC) networks is a complex challenge due to highly variable vehicle mobility, fluctuating network conditions, and limited resource availability. To address this, we propose a real-time task offloading framework that integrates DT modeling, Graph Attention Networks (GAT), and Proximal Policy Optimization (PPO). The DT creates a real-time virtual replica of the physical environment, enabling predictive monitoring of vehicles, edge servers, and network links. GAT models spatial dependencies among nodes, while PPO learns adaptive, mobility- and QoS-aware offloading strategies. Simulation results show that the proposed framework consistently outperforms baseline schemes, achieving higher task success rates, lower latency, reduced energy consumption, and faster decision-making. The framework demonstrates strong robustness under dynamic conditions and scalability to large VEC networks, highlighting its potential for practical deployment in intelligent transportation systems.

Keywords: Vehicular Edge Computing · Digital Twin · Graph Attention Network · Proximal Policy Optimization · Task Offloading

1 Introduction

The rapid growth of Intelligent Transportation Systems (ITS) has transformed vehicular networks into advanced computing platforms, where vehicles process complex workloads such as autonomous driving, real-time HD mapping, and multi-sensor fusion. Traditional cloud architectures struggle to meet the stringent latency and reliability requirements of these applications, motivating the

H. R. Arabnia et al. (Eds.): CSCI 2025, CCIS 2974, pp. 55–67, 2027.
https://doi.org/10.1007/978-3-032-28475-4_4

adoption of VEC. VEC enables task offloading to Road Side Units (RSUs), MEC servers, and neighboring vehicles via V2V links [6]. However, high vehicle mobility introduces significant challenges for conventional offloading strategies, including dynamic topologies, intermittent connectivity, and heterogeneous resource capacities.

In VEC, efficient task offloading requires maintaining low latency, energy efficiency, load balancing, and effective management of inter-task dependencies [15]. Conventional Deep Reinforcement Learning (DRL) methods often lack spatial and temporal awareness, and many DT implementations rely on oversimplified models. DTs provide synchronized virtual replicas that predict connectivity, simulate task execution, and enable proactive resource allocation. Recent DRL-based approaches have advanced from DQN methods with discrete actions and limited spatial awareness [10], to graph-based methods capturing topological context [19], and attention-driven GAT models that prioritize links based on connectivity [16]. Despite these improvements, critical challenges remain, including modeling inter-task dependencies, predicting mobility, enabling V2V offloading, and seamless integration with DT systems [7].

To address these limitations, this paper proposes a DT-enhanced GAT-PPO framework for proactive, context-aware task offloading. The DT layer constructs dynamic graph representations encoding physical (capacity, energy) and predicted (mobility, load) states. The GAT model captures spatiotemporal dependencies among nodes, while the PPO agent optimizes a multi-objective reward considering latency, energy consumption, task success, and load balancing. Experimental results demonstrate significant improvements in latency (28–45 ms), energy efficiency (30–35% savings), task success (93–97%), and resource utilization (20–25% higher) compared to baseline methods.

The rest of the paper is organized as follows: Section 2 reviews related work on VEC task offloading, Section 3 presents the system model, Section 4 describes the proposed scheme, Section 5 evaluates performance, and Section 6 concludes the paper.

2 Related Work

VEC enables low-latency and efficient task offloading, and recent DRL-based methods have significantly advanced task scheduling and resource allocation. For instance, You et al. [20] proposed a hierarchical RL framework that leverages DAGs and GATs for V2I offloading, effectively reducing delay and communication overhead. Shen et al. [13] and Liu et al. [11] introduced MGARL and SATA-DRL for content caching and dynamic task scheduling, achieving improved efficiency and lower deadline violations. Other DRL approaches incorporate federated learning, multi-agent coordination, and trajectory prediction to enhance latency, energy efficiency, and load balancing [3,4,8,12,14,18].

DT-assisted frameworks further improve adaptability and predictive capabilities in edge-cloud systems. DTDRLMO [1] utilizes DTs for microservice offloading in collaborative edge environments. Dong et al. [2] applied DTs for predictive user association and resource allocation, while Hu et al. [5] and Zhang et al. [21] focus on collaborative DNN inference and heterogeneous edge-cloud offloading optimization. UAV-assisted DT frameworks, such as FlexEdge [9], exploit mobility-aware DRL, and multi-agent or SAC-based DT methods enhance system reliability, adaptability, and multi-objective optimization [17,22].

Despite these advances, most existing DRL- and DT-based approaches target specific aspects, such as mobility prediction, microservice offloading, or multi-agent cooperation. Few integrate real-time system monitoring with graph-based embeddings and attention mechanisms to prioritize tasks. Many also depend on offline training, heuristics, or game-theoretic methods, limiting adaptability, scalability, and convergence speed. To address these limitations, our proposed framework combines real-time DT monitoring, graph attention embeddings, and an attention-augmented PPO agent. This unified approach simultaneously optimizes latency, energy consumption, QoS, and load balancing in complex vehicular edge-cloud systems.

3 System Model

We model a VEC environment using DTs to represent vehicles, edge servers, and the overall network state. The DT provides a virtual replica of the physical system for monitoring, prediction, and state representation, capturing both real-time and historical information for proactive management. The system consists of three layers of DTs:

Vehicle Digital Twins (VDTs): Each vehicle $v_i \in \mathcal{V}$ maintains a VDT, denoted as DT_{v_i}, which continuously monitors sensor readings such as LiDAR and camera data, mobility state including position, speed, and trajectory, local computation capacity, and the current task queue. This virtual representation allows the system to maintain real-time awareness of vehicle states and task characteristics, forming a foundation for predicting task execution and communication performance.

Edge Server Digital Twins (EDTs): Similarly, each roadside unit (RSU) $e_j \in \mathcal{E}$ is associated with an EDTs, DT_{e_j}, which monitors available CPU cycles, memory, network bandwidth, latency, and the edge task queue. The EDT provides a virtual view of edge resources, enabling estimation of processing capabilities and network conditions to support low-latency and efficient offloading.

Central Digital Twin (CDT): At the global level, the CDT aggregates data from all VDTs and EDTs to maintain a synchronized view of the overall system. The CDT records network topology, traffic density and patterns, and the availability of edge and cloud resources, offering a holistic snapshot of the VEC environment that is crucial for system-wide monitoring and planning.

3.1 Task Representation in DT

Each vehicle generates tasks T_{vi}, which are represented in its DT as

$$T_{vi} = (D_{vi}, C_{vi}, \tau_{vi}, \gamma_{vi}), \tag{1}$$

where D_{vi} denotes the input data size, C_{vi} is the required CPU cycles, τ_{vi} is the task deadline, and γ_{vi} indicates task priority. The DT maintains these task attributes in synchronization with the physical system to ensure accurate representation of task execution requirements.

3.2 Communication State

The communication state of the VEC environment is also captured in the DTs. The instantaneous vehicle-to-infrastructure (V2I) data rate between vehicle v and RSU e is modeled as

$$r_{ve}^{DT}(t) = B_{ve} \log_2 \left(1 + \frac{P_v h_{ve}(t)}{N_0 + I_{ve}(t)}\right), \tag{2}$$

where B_{ve} represents the allocated bandwidth, P_v is the vehicle transmit power, $h_{ve}(t)$ is the channel gain, N_0 is noise power, and $I_{ve}(t)$ captures interference. Similarly, V2V link conditions are monitored in the DT as metadata to represent vehicle-to-vehicle connectivity.

The complete DT state of the VEC environment at time t is represented as

$$\mathcal{S}^{DT}(t) = \{\text{VDTs}, \text{EDTs}, \text{CDT}\}, \tag{3}$$

where each component contains both physical and virtual attributes. VDTs include vehicle position, speed, CPU frequency, and task queue information. EDTs include edge CPU frequency, bandwidth, task queue length, and latency. The CDT captures the global network topology, traffic density, and resource availability of edge and cloud servers. This unified DT state provides a structured representation of the VEC environment, suitable for conversion into graph-based embeddings for GAT input and subsequent intelligent task offloading.

In summary, the DT framework provides a real-time, synchronized representation of the entire VEC environment, capturing vehicle mobility, task attributes, edge resource availability, and network conditions. By encoding all these states into DTs, the system forms a rich and structured representation that can later serve as input for learning agents such as GAT-PPO for proactive and context-aware task offloading.

4 Proposed Scheme

To optimize task offloading in DT-enabled VEC, we propose a framework that integrates DT modeling, GAT, and PPO, as illustrated in Fig. 1. The DT captures real-time network states of vehicles, edge servers, and cloud nodes, including task attributes, resource availability, mobility, and network conditions. Using these synchronized DT states, we dynamically construct a heterogeneous, attributed graph $\mathcal{G}_t = (\mathcal{N}_t, \mathcal{E}_t)$ at each time step t.

The node set $\mathcal{N}_t$ consists of task nodes $\mathcal{N}_T$ and computational nodes $\mathcal{N}_C$, where $\mathcal{N}_T = \{T_{vi}\}$ represents individual tasks, and $\mathcal{N}_C = \mathcal{N}_V \cup \mathcal{N}_E \cup \mathcal{N}_K$ includes

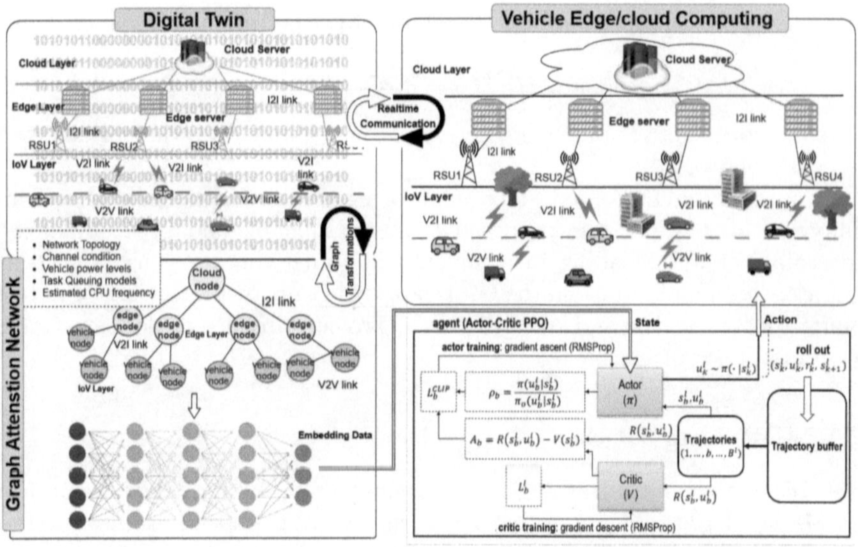

Fig. 1. Architecture of the Proposed Task Offloading Framework.

vehicles, edge servers, and cloud resources. Each node i is associated with a feature vector $\boldsymbol{x}_i$ that encodes relevant attributes such as task size, priority, vehicle location and velocity, available CPU cycles, memory, and edge latency. The edge set $\mathcal{E}_t$ represents both computation and communication links, and each edge (i, j) carries features $\boldsymbol{e}_{ij}$ including bandwidth, latency, and reliability. Formally, the graph can be expressed as

$$\mathcal{G}_t = \left\{ \{\boldsymbol{x}_i\}_{i \in \mathcal{N}_t}, \{\boldsymbol{e}_{ij}\}_{(i,j) \in \mathcal{E}_t} \right\}. \tag{4}$$

This graph is then fed into a Graph Attention Network, which computes node embeddings by aggregating information from neighboring nodes weighted according to their importance. For each node i, the GAT embedding is computed as

$$\boldsymbol{h}_i^{(l+1)} = \sigma\Big(\sum_{j \in \mathcal{N}_i} \alpha_{ij}^{(l)} \mathbf{W}^{(l)} \boldsymbol{h}_j^{(l)} \Big), \tag{5}$$

where $\boldsymbol{h}_i^{(l)}$ is the node feature at layer l, $\alpha_{ij}^{(l)}$ is the attention coefficient for neighbor j, $\mathbf{W}^{(l)}$ is a learnable weight matrix, and σ is a non-linear activation function. The attention mechanism ensures that nodes with more critical connectivity or resource significance have higher influence on the embedding.

The output node embeddings $\boldsymbol{H}^{GAT} = \{\boldsymbol{h}_i^{(L)}\}$ from the final GAT layer capture both the structural information of the network and the dynamic state of tasks and resources. These embeddings form a rich and compact representation of the environment, which is then used as the input to a PPO agent. The PPO agent interprets these embeddings to make context-aware and resource-aware task offloading decisions, considering latency, energy consumption, load balancing, and task success probability. Formally, the PPO policy can be defined as

$$\pi_\theta(a_t|\boldsymbol{H}_t^{GAT}) = \mathrm{PPO}(\boldsymbol{H}_t^{GAT}), \tag{6}$$

where a_t represents the offloading action at time t, and θ are the parameters of the policy network.

By integrating DT-based state monitoring, graph-based embeddings, and reinforcement learning, the proposed framework enables adaptive, predictive, and scalable task offloading in complex VEC environments. The DT ensures accurate state representation, the GAT captures the network topology and task-resource dependencies, and the PPO agent provides intelligent decision-making for optimal task allocation.

4.1 MDP Formulation

To enable reinforcement learning-based task offloading, we formulate the problem as a Markov Decision Process (MDP) using the GAT embeddings derived from the DT representation of the VEC environment.

State Space. The system state at time t, $\boldsymbol{s}_t \in \mathcal{S}$, is constructed from the output embeddings of the GAT. These embeddings capture both the network topology and the dynamic status of tasks and resources. Formally, for each node i, the embedding is

$$\boldsymbol{h}_i^{(l+1)} = \sigma\left(\sum_{j\in\mathcal{N}_i} \alpha_{ij}^{(l)} \mathbf{W}^{(l)} \boldsymbol{h}_j^{(l)}\right), \tag{7}$$

where $\alpha_{ij}^{(l)}$ is the attention coefficient between node i and neighbor j, $\mathbf{W}^{(l)}$ is a learnable weight matrix, and σ is a non-linear activation function. The global system state is then obtained by a permutation-invariant readout function $R(\cdot)$ over all nodes:

$$\boldsymbol{s}_t = R(\{\boldsymbol{h}_i^{(L)} \mid i \in \mathcal{N}_t\}), \tag{8}$$

where L is the final GAT layer. The attention coefficient can incorporate both network connectivity and temporal relevance:

$$\alpha_{ij}^t = \frac{\exp(\mathrm{LeakyReLU}(\boldsymbol{a}^T[\mathbf{W}\boldsymbol{h}_i \| \mathbf{W}\boldsymbol{h}_j]))}{\sum_{k\in\mathcal{N}_i}\exp(\cdot)} \cdot e^{-\lambda\|t-t_{\mathrm{last}}\|} \cdot \mathbb{I}_{\mathrm{conn}}(i,j), \tag{9}$$

where $\mathbb{I}_{\mathrm{conn}}(i,j)$ indicates whether a communication link exists, and the exponential term models decay based on time since last interaction.

Action Space. At each time step, the agent selects an offloading decision for each task node:

$$a_t^{vi} \in \{\mathrm{local}, \mathrm{edge}, \mathrm{cloud}\}, \quad \boldsymbol{a}_t = (a_t^{v_1 i_1}, \ldots, a_t^{v_{|\mathcal{N}_T|} i_{|\mathcal{N}_T|}}). \tag{10}$$

The policy network π_θ maps the state $\boldsymbol{s}_t$ to a distribution over possible actions for all tasks:

$$\pi_\theta(\boldsymbol{a}_t \mid \boldsymbol{s}_t) = \prod_{T_{vi} \in \mathcal{N}_T} \pi_\theta(a_t^{vi} \mid \boldsymbol{s}_t), \tag{11}$$

enabling independent yet coordinated offloading decisions across tasks.

Reward Function. The reward function is designed to reflect key objectives in VEC task offloading, including latency, energy consumption, QoS, reliability, and load balancing:

$$R_t = \sum_{T_{vi} \in \mathcal{N}_T} \Big(-\alpha L_{vi} - \beta E_{vi} + \gamma Q_{vi} - \eta D_{vi} - \zeta B_{vi} - \xi R_{vi} + \rho(a_{vi}) \Big), \tag{12}$$

The MDP formulation directly connects as, the DT captures real-time states, the GAT generates node embeddings representing structural and resource aware information, and the PPO agent uses these embeddings to select optimal offloading actions according to the defined reward. This design ensures that task offloading decisions are both context-aware and adaptive to dynamic vehicular environments.

4.2 Learning Policy with PPO

To make intelligent and adaptive task offloading decisions, we employ PPO, a state-of-the-art reinforcement learning algorithm suitable for high-dimensional and continuous action spaces. PPO learns a stochastic policy $\pi_\theta(a_t \mid s_t)$ that maps the DT-GAT embeddings of the VEC environment (the MDP state) to a probability distribution over offloading actions for each task. The input to the PPO policy network is the node embeddings generated by the GAT. The policy network outputs a probability distribution over offloading actions for each task. $\pi_\theta(a_t \mid s_t)$. To estimate expected returns, PPO uses a separate value network $V_\phi(s_t)$. which predicts the cumulative discounted reward from the current state. This value estimate stabilizes training by reducing variance in the policy gradient updates.

PPO improves the policy while preventing large updates that may destabilize learning. The clipped surrogate objective is:

$$\mathcal{L}(\theta) = \mathbb{E}_t \Big[\min \big(r_t(\theta) \hat{A}_t, \text{clip}(r_t(\theta), 1-\epsilon, 1+\epsilon) \hat{A}_t \big) \Big], \tag{13}$$

where $r_t(\theta) = \frac{\pi_\theta(a_t|s_t)}{\pi_{\theta_{\text{old}}}(a_t|s_t)}$ is the probability ratio, $\hat{A}_t$ is the advantage estimate, and ϵ is a hyperparameter controlling the clipping range.

The advantage function $\hat{A}_t$ can be computed using Generalized Advantage Estimation (GAE):

$$\hat{A}_t = \sum_{l=0}^{\infty} (\gamma\lambda)^l \delta_{t+l}, \quad \delta_t = R_t + \gamma V_\phi(s_{t+1}) - V_\phi(s_t), \tag{14}$$

where γ is the discount factor and λ controls bias-variance tradeoff.

Table 1. Computational Complexity and Overhead Comparison

Model	Training Complexity	Online Inference Cost	Decision Latency
PPO	Low	Very Low	Low
GPPO	Medium	Medium	Medium
DGPPO	High	Medium	Acceptable

The computational complexity of the models increases from PPO to GPPO and DGPPO (Table 1). PPO uses a flat state representation, resulting in the lowest training and inference cost. GPPO introduces graph-based modeling to capture spatial dependencies, increasing overhead from message passing and attention. DGPPO adds a DT layer for predictive system states, with extra cost from state updates and mobility prediction, while training remains offline. During online deployment, all models generate decisions via a single forward pass of the PPO actor, ensuring practical latency. Despite higher overhead, DGPPO's improvements in task success, latency, energy efficiency, and robustness justify its use in dynamic vehicular

5 Performance Evaluation

To evaluate the proposed DT-enabled task offloading scheme in VEC networks, we conducted simulations on a desktop equipped with an Intel Core i7-4790 CPU (3.6 GHz) and 8 GB of RAM. The environment was implemented in Python 3.10, using TensorFlow 2.14 for general computation, Stable-Baselines3 for reinforcement learning, and PyTorch Geometric for the GAT implementation. The simulated network comprises vehicular nodes, RSUs serving as Mobile Edge Servers (MES), UAVs, and cloud servers. Vehicles are randomly distributed over a 1000×1000 m urban grid, while RSUs are fixed at intersections. UAVs dynamically reposition according to task density to support offloading. Each vehicle generates computational tasks that can be offloaded to nearby RSUs, UAVs, or remote cloud servers. For learning, the PPO actor-critic algorithm uses a GAT encoder to process graph-structured states, capturing server load, link quality, and task demands. Both the policy and value networks adopt a four-layer architecture with two hidden layers of 18 neurons each, and the networks are optimized using the Adam optimizer. Table 2 shows the simulation and training parameters.

Figure 2 shows the average rewards over 50,000 episodes for DGPPO (proposed), GPPO (baseline), and PPO (baseline). DGPPO achieves the highest rewards consistently, rising rapidly after 10,000 episodes and stabilizing, indicating efficient policy learning and balanced exploration-exploitation. GPPO performs moderately, benefiting from graph-attention state representation but plateauing earlier and converging slower than DGPPO. PPO has the lowest rewards, with gradual growth and higher variance, reflecting less effective

Table 2. Simulation and Training Parameters for DGPPO in VEC (Enhanced Training)

Parameter	Value/Description
Number of Vehicles	50–200 (random mobility)
Vehicle Speed	20–120 km/h
Number of Edge Servers	5–20
Edge Server Capacity	10–50 tasks/sec
Network Bandwidth	10–100 Mbps
Task Arrival Rate	Poisson, 1–10 tasks/sec per vehicle
Task Size	1–10 MB
Task Deadline	50–500 ms
Energy Constraint (Edge)	100–500 units/server
DT Prediction Horizon	5–10 time steps
GAT Hidden Layers	[64,64]
GAT Attention Heads	4
PPO Hidden Layers	[256,256,128]
Learning Rate (PPO)	0.0003
Discount Factor (γ)	0.99
Batch Size	256
Policy Epochs (PPO)	20
Training Episodes	5000
Activation Function	ReLU (PPO), LeakyReLU (GAT)
Exploration Strategy	PPO deterministic/stochastic policy

policy optimization. Shaded areas indicate reward variability; DGPPO stabilizes quickly after initial fluctuations. Integrating DT information with GAT in DGPPO significantly improves offloading decisions and accelerates reward accumulation.

Figure 3 presents training loss curves over 50,000 episodes. All three schemes show decreasing loss, indicating effective learning. DGPPO achieves the lowest loss, converging faster and more smoothly than the baselines, reflecting efficient gradient updates and better policy optimization. GPPO reduces loss moderately, outperforming PPO due to graph-attention-based state representation, but its convergence is slower. PPO shows the highest loss with slower decrease and more variability. Shaded regions show variance; DGPPO stabilizes quicker with lower uncertainty. These results confirm that DGPPO not only achieves higher rewards but also ensures more stable and efficient training.

Figure 4 illustrates task deadline adherence as the task load increases. DGPPO consistently performs best, maintaining a high rate of on-time task completion even with 1000 tasks. GPPO performs moderately but drops under

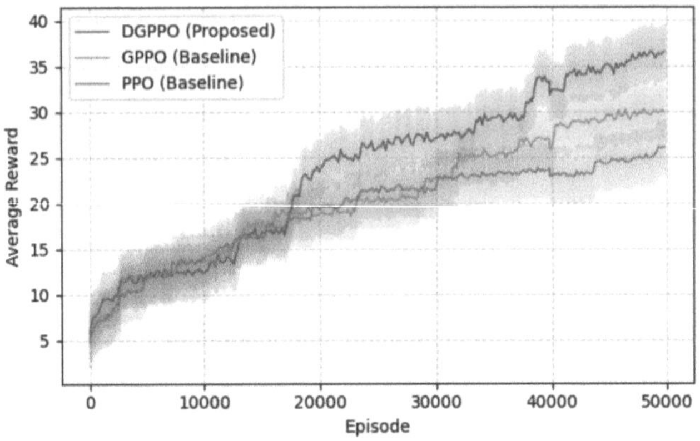

Fig. 2. Training reward comparison for DGPPO, GPPO, and PPO.

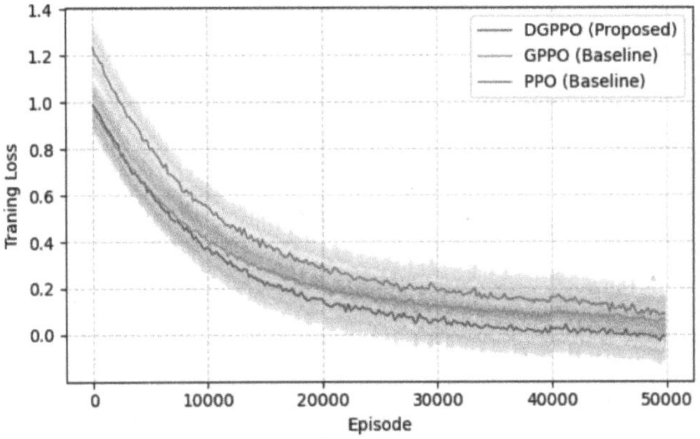

Fig. 3. Training loss comparison for DGPPO, GPPO, and PPO.

heavier loads, showing limitations in handling high demand. PPO performs the worst, with adherence decreasing sharply beyond 600 tasks. DGPPO's dual-graph policy enables more effective use of resources in both time and space, leading to superior scheduling performance.

Figure 5 shows average task completion time (in seconds) as the task load increases from 200 to 1000 tasks. All schemes show increased completion time with heavier workloads. DGPPO consistently achieves the lowest completion time, demonstrating efficient resource usage. GPPO performs moderately but lags behind DGPPO. PPO shows the highest completion times, reflecting limited scalability and less effective task allocation under load. Combined with the deadline results, this confirms that DGPPO is the most effective method for both timely execution and meeting deadlines.

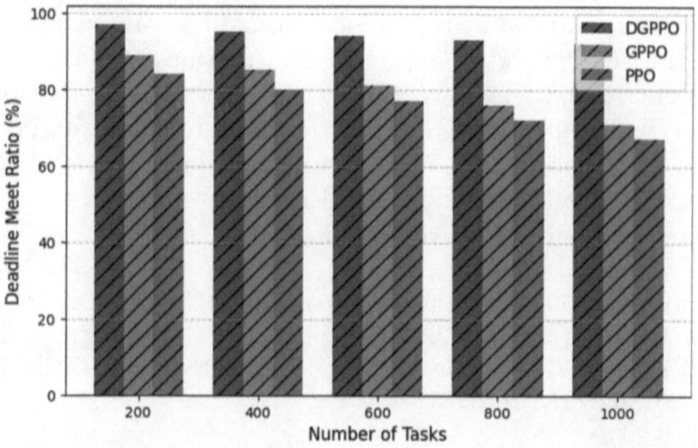

Fig. 4. Task deadline adherence comparison for DGPPO, GPPO, and PPO.

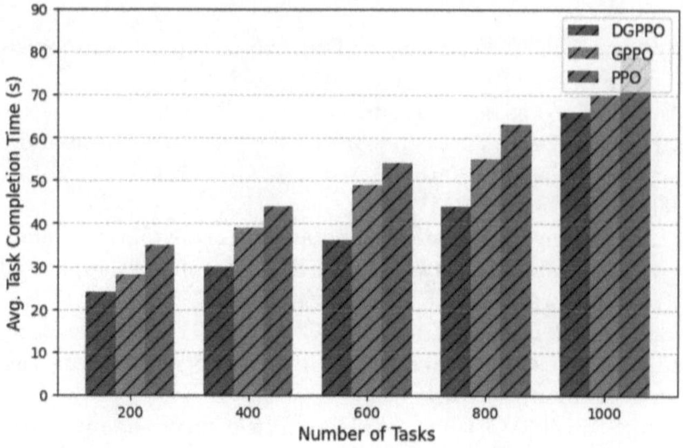

Fig. 5. Average task completion time comparison for DGPPO, GPPO, and PPO.

6 Conclusion and Future Work

We proposed DGPPO, a DT-assisted task offloading framework for dynamic MEC environments. By integrating GAT for topology-aware representation and PPO for adaptive policy learning, the framework enables intelligent offloading under non-stationary network conditions. The DT provides a real-time virtual replica of the MEC system, synchronizing server availability, network quality, and workload distribution to guide decisions. A blockchain-based mechanism ensures data consistency and trust among MES candidates. Task offloading is modeled as an MDP, capturing stochastic network behavior and multi-dimensional costs, with sub-processes optimized via DRL within the DT environment. Simulations show DGPPO improves energy efficiency, reduces latency, and enhances task success rate and decision accuracy compared to baselines. Future work includes integrating energy-constrained UAV edge servers, cross-layer optimizations for offloading, caching, and bandwidth allocation, and advanced DT models with graph forecasting and federated learning. In future work, we plan to enhance the blockchain layer with efficient consensus mechanisms and adaptive verification to reduce latency, while integrating metrics such as throughput and delay to evaluate trust versus responsiveness. We also aim to explore cooperative offloading, task replication, model compression, and security- and privacy-aware mechanisms, validated through DT-driven co-simulations and V2X testbeds.

Acknowledgments. This work was supported by the National Science and Technology Council of Taiwan (NSTC114-2221-E-155-021) and National Research Foundation of Korea (NRF) funded by the Ministry of Education (No. NRF-2023R1A2C1003143 and No. NRF-2018R1A6A1A03025526).

References

1. Chen, X., Wang, L., Liu, Y., Zhao, J.: Dtdrlmo: A digital twin-assisted deep reinforcement learning algorithm for microservice offloading. IEEE Trans. Mob. Comput. (2024). https://doi.org/10.1109/TMC.2024.1234567
2. Dong, M., Li, K., Zhang, Y.: Deep learning for user association and resource allocation via digital twin in mec. IEEE Access **7**, 151314–151326 (2019). https://doi.org/10.1109/ACCESS.2019.2947985
3. Gao, Y., Li, X., Zhang, Y.: Spatio-temporal graph convolutional networks for task migration in vehicular edge networks. IEEE Trans. Mob. Comput. **23**(5), 345–356 (2024). https://doi.org/10.1109/TMC.2024.1234567
4. Han, M., Sun, X., Wang, X., Zhan, W., Chen, X.: Transformer-based distributed task offloading and resource management in cloud-edge computing networks. IEEE J. Sel. Areas Commun. , (2025)
5. Hu, Y., Zhang, T., Liu, W., Chen, M.: Digital twin-driven adaptive dnn inference in aiot. IEEE Internet Things J. (2024). https://doi.org/10.1109/JIOT.2024.1234567
6. Huang, W., Zhao, Z., Min, G., Wang, Y., Chang, Z.: Mmto: Multi-vehicle multi-hop task offloading in mec-enabled vehicular networks. IEEE Trans. Mob. Comput. , (2025)
7. Jeremiah, S.R., Yang, L.T., Park, J.H.: Digital twin-assisted resource allocation framework based on edge collaboration for vehicular edge computing. Futur. Gener. Comput. Syst. **150**, 243–254 (2024)
8. Li, H., Xiong, K., Lu, Y., Chen, W., Fan, P., Letaief, K.B.: Collaborative task offloading and resource allocation in small-cell mec: A multi-agent ppo-based scheme. IEEE Trans. Mob. Comput. , (2024)
9. Li, J., Zhang, X., Zhou, F., Wu, D.: Flexedge: Dt-enabled uav-assisted task offloading for vehicular edge computing. IEEE Trans. Veh. Technol. **72**(5), 5623–5636 (2023). https://doi.org/10.1109/TVT.2023.3245678
10. Li, X., Zhang, Y., Wang, Z.: Deep learning-based task offloading in vehicular edge computing: A survey. IEEE Access **12**, 7890–7905 (2024). https://doi.org/10.1109/ACCESS.2024.1234567
11. Liu, Y., Chen, X., Wang, H.: Sata-drl: A task graph-based offloading mechanism in mobile edge computing. IEEE Trans. Parallel Distrib. Syst. **34**(7), 1234–1245 (2023). https://doi.org/10.1109/TPDS.2023.1234567

12. Mustafa, E., Shuja, J., Rehman, F., Namoun, A., Bilal, M., Bilal, K.: Deep reinforcement learning and sqp-driven task offloading decisions in vehicular edge computing networks. Comput. Netw. **262**, 111180 (2025)
13. Shen, J., Lin, Y., Zhang, Y., Zhang, W., Shu, F., Li, J.: Content caching-assisted vehicular edge computing using multi-agent graph attention reinforcement learning. arXiv preprint arXiv:2410.10071 (2024)
14. Sonam, S., Gupta, P., Sapra, S.: Deep reinforcement learning-based task scheduling in edge computing. In: Optimizing Edge and Fog Computing Applications with AI and Metaheuristic Algorithms, pp. 173–205. Auerbach Publications (2025)
15. Tan, X., Wang, M., Wang, T., Zheng, Q., Wu, J., Yang, J.: Adaptive task scheduling in digital twin empowered cloud-native vehicular networks. IEEE Trans. Veh. Technol. **73**(6), 8973–8987 (2024)
16. Uddin, M., Rahman, A., Chowdhury, M.: Task offloading strategies in vehicular edge computing: A deep reinforcement learning approach. IEEE Transactions on Intelligent Vehicles **10**(1), 56–67 (2025). https://doi.org/10.1109/TIV.2025.1234567
17. Ullah, I., Singh, S.K., Adhikari, D., Khan, H., Jiang, W., Bai, X.: Multi-agent reinforcement learning for task allocation in the internet of vehicles: Exploring benefits and paving the future. Swarm Evol. Comput. **94**, 101878 (2025)
18. Wu, Z., Jia, Z., Pang, X., Zhao, S.: Deep reinforcement learning-based task offloading and load balancing for vehicular edge computing. Electronics **13**(8), 1511 (2024)
19. You, J., Kim, S., Lee, D.: Hierarchical reinforcement learning-based task offloading in v2i networks using graph attention networks. IEEE Trans. Veh. Technol. **73**(2), 234–245 (2024). https://doi.org/10.1109/TVT.2024.1234567
20. You, J., Kim, S., Lee, D.: Hrl-gat: Hierarchical reinforcement learning with graph attention networks for task offloading in vehicular networks. IEEE Trans. Mob. Comput. **23**(3), 345–356 (2024). https://doi.org/10.1109/TMC.2024.1234567
21. Zhang, H., Sun, Y., Chen, L., Zhao, Q.: Distributed dl-based task offloading via digital twin. Futur. Gener. Comput. Syst. **143**, 85–98 (2023). https://doi.org/10.1016/j.future.2023.01.012
22. Zhang, Y., Zhao, R., Huang, X., Xu, W.: Multiobjective task offloading with dt in vehicular edge computing. IEEE Trans. Veh. Technol. **72**(9), 9845–9858 (2023). https://doi.org/10.1109/TVT.2023.3256789

Assessing the Operational Effectiveness of e-Highways in Germany Through Simulation

Nicole Müller[1], Suchithra Rajendran[2(✉)], and Adithya Gupta[3]

[1] Department of Production, Operations and Logistics Management, University of Passau, Innstraße 41, 94032 Passau, Germany

[2] Department of Industrial and Manufacturing Systems Engineering, University of Missouri, Columbia, MO 65211, USA

RajendranS@missouri.edu

[3] Rock Bridge High School, Columbia, MO 65203, USA

Abstract. In the forthcoming decade, it is expected that a significant proportion of trucks will operate on electric highways worldwide. While existing research has primarily emphasized infrastructure design, limited attention has been devoted to the operational aspects of e-Highway networks. This study addresses this gap by examining network-level operations along the A5 corridor in Germany, between Langen/Mörfelden and Weiterstadt. A structured Define, Measure, Analyze, Design, and Verify (DMADV) methodology is applied to (i) characterize the operational problem, (ii) acquire and process data from multiple sources, (iii) develop a simulation model, and (iv) analyze system performance under varying conditions. Sensitivity analyses are conducted with respect to truck speed, fleet size, and highway lane capacity. The findings demonstrate that truck speed and fleet size substantially influence both the number of operating trucks and the average time vehicles spend in the system, whereas lane capacity exerts a comparatively limited effect. These insights provide a foundation for data-driven decision support in the design and management of e-Highway operations.

Keywords: Electric Highways · Define, Measure, Analyze, Design and Verify (DMADV) · Network operations · Simulation modeling

1 Introduction

1.1 On-Road Logistics

Road freight transportation in Europe has exhibited a consistent upward trend, reaching its peak in 2017 with 1,913,116 million tonne-kilometers (tkm), the highest volume recorded in the preceding five years (Eurostat-1, 2018). A wide spectrum of commodities is transported by road, including food and beverages, agricultural products, metals, and hazardous materials. In terms of tonnage, construction-related materials such as metal ores and other mining and quarrying products represented the largest share in 2017, accounting for 24.8% of the total. When transport distances are taken into account, food products, beverages, and tobacco emerge as the most significant category, contributing 17.2% of the total tonne-kilometers transported in the European Union (Eurostat-2, 2018).

H. R. Arabnia et al. (Eds.): CSCI 2025, CCIS 2974, pp. 68–90, 2027.
https://doi.org/10.1007/978-3-032-28475-4_5

At the regional level, road freight activity is heavily concentrated in Germany, Poland, Spain, and the United Kingdom. Figure 1 illustrates the distribution of road freight across these major economies. In contrast, the lowest levels of road freight are observed in rural areas, particularly in eastern and southern Europe, as well as in smaller island regions, where geographical and infrastructural constraints limit freight movement (Eurostat-3, 2019).

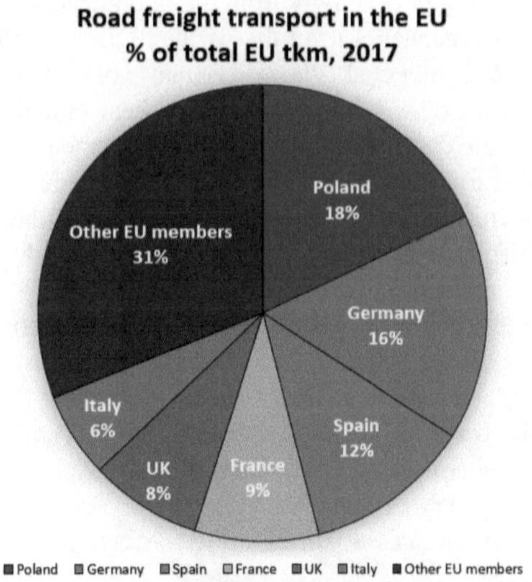

Fig. 1: Distribution of road freight transportation in the EU in 2017

According to Eurostat, road freight transport still keeps its leading position in the modal split of road, rail and inland waterways. When comparing the share of inland freight transported within the EU in 2017 by the different transport modes, road transportation (76.7%) was more than four times higher as the transport by rail (17.3%) and significantly higher than inland waterways (6.0%) (Eurostat-4, 2019). Similarly, in the U.S., road freight transport is also the number one of moving goods between cities: nearly 70% of all freight is transported by road (Truckinfo, 2018). The primary reason for road transport being the most commonly used shipment mode is that freight can be delivered door-to-door, which is crucial for day to day items such as groceries, fruits, flowers, etc. Other advantages are the flexibility from a time, route and capacity perspective and the cost-effectiveness (Rodrigue et al. 2013; McKinnon et al., 2008).

On the EU's roads, 6.3 million trucks were operating in 2016, with Poland possessing the largest truck fleet, followed by Germany and Italy (Fig. 1). Nearly all of the existing trucks in the European Union run on diesel (96.1%) (ACEA-1, 2018). Although road freight transportation in Europe is fundamental for the development of the economy, it is adverse for the environment, and it affects human health. For 23 out of 38 Member States of the EU, the air quality standards exceeded the recommended limit (European Commission, 2016). Pollution caused by transport and noise was linked to several increasing health problems, such as respiratory diseases, lung cancer, or cardiovascular diseases (Xia et al., 2015).

1.2 Climate Change and Measures to Ensure Sustainability

Earth's climate has been changing with a constant increase in atmospheric CO_2, ever since the industrial revolution (NASA, 2019a, b). More than 95% of global warming is due to the consequence of human activity from the mid 20th century. More specifically, the GHG emissions have reached a level that is the highest of all times (IPCC, 2014; Ramaswamy et al., 2006; Santer et al., 2003). The cause of global warming, i.e., "greenhouse effect", is the result of atmosphere confining heat radiating from Earth toward space. Certain gases like CO_2, Nitrous oxide (N_20), Methane (CH_4), and Water vapor (H_20) contribute to the greenhouse effect and block heat from escaping. The concentration of GHG has increased as a result of human activities like industry, clearing of agriculture land, etc. To a more considerable extent, burning fossil fuels like oil and coal has increased the concentration of CO_2 (NASA, 2019a, b). One of the primary implications of the greenhouse effect is the global temperature rise: Since the late 19th century, the temperature of the planet's average surface has risen about 0.9 °C (1.62 °F), caused by increased CO_2 and other emissions due to human activities (Climatic Research Unit, 2016; GISTEMP Team, 2019; NOAA, 2019). Several policies, such as the Paris Agreement for climate change, attempt to limit the relative global temperature rise to less than 2 °C in this century (United Nations, 2015). Further significant impacts of climate change are warming oceans, shrinking ice sheets, glacial retreat, among other things (NASA, 2019a, b).

Transport is one of the biggest contributors to CO_2 emissions in Europe, calculating 32% of total emissions. In road transportation, heavy-duty trucks (HDV) represent approximately 25% of CO_2 emissions (ICCT, 2018). Since the freight demand and the stagnating HDV fuel efficiency, emissions are expected to increase by 10% within the next decade (EUR-Lex, 2016). Considering that road transportation in the EU accounts for 76.7% of total inland freight transport, the EU currently has the third biggest CO_2 footprint of the major markets regarding HDVs, behind China and the U.S. (Facanha et al. 2012; Eurostat-4, 2019). The European Commission targets to reduce GHG emissions from all sources by 80–95% before 2050 (European Commission, 2011). Whereas, a two-step target is established for CO_2 emission reduction, expecting a decline of 15% and 30% before 2025 and 2030, respectively (European Commission, 2019).

These conditions lead to a situation that forces countries to deal with the impacts of transportation operations. In an attempt to achieve a sustainable solution, the automotive industry is forced to consider alternative technologies. Given the critical contributions of HDVs towards CO_2 emissions, alternative technologies are particularly essential for fulfilling climate stabilization goals (ICCT, 2017).

1.3 Electric and Hybrid Electric Trucks

The pressure of HDV to act and reduce CO_2 emission in freight transport to achieve the targets of climate protection is highly relevant. For the reduction of emissions, different alternative drive trains of fuel options are required. In contrast to the above mentioned alternative fuels and powertrains, potential technologies for heavy-freight transport are disputed, as in particular, the high energy consumption means a challenge in terms of range. For HDV, the following are alternative technologies or fuel options: electric plug-in), electric hybrid (catenary or in-road charging), hydrogen-powered fuel cell and ICE running on synfuel.

For this paper, we focus on electric and hybrid technologies. Battery electric drive (plug-in) consists of an electric motor and battery as energy storage. It increases energy efficiency, reduces fuel and maintenance costs, reduces GHG emissions, and eliminates local air pollution. Due to the limited electric range and the long charging time of the battery, it actually is more convenient for shorter distances (Fraunhofer Institute, 2018). The fuel cell works with an electric motor and fuel cell with hydrogen that act as energy storage. This technology enables quick refueling time, longer ranges, and is also reducing GHG and eliminating air pollution, but the challenges are the infrastructure development and renewable hydrogen cost.

Battery electric vehicles that run in-road or with overhead catenary, enable zero-emission transport, increased energy efficiency and reduced maintenance and fueling costs. Trucks that are powered by an overhead catenary system are considered as hybrid vehicles. The cost for the overhead catenary wires and the construction of underground infrastructure and the standardization across regions is a key challenge and a market entry barrier of this technology. Currently, the total cost of ownership for alternative vehicles is above the cost of diesel vehicles. By 2030, it is expected that the total costs for all alternative vehicles are going to be less than diesel, due to benefits from economies of scale, production optimization, and technology improvements. For example, the cost of overhead catenary HDV would be 25–30% less than diesel vehicles. This illustrates that the alternative electric technologies for HDV represent promising and cost-effective possibilities for an emissions-reduced future (ICCT, 2017).

1.4 Electric Highway

Electric highways or so-called “e-roads”, provide vehicles with a source of power by charging onboard batteries or transferring electricity to an electric motor. In other words, electric vehicles have the option to use overhead catenary electricity or inductive dynamic charging that are integrated into roadways. When these overhead catenary wires are used on different sections of a highway, they enable greater range and smaller batteries for vehicles. The vehicles used on electric highways are either a hybrid drivetrain with an ICE or they are equipped with a fully electric drivetrain, where power electronics, pantographs, or induction coils are combined with a battery pack. The hybrid powertrain or added battery capacity allows the vehicle to achieve a greater range from the roads (ICCT, 2017). These kinds of projects require high upfront investments due to high infrastructure costs, i.e., the construction of overhead wires.

Despite the high costs, a study found that hybrid trucks are competitive to conventional diesel trucks since their maintenance and fuel costs are low (Den Boer et al., 2013). Other models of e-roads, e.g., conductive on-road or inductive charges, have the possibility of a wider variety of vehicles that can be used and lower infrastructure costs, but the development of their technology is not as mature as the catenary system (Connolly,

2016). The inductive charging technology is based on wireless inductive power, either dynamic inductive charging or stationary inductive charging technologies.

For stationary wireless inductive charging, vehicles are usually parking for a long time, e.g., in a bus station, parking lot, etc. or it stops during the way for a short time, e.g., bus stop, traffic light. This technology can be seen as mature, particularly focusing on passenger cars. On the contrary, for dynamic charging, the vehicle is allowed to drive during the loading process, but on the related road or the special lane. Dynamic wireless charging creates the opportunity for electric vehicles to have smaller batteries and to avoid charging time. It is already demonstrated in many prototypes of buses and rail projects (e.g., KAIST, Bombardier, Oak Ridge National Laboratory, the University of Auckland, etc.) (Zhao et al., 2018). Still, the technology is in its early stages, and it faces more challenges compared to stationary wireless charging due to the in-road charging.

The cost of the dynamic inductive charging infrastructure strongly varies with the vehicle type, peak traffic flow, road electrification coverage, and charging demands. When breaking down the costs into road installation cost and power supply installation cost, one can see that the road installation makes up most of the infrastructure cost, and it depends on road electrification coverage and power demands. For the power supply installation, the peak power demands are crucial to establish the costs. Overhead catenary technology is well developed and can be determined as mature compared to wireless charging. The combination of overhead power supply with the flexibility of road transport could increase the efficiency of freight transport with electric trucks. These trucks can connect/disconnect to/from overhead catenary wires by using a pantograph.

When looking at the cost of these technologies, we divide into truck capital cost and infrastructure cost. The vehicle cost varies with the road electrification coverage. For an electrified highway with an electrification coverage of 50%, the costs of a dynamic charging truck and a catenary truck are competitive with the costs of a diesel truck. Costs to operate are lower for the dynamic charging truck and the catenary electric truck compared to the diesel truck, due to low maintenance cost and the low electricity price. So when comparing the overall total cost of a mile, the costs of these electric operated trucks are similar to those of a diesel truck. Nevertheless, the infrastructure costs of the dynamic charging technology are high, likewise the costs for catenary charging technology (Zhao et al., 2018).

1.5 Operation of Electric Highway

1.5.1 Sweden's e-Highway

In 2016 Sweden tested the first electric highway with a catenary system for HDV. A two-kilometer strip of the highway was electrified with a 750 V DC overhead catenary system designed by Siemens and the trucks provided by Scania. The trucks on the road are hybrids, with 9-L diesel engines, that run on biofuel and as well on electric motors. They pull electrical power from a pantograph that can freely connect and disconnect with the overhead catenary wires at a speed of 90 km/h. The hybrid powertrain enables the trucks to drive in all-electric mode while they are connected to the catenary system. When the truck is outside the electrically-powered lane, it is powered by hybrid electric mode, i.e., by the electric motor or the combustion engine (Zhao et al., 2018).

1.5.2 California's e-Highway

The ports of Los Angeles and Long Beach are also operating with a similar catenary system to reduce the pollution of diesel trucks between the ports and inland warehouses. The short-term goal is to minimize pollutant emissions around the dock rail yards. For the long-term, the goal is to achieve a zero-emission truck corridor between the ports and Los Angeles along the I-710 and CA-60 freeways. The strip is one kilometer long in two directions, and the system is also using 750 V DC power. In California, four demonstration trucks are planned: a diesel hybrid truck, a battery-electric truck, and two-hybrid trucks running on natural gas (Zhao et al., 2018).

1.5.3 Germany's e-Highway

In 2018, Germany opened his first electric highway for overhead hybrid trucks on a 10 km strip of the A5 highway, between Frankfurt and Darmstadt. The e-Highway aims to achieve a tone of neutral emissions by the technology of catenary overhead electricity, that is developed by Siemens. The trucks will use a hybrid powertrain so that they can operate without being connected to the catenary wires (Carrilero et al., 2018).

1.5.4 KAIST the Online Electric Vehicle (OLEV)

KAIST demonstrated a dynamic inductive charging project called OLEV in South Korea, in 2012. This project included two wireless recharged buses running on a 15-mile line, which is electrified by 10–15% and covered with wireless recharging pads. The OLEV buses received 100 kW power from coils that were installed at the underbody of the bus and placed under the road surface of the route. Since the electrified road was designed to differentiate regular cars from the OLEV buses, the power supply was only activated when the buses were driving on the road; otherwise, it was turned off to prevent unnecessary power consumption (Zhao et al., 2018).

1.5.5 Bombardier PRIMOVE

Bombardier has already demonstrated its first inductive power transfer system with a light rail and an e-bus project in 2012. The PRIMOVE technology was also adopted in a research project in Belgium, and in 2013, Scania and Bombardier jointly presented the dynamic inductive charging technology as part of a project. A Scania diesel hybrid truck with the Bombardier inductive charging technology was tested on a 300-m test road. The charging system was able to transfer 200 kW between the road and trucks. The purpose of this project and the test track was to validate the functionality of the dynamic inductive energy transmission for HDVs and to learn more about infrastructure costs (Zhao et al., 2018).

1.6 DMADV Approach

Define, Measure, Analyze, Design, and Verify (DMADV) is a six sigma framework for the development of new processes, products, or services. It is especially useful when implementing a whole new initiative or strategy, but also there is a need for redesigning an existing process, product, or service. The DMADV approach is used to develop a simulation model, and it is structured in a five-step procedure. The identification of the purpose, goals, and risks of the process, project or service and a clear definition of the project is performed in the *Define* phase. In the *Measure* phase, data is collected and pre-processed to achieve a summary and better understanding of the input data that is needed to design the model. The third step *Analyze* includes the examination of the data and the development of possible alternatives to a baseline model. In the *Design* phase, the selected alternative is drafted and developed in detail. In the final stage *Verify*, the proposed model is validated, i.e. it has to pass a series of test runs to ensure high quality (Celano et al., 2012; Six Sigma Daily, 2012).

The remaining of the paper is organized as follows. Section 2 details the review of the literature in electric highways and logistic operations. Section 3 provides the methodology used in this study, while Sect. 4 describes the data used for this research in detail. The results are given in Sects. 5, and 6 overviews the conclusions and potential future work.

2 Literature Review

2.1 Electric Highways

Lipp (2007) underlined the difference between the progress of achieving full electrification of different countries. While many countries are moving towards the inevitable future of electric mobility, some are still far away from achieving full electrification. Another article by Krok (2019) specifically highlighted the benefits of electric truck operations on e-Highways in Germany., i.e., trucks driving without contribution to air pollution. Similarly, Kottasová (2019) described the environmental boost as well as the road benefits when using the system of Siemens for trucks on electric highways in Germany. The author also mentioned that the demonstrations and tests of this e-Highway technology have also been carried out near the ports of Los Angeles and Long Beach as well as in Sweden.

Plötz et al. (2019) analyzed electric trucks powered by overhead lines (trolley trucks) in the EU and their impact on CO_2 emissions and the electricity system. The results demonstrated that CO_2 emissions from heavy road transport could be reduced, even after considering the additional CO_2 emissions that can be produced in the electricity sector. The authors concluded that for heavy road transport, electric trucks constitute an attractive alternative for reducing CO_2 emissions.

Taljegard (2019) investigated the impacts of road transport electrification on the electricity generation system, in terms of energy demand and power on different temporal and geographical scales. After developing a vehicle energy consumption model, the results showed that electrification could affect the electrical system in different ways, depending on the scale of electrification, the electrification strategy used, and the time of consumption. When comparing indirect electrification (e.g., electrofuels or hydrogen) and direct electrification of passenger cars (e.g., involving static charging), both technologies can increase the electricity demand, but indirect electrification can increase the demand four times as high as direct electrification.

Benato et al. (2018) examined a High Voltage Direct Current (HVDC) interconnection between Italy and France, which is a modern example of the interaction between electric energy and highway infrastructures. The main characteristics of the link, i.e., the power system choices (geometrical and electrical characteristics, type of cable, converter station technologies), and the main drivers for the installation design, are described.

2.2 Simulation Study for Analyzing Highway Operations

While discrete-event simulation has been widely used for studying different modes of transportation (Rajendran et al., 2015; Murarvev et al., 2016; Smith and Srinivas, 2019; Rajendran, Harper, 2020) the application of these tools to analyzing highway operations is limited (Polat et al., 2009; Oh et al., 2017; Motro et al., 2016).

In the study presented in Polat (2009), the main objective was scheduling a real-life asphalt highway construction project of on-site and off-site operations, using a combination of discrete-event simulation techniques and line-of-balance. The paper used the combination of project scheduling problems and did not only consider stochastic features of activities and constraints on resources, but it also allows the contractor to achieve workflow benefits between on-site and off-site operations. It is advantageous to the construction industry as it enables them to perceive the potential benefits of combining balance lines and discrete event simulation techniques to solve the resource allocation and project scheduling problems of linear construction projects.

Oh et al. (2017) reviewed different model-based approaches that are predicting travel time on highways. After dividing the models into four categories, they were evaluated from five perspectives, such as accuracy, prediction range, efficiency, robustness, and applicability. The results show that contrary to data-driven approaches, model-based approaches tend to be relatively resistant against changes in experimental circumstances, e.g., changing geographic conditions. However, the models also present some drawbacks (e.g., capacity, security gap, the density of traffic jams) that restrict the predictive performance for these limits.

Motro et al. (2016) use a microscopic traffic simulator to evaluate the effectiveness of an overtaking assistant, a wireless communication system based on short-range communication from vehicle to vehicle, which is intended to improve safety when overtaking on highways. After performing 18,000 overtaking scenarios, the results show that the overtaking assistant predicted 67% of the performed collisions successfully. What could not be detected was the communication range, because vehicles were beyond this range while overtaking. It is suggested to improve the short-range communications power settings to broaden their communication range for enhancing the effectiveness and proving safety on highways. However, the authors could have improved the research in several respects, e.g., increase the complexity of the performed overtaking scenarios to make them more similar to real-life or using collected data from the field instead of simulated data.

2.3 Application of DMAIC/DMADV to Improve Logistic Operations

Several papers have leveraged six sigma tools such as DMAIC (Define, Measure, Analyze, Improve and Control) and DMADV for improving logistic operations. For instance,

Smith and Srinivas (2019) evaluated the problem of long waiting time by improving inbound logistics operations, with the objectives to lowering the detention fees that have to be paid to the carrier and to minimize CO_2 emissions. They use DMAIC, a methodology to improve data-driven processes, to create a simulation of alternative check-in policies and their impact on environmental pollution and traffic congestion. The authors propose the construction of a staging area with dynamic dispatching rule as a short-time solution since it achieves lower CO_2 emissions and detention fees, and the capital investment is minimal. The authors concluded that the automated technology for the check-in process is a superior alternative because it reduces CO_2 emissions of trucks by 80% and removes detention fees, but it is more cost-intensive.

Another paper by Rajendran and Harper (2020) focused on the hyperloop technology by studying the operational perspective based on simulation models to provide insights on the hyperloop network operations. The author developed a DMADV approach and performed a sensitivity analysis by varying parameters, such as the number of Hyperloop capsules in the system, pod capacity, impact on the performance, etc. The results show that for the parameters, intense market research is recommended due to the exponential influence of these factors on the model output. For any company with a focus on logistics that is interested in using the hyperloop technology, the proposed simulation model can be used as a reference.

2.4 Contributions to the Literature

This research documents various contributions to the existing literature. Several recent studies examined the design of e-Highway systems, but to the best of our knowledge, this research is the first to concentrate on network operations. For this research, a two-phased approach was applied. In the first phase, the demand for the e-Highway network in highway A5 in Germany for every hour of the day and all days of the week using freely accessible data sources was estimated. The second step is to develop a simulation model using the input data of phase 1. Another key contributor to this research is the usage of the DMADV approach (Define, Measure, Analyze, Design, Verify), to gather a logical overview, understand the input data and their processing, and develop the simulation model.

3 Methodology

3.1 Application of DMADV Approach to the Current Study

In this study, we adopt the well-established data-driven methodology (DMADV) to examine the effectiveness of electric highways based on an e-Highway project that is currently operating on a test track in Germany (ELISA eHighway Hessen, 2020). The DMADV approach is used to develop a simulation approach, and it constitutes five stages: Define, Measure, Analyze, Design, and Verify. In the *Define* Phase, the highway structure and the system is observed, and to gather a logical overview, it was illustrated by a flow chart. Research goals and related challenges are also determined in this first phase.

The *Measure* phase serves to ensure the collection and the pre-processing of the required data to get an understanding of the input data that is needed to develop the model, e.g., the average number of trucks per day, truck speed per hour, etc. In this phase, we also generated several statistics and graphs to deep dive into the required data and to get a better overview. For example, we contemplated not only the average number of trucks per day but the average truck traffic per hour, as well as the distribution of truck traffic for each hour of the day and each day of the week. For the *Analyze* stage, we used the discrete event simulation, which is widely spread for supply chain and logistics operations (Celano et al., 2012; Jones et al., 2010). This phase includes the first development and the constant verification of a baseline model employing the collected and pre-processed data.

The *Design* phase is responsible for establishing the model and for the design and the evaluation of different alternate scenarios the model is running through. For our model, we used settings, such as different variations in truck speed, number of trucks, etc. The findings and estimations, in our case the positive and negative impacts of electric highways are presented and verified in the last phase, the *Verify* stage.

3.2 Terminology

Table 1 presents the list of commonly used terminology in the e-Highway operations.

Table 1: Terms used in the proposed e-Highway operations study

Terms	Definitions
E-Highway	Electric highways that provide vehicles with a source of power by charging onboard batteries or transferring electricity to an electric motor
Overhead truck	Trucks, that are powered by an overhead catenary system
Catenary system/ Overhead line	The overhead line supplies the electric motor of the truck with power, i.e., the batteries of the truck are charged
Pantograph	The installed pantograph of a truck connects to the overhead line and supplies the electric motor of the OH-truck with power
Time in System	Travel time of the trucks for the specific section of the highway based on the determined speed

3.3 Sequence of Events in e-Highway Operations

The operation of a truck on an electric highway follows a defined sequence of events, as illustrated in Fig. 2. Initially, the truck travels on the highway either in electric mode, powered by its battery, or through its onboard diesel generator. Upon approaching the overhead line, roof-mounted sensors detect the infrastructure, prompting the pantograph to extend and establish a connection with the line. Once connected, the overhead line supplies power directly to the truck's electric motor while simultaneously charging the onboard battery. At the end of the electrified section, the system automatically disengages the pantograph, thereby disconnecting the truck from the overhead line. Following disconnection, propulsion is maintained by the battery. If the battery becomes depleted after extended travel or if its charge is reserved for emission-free operation in urban areas, the diesel generator is activated to sustain continued movement.

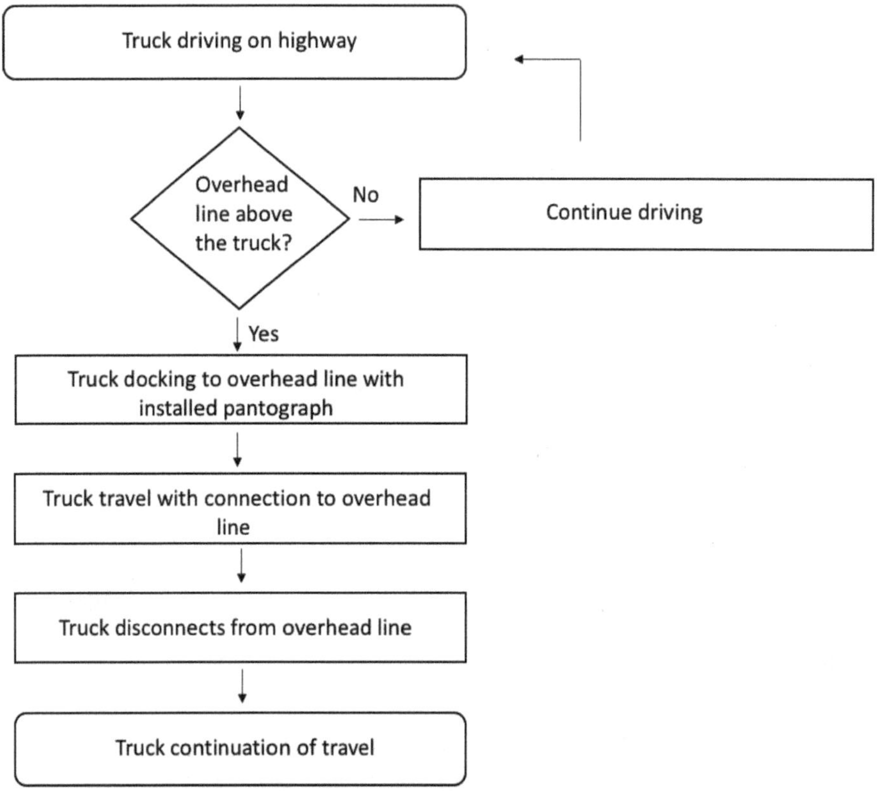

Fig. 2: Overview of the steps of an e-Highway operation

3.4 Model Overview

3.4.1 Assumptions

To establish the scope of analysis, several assumptions are adopted in modeling the e-Highway system. First, the number of electric trucks (e-trucks) that can connect to the overhead line at any given time is restricted, reflecting system capacity constraints. Second, the maximum travel speed is assumed to be identical for all e-trucks, ensuring uniform traffic flow and simplifying comparative analysis. Third, the highway is accessible to both e-trucks and conventional trucks. The overhead line is only activated when the pantograph of an e-truck establishes contact, which allows conventional trucks to operate on the same lanes without disruption. These assumptions provide a controlled framework for evaluating network operations and assessing system performance.

3.4.2 Simulation Model

For the simulation of the electric highway, we used the simulation software SIMIO®. In order to gain an overview of the different steps that are involved in the process, we are using an illustration of the e-Highway, which can be seen in Fig. 3.

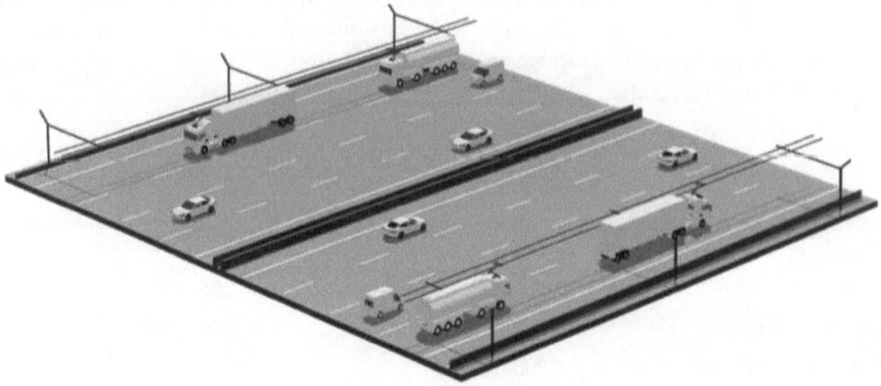

Fig. 3: Illustration of the e-Highway system (adapted from Riegelhuth, 2017)

4 Data

4.1 Data Description

The data employed in this study is obtained from the official reports of the Federal Highway Research Institute in Germany (Berichte der Bundesanstalt für Straßenwesen, BAST, 2018). The institute conducts annual analyses of traffic developments on federal highways, based on records collected from permanent automatic counting stations. For the purpose of developing the simulation model, the average daily traffic volume of vehicles and trucks on the selected test road was examined. Accordingly, the most recent dataset available from BAST, corresponding to the year 2018, was utilized.

After determining the total number of trucks operating on the highway per day, it was necessary to estimate the hourly distribution of truck flows across each day of the week. Since the BAST dataset reports hourly distributions for overall traffic volumes without distinguishing between vehicle types, the daily traffic pattern for all vehicles was applied to trucks. The hourly distribution of truck traffic was therefore derived using the available aggregate parameters, under the assumption that the temporal patterns of trucks are consistent with those of the broader vehicle population.

4.2 Steps Involved in the Data Estimation Process

Step 1: To calculate the traffic volume of trucks per hour and for each day of the week, we used the number of trucks traveling on the highway from the BAST. This number specifies the daily volume of truck traffic, i.e., the number of trucks that are driving on this highway per day.
Step 2: Secondly, we looked at the general daily traffic flow curves for each day of the week (Ruhr-Universität Bochum, 2007). These show the percentage of traffic per hour throughout the whole day and for each day of the week.
Step 3: Based on this data, we were able to calculate the number of trucks that are driving on the highway, for each day of the week and each hour of the day (as shown in Table 2). Therefore, we multiplied the percentage of traffic volume per hour for any day with the volume of daily truck traffic.

Table 2: Volume of trucks per hour for each day of the week

Time Interval	Weekdays (Mon–Fri)	Saturday
12:00–6:00 AM	200–400	250–330
6:00–9:00 AM	1,200–1,750	440–1,030
9:00 AM–12:00 PM	1,000–1,400	1,400–1,500
12:00–3:00 PM	950–1,200	1,160–1,200
3:00–6:00 PM	1,050–1,250	1,030–1,090
6:00–9:00 PM	740–1,020	610–890
9:00 PM–12:00 AM	185–425	370–500

4.3 Data Analysis

Figure 4 shows the daily truck traffic for every day of the week. In general, the volume of traffic is quite similar for each day of the week except for Sundays, since there is a driving ban for trucks in Germany on this day. On Saturdays, the number of traffic is the highest with 18,597 trucks/day, followed by Monday (18,487 trucks/day) and Friday (18,400 trucks/day). Interestingly, the volume of truck traffic is the same from Tuesday - Thursday, with 18,379 trucks/day.

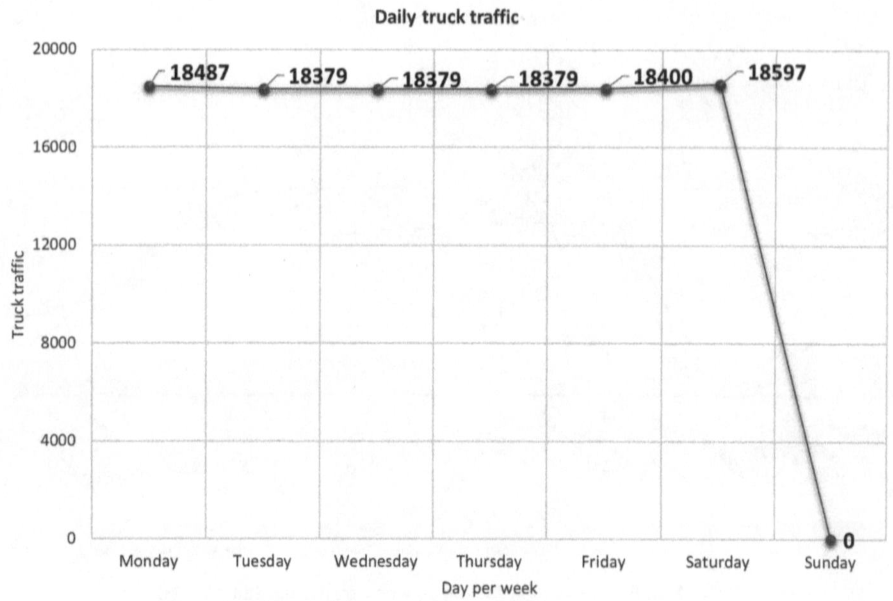

Fig. 4: Daily truck traffic for each day of the week

The daily distribution of the truck traffic for each hour of the day and every day of the week are shown in Fig. 5. On Monday, the graph shows an above-average percentage early in the morning, and it is falling steadily after the morning peak. For Tuesday to Thursday, we can observe a morning peak and relatively lower traffic in the afternoon. The graph for Friday has a strongly pronounced, broad maximum in the afternoon. Saturday has a relatively balanced distribution of traffic throughout the day.

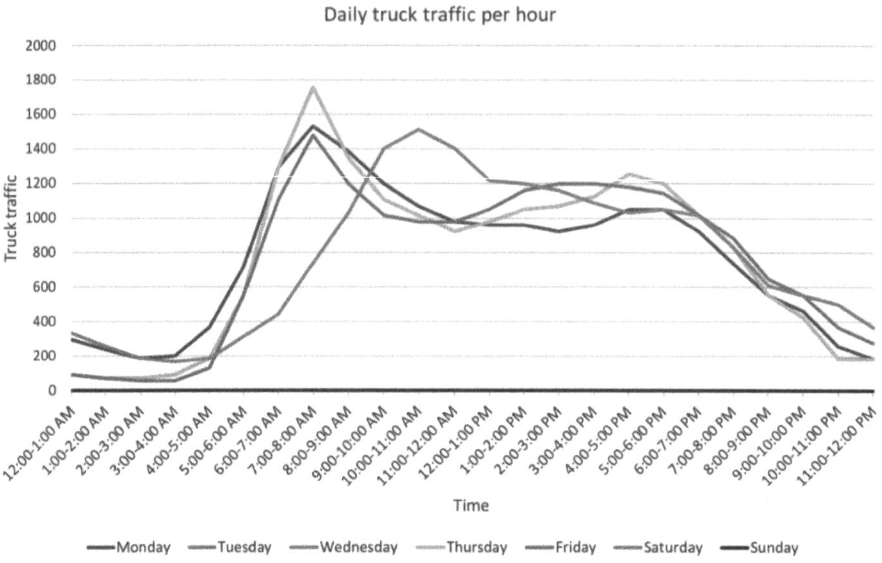

Fig. 5: Daily distribution of truck traffic per hour

Figure 6 shows the distribution of truck traffic for different periods of the day (morning, afternoon, evening, and night) and each day of the week. The traffic is highest in the morning where it is about twice times higher than evening traffic and approximately six times higher than night traffic. Afternoon traffic is also very pronounced and is in second place after morning traffic.

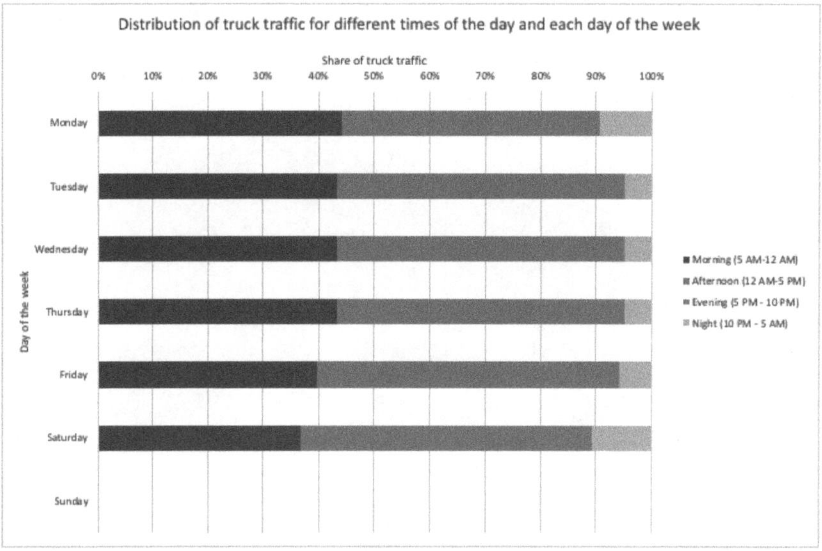

Fig. 6: Distribution of truck traffic for different time periods of the day and each day of the week

5 Results

5.1 Base Case

5.1.1 Base Case Model Parameter Settings

In Table 3, we have listed all input parameters that are used in the simulation model. These parameters are based on the framework conditions of the e-Highway project to be simulated. The variation of the parameters and their impact are investigated in a sensitivity analysis in Sect. 5.3.

Table 3: Input parameters

Parameter	Value
Length of the e-Highway	5000 m (Riegelhuth, 2017)
Speed of e-trucks	90 km/h (Siemens, 2012)
Average travel time of e-trucks	3.33 min (calculated based on the maximum truck speed and e-Highway length)
Length of e-trucks	9.75 m (SCANIA, 2018)
Average daily number of trucks	18,439 (BAST, 2018)

5.1.2 Base Case Results

For our base scenario, we assume that on average, five trucks per section can drive on the highway and use the overhead line at the same time (eHighway.SH). If we run five trucks at the same time with an average time in the system, we get a number of 90 trucks/h. For example, according to Table 1, the average number of trucks per hour is 771 on Monday. The maximum number of trucks per hour is 1531, the minimum is 185. If we assume that in the future all truck traffic will be electrified, a catenary overhead line that allows more than five trucks at the same time, as well as more than one line, is required. These are realistic scenarios and assumptions for the future, as the highway is a pilot project and started only in autumn 2019. It will be examined whether overhead line technology proves itself in practical operation and whether goods can be transported on the road in a climate- and noise-friendly manner.

5.2 Comparison of the e-Highway

“After the first thousand kilometers that were evaluated, we were able to record a fuel saving of diesel of about 10 percent. This is a pleasingly positive result even after the first weeks of the test phase” says Stefan Ziegert, Product Manager for Scania Germany Austria (emobilitaet.online, 2019). This is the first conclusion of the truck manufacturer of the e-Highway practical test after a few months.

Within a research project by the institute of energy- and environment research Heidelberg (Jöhrens et al., 2018) a potential analysis was carried out to examine the short-term realizable potential for the economic operation of overhead trucks (OH-truck). According to the research, it can be assumed, that for the initial phase of an overhead line system, the achievable cost advantages on the part of the operators will be small and will not be the decisive driver for switching to OH-trucks. The main reason for this is that investments are necessary for the infrastructure and the vehicle prices are very high.

In addition to that, the fuel prices are currently comparatively low, which limits the cost advantage of OH-trucks in electric operation compared to their diesel counterparts. However, it must be taken into account that diesel prices can also change noticeably at short notice. Until now, the Federal Environment Ministry has financed the project e-Highway in Germany with almost 14.6 million euros. A further 15 million euros will be spent on data collection and evaluation (Riegelhuth, 2017).

If an overhead line system is nevertheless successfully launched, a significant decline in vehicles cost could be expected. This will make the overhead truck economically attractive for many applications, provided that the infrastructure allows sufficient electric driving shares. If the establishment of a comprehensive overhead line network is successful, up to 65% of the mileage could be provided by OH-trucks in Germany. This would correspond to a CO_2 reduction of 3.7 million tonnes based on the number of travels that can be replaced by overhead trucks. This represents a reduction of 26%, which would make a relevant contribution to achieving the sector target for transport in 2030 (Jöhrens et al., 2018).

5.3 Sensitivity Analysis

For the sensitivity analysis, the different parameter settings are summarized in Table 4 and the related results in Table 5.

5.3.1 Impact of Truck Speed

In the base setting TS-1, the maximum truck speed is 90 km/h. For the different settings, we reduced the truck speed by 10 km/h, which yields a speed of 80 km/h for setting TS-2 and 70 km/h for setting TS-3. In the base setting TS-1, the average time in the system is 3.34 min with a number of 57 trucks per hour driving on the highway. For setting TS-2 the average time in the system is higher. However, one would expect the number of trucks using the highways to decrease with the decline in speed, but it remains almost the same. Similarly, setting TS-3 has a number of trucks per hour with 65, and an average time amounts to 5.06 min (7.2% increase from setting TS-2). In general, although the speed of the trucks has a linear impact on the average time in the system, the number of trucks utilizing the system remains almost the same, and thus the current system is robust to the change in truck speed.

Table 4: Parameter settings for conducting sensitivity analysis

Scenario	Truck Speed (TS)	Number of Lanes (NL)	Trucks Connected at the Same Time (TC)
Base	90 km/h	1	5
TS-1 (Base)	90 km/h	1	5
TS-2	80 km/h	1	5
TS-3	70 km/h	1	5
NL-1 (Base)	90 km/h	1	5
NL-2	90 km/h	2	5
NL-3	90 km/h	3	5
TC-1 (Base)	90 km/h	1	5
TC-2	90 km/h	1	10
TC-3	90 km/h	1	15

5.3.2 Impact of Number of Lanes

In this setting, the number of lanes in the system is increased from one lane to three lanes with an increment of 1. In the base setting NL-1, the average time in the system amounts to 3.34 min, and the number of trucks per hour is 57. When increasing the number of lanes in setting NL-2 by one lane, both the number of trucks and the average time in the system remain the same as in the base case (i.e., 57 trucks/h and 3.34 min respectively). A similar pattern is observed for setting NL-3, in which the number of lanes is increased to three. Thus, the number of lanes is observed to have no impact on the performance measures. In other words, the current system with a single lane is sufficient to accommodate the estimated truck traffic.

5.3.3 Impact of Truck Traffic

In the base case, it is assumed that only five trucks can be connected to the overhaul electric channel (according to the information provided by eHighway.SH, 2020). However, in the future, with better facilities, it is possible to have more trucks operating at any point in time. Therefore, in order to understand the impact of the number of trucks operating at any time, we varied the truck traffic from 5 to 15, in increments of 5 (settings TC-1 - TC-3). With an increasing number of trucks that are allowed to be connected at the same time, the average time in the systems remains the same, but the number of trucks per hour is increasing. Starting from 57 trucks per hour in TC-1, to 114 trucks per hour in TC-2 and 171 trucks per hour in TC-3.

Table 5: Results summary

Scenario	Average time in system (minutes)	Trucks per hour
Base (calculated)	3.33	57
TS-1 (Base)	3.34	57
TS-2	4.72	61
TS-3	5.06	65
NL-1 (Base)	3.34	57
NL-2	3.34	57
NL-3	3.34	57
TC-1 (Base)	3.34	57
TC-2	3.34	114
TC-3	3.34	171

6 Conclusions

Various climate agreements with the aim of reducing the greenhouse gas emissions and to limit the global temperature rise intensify the pressure for the transportation sector to reduce emissions steadily. Transportation is one of the biggest contributors to CO_2 Emissions in Europe, calculating 32% of total emissions. Especially in road transportation, HDV represents approximately 25% of CO_2 emissions (ICCT, 2018). One option to achieve the goals is the introduction of the overhead line hybrid trucks. The OH-truck draws its traction from an overhead line system, which enables extremely efficient use of electrical energy and great potential for reducing GHG emissions if renewable electricity is used (Jöhrens et al., 2018). The catenary system designed by Siemens and the trucks provided by Scania was implemented in Sweden, and thus, for the first time in Europe. Following other e-Highway system operations in the U.S. and Germany (Mobility Siemens).

In this research, we study and understand the effectiveness of e-Highway operations using the Define, Measure, Analyze, Design and Verify (DMADV) approach. To gather a logical overview, the highway structure is observed, and the research goals, as well as the challenges, are determined in the define phase. The measure phase is responsible for getting an understanding of the input data and their preprocessing, as well as the generation of several graphs and statistics to gain a better overview. Afterward, the development and constant verification of the baseline model takes place in the analyze phase, followed by the design phase, which is responsible for the design and the establishment of the model. In the verify stage, different parameter settings and key factors, such as truck speed, the impact of lanes, and truck traffic are varied, and their impact is studied and presented.

Results show that the system is affected by the truck speed and truck traffic. When varying these parameters, there is a significant impact on the number of operating trucks as well as on the time in the system. If the overhead lines are successfully introduced despite high vehicle and infrastructure costs at the beginning, the decrease in vehicle cost could be expected. Likewise, with increasing infrastructure development, the number of truck journeys by using a high number of overhead lines will increase. Within a research project of Fraunhofer ISS, PTV Group, and the Institure for Energy and Environmental Research Heidelberg, it was determined that both factors lead to a strong increase in the potential total mileage of OH trucks to around 65% in Germany. If this potential is fully exploited, about 3.7 Mt. CO_2 could be saved annually in 2030. Thus, OH trucks could make a relevant contribution to achieving the climate target for 2030 (Jöhrens et al., 2018).

One of the limitations of this study is that the e-Highway is currently still in the test phase, and the numbers we present have been estimated based on the current publicly-available truck traffic records. For future work, the actual testing results of the highway can be studied and analyzed. In this study, only the e-Highway on A5 in Germany is being considered. However, the proposed simulation model is generic and can be used to analyze the e-Highway operations for any expressway globally.

References

ACEA-1: European Automobile Manufacturers Association. Report vehicles in use Europe 2018. https://www.acea.be/statistics/article/report-vehicles-in-use-europe-2018 (2018). Accessed 10 Dec 2019

ACEA-2: European Automobile Manufacturers Association. New passenger car registrations by fuel type in the European Union. https://www.acea.be/uploads/press_releases_files/20191107_PRPC_fuel_Q3_2019_FINAL.pdf (2019). Accessed 10 Dec 2019

ACEA-3: European Automobile Manufacturers Association. Making the transition to zero-emission mobility. https://www.acea.be/uploads/publications/ACEA_progress_report_2019.pdf (2018). Accessed 10 Dec 2019

Auto.ndtv: Germany Gets Its First Electric Highway For Hybrid Trucks. https://auto.ndtv.com/news/germany-gets-its-first-electric-highway-for-hybrid-trucks-2035284 (n.d.). Accessed 4 Feb 2020

Brilon, W.: Modellierung von Dauerlinien und Ganglinien der Verkehrsnachfrage im Autobahnnetz, Ruhr-Universität Bochum. https://www.ruhr-uni-bochum.de/verkehrswesen/download/literatur/Modellierung_von_Dauerlinien_und_Ganglinien_der_Verkehrsnachfrage_im_Autobahnnetz.pdf (2007)

Carrilero, I., Gonzalez, M., Anseán, D., Viera, J.C., Chacón, J., Pereirinha, P.G.: Redesigning European Public Transport: Impact of New Battery Technologies in the Design of Electric Bus Fleets, XIII Conference on Transport Engineering: Transportation Research Procedia, p. 201 (2018)

Celano, G., Costa, A., Fichera, S., Tringali, G.: Linking six sigma to simulation: a new roadmap to improve the quality of patient care. Int. J. Health Care Qual. Assur. **25**(4), 254–273 (2012)

Climatic Research Unit: Data. https://crudata.uea.ac.uk/cru/data/temperature/ (2016). Accessed 15 Dec 2019

CNN: Germany is opening its first electric highway for trucks. https://edition.cnn.com/2019/05/07/tech/e-highway-a5/index.html (n.d.). Accessed 12 Dec 2019

Connolly, D.: eRoads: A Comparison between Oil, Battery Electric Vehicles, and Electric Roads for Danish Road Transport in Terms of Energy, Emissions, and Costs. Aalborg University. http://vbn.aau.dk/les/237844907/David_Connolly_eRoads_2016.pdf (2016). Accessed 19 Dec 2019

Delgado, O., Rodríguez, F.: CO2 emissions and fuel consumption standards for heavy-duty vehicles in the European Union, International Council on Clean Transportation (ICCT). https://theicct.org/sites/default/files/publications/Efficiency_standards_HDVs_EU_Briefing_20180504.pdf (2018). Accessed 3 Jan 2020

Den Boer, E., Aarnink, S., Kleiner, F., Pagenkopf, J.: Zero emission trucks. An overview of state-of-the-art technologies and their potential. http://www.theicct.org/sites/default/les/publications/CE_Delft_4841_Zero_emissions_trucks_Def.pdf (2013). Accessed 9 Dec 2019

Ehighway.SH. https://www.ehighway-sh.de/de/faq.html (2020). Accessed 12 Feb 2020

Emobilitaet.online: Das Anwenderportal. https://emobilitaet.online/news/forschungsprojekte/5608-scania-e-highway-a5-hybrid-lkw (2019). Accessed 10 Jan 2020

EUR-Lex: Access to European Union Law. Communication from the commission to the european parliament, the council, the euopean economic and social committee and the committee of the regions. https://eur-lex.europa.eu/legal-content/en/TXT/?uri=CELEX%3A52016DC0501 (2016). Accessed 18 Jan 2020

European Comission: Environmental Implementation Review: new way to help Member States apply EU rules benefits citizens, administrations and economy. https://ec.europa.eu/commission/presscorner/detail/en/IP_17_197 (2016). Accessed 16 Jan 2020

European Commision-2: Alternative Fuels. Expert group report. https://ec.europa.eu/transparency/regexpert/index.cfm?do=groupDetail.groupDetailDoc&id=34592&no=1 (2017). Accessed 28 Dec 2019

European Commission: A roadmap for moving to a competitive low carbon economy in 2050. Communication from the Commission to the European Parliament, the Council, the European Economic and Social Committee and the Committee of the Regions. http://eur-lex.europa.eu/legal-content/EN/TXT/?uri=CELEX:52011DC0112 (2011). Accessed 15 Jan 2020

European Commission: Alternative Fuels - Expert group report. https://ec.europa.eu/transparency/regexpert/index.cfm?do=groupDetail.groupDetailDoc&id=34592&no=1 (2017). Accessed 13 Jan 2020

European Commission-1: CO2 standards for heavy-duty vehicles in the European Union. https://theicct.org/publications/co2-stds-hdv-eu-20190416 (2019). Accessed 22 Dec 2019

European Environment Agency: National emissions reported to the UNFCCC and to the EU Greenhouse Gas Monitoring Mechanism, Directorate-General for Environment, United Nations Framework Convention on Climate Change. https://www.eea.europa.eu/data-and-maps/data/national-emissions-reported-to-the-unfccc-and-to-the-eu-greenhouse-gas-monitoring-mechanism-13 (2017). Accessed 13 Jan 2020

Eurostat-1: Road freight transport statistics. https://ec.europa.eu/eurostat/statistics-explained/index.php?title=Road_freight_transport_statistics#Growth_in_EU_road_freight_transport_for_the_fifth_consecutive_year (2018). Accessed 12 Jan 2020

Eurostat-2: Road freight transport by type of goods. Road freight transport by type of cargo. https://ec.europa.eu/eurostat/statistics-explained/index.php/Road_freight_transport_by_type_of_goods (2018). Accessed 12 Jan 2020

Eurostat-3: Transport statistics at regional level. https://ec.europa.eu/eurostat/statistics-explained/pdfscache/14273.pdf (2019). Accessed 11 Jan 2020

Eurostat-4: Freight transport statistics. https://ec.europa.eu/eurostat/statistics-explained/index.php?title=Freight_transport_statistics#Modal_split (2019). Accessed 12 Jan 2020

Facanha, C., Miller J., Blumberg K.: Global Transportation Energy and Climate Roadmap: The International Council on Clean Transportation, p. 31. https://theicct.org/sites/default/files/publications/ICCT%20Roadmap%20Energy%20Report.pdf (2012). Accessed 19 Jan 2020

GISTEMP Team: GISS Surface Temperature Analysis (GISTEMP), version 4: NASA Goddard Institute for Space Studies. https://data.giss.nasa.gov/gistemp/ (2019)

Hessen Mobil - Straßen- und Verkehrsmanagement: ELISA eHighway Hessen. https://mobil.hessen.de/ELISA (2020). Accessed 3 Jan 2020

IPCC Fifth Assessment Report: Summary for Policymakers. https://www.ipcc.ch/site/assets/uploads/2018/02/AR5_SYR_FINAL_SPM.pdf (2006). Accessed 5 Jan 2020

Jöhrens, J. et al. : Roadmap OH- LKW: Potentialanalyse 2020–2030, Fraunhofer ISS, PTV Group, Institut für Energie- und Umweltforschung Heidelberg. https://www.ifeu.de/wp-content/uploads/2018-12-20-ifeu-PTV-SCS-Potentialanalyse-Roadmap-OH-Lkw.pdf (2018). Accessed 25 Feb 2020

Lipp, J.: Lessons for effective renewable electricity policy from Denmark, Germany and the United Kingdom. Energy Policy. **35**(11), 5481–5495 (2007)

Maxwell, T.T., Jones, J.C.: Alternative Fuels: Emissions, Economies and Performance. SAE (1995)

McKinnon, A., Palmer, A., Edwards, J., Piecyk, M.: Reliability of Road Transport from the Perspective of Logistics Managers and Freight Operators. Logistics Research Centre Heriot-Watt University, Edinburgh, UK (2008)

Mobility Siemens. https://www.mobility.siemens.com/global/de/portfolio/strasse/ehighway.html (n.d.). Accessed 4 March 2020

Motro, M. et al.: Vehicular Ad-Hoc Network Simulations of Overtaking Maneuvers on Two-Lane Rural Highways, pp. 60–76 (2016)

Moultak, M., Lutsey, N., Hall, D.: Transitioning to zero-emission heavy-duty freight vehicles, International Council on Clean Transportation (ICCT). https://theicct.org/sites/default/files/publications/Zero-emission-freight-trucks_ICCT-white-paper_26092017_vF.pdf (2017). Accessed 3 Jan 2020

Murarvev, D., Aksoy, S., Rakhmangulov, A., Aydogdu, Y.V.: Comparing model development in discrete event simulation on Ro-Ro terminal example. Int. J. Log. Syst. Manag. **24**(3), 283–297 (2016)

Nasa: Global climate Change. Climate Change: How do we know? https://climate.nasa.gov/evidence/ (2019a). Accessed 9 Jan 2020

Nasa: Global climate Change. Vital Signs of Planet. https://climate.nasa.gov (2019b). Accessed 9 Jan 2020

NOAA: National Oceanic and atmospheric administration. https://www.ncdc.noaa.gov/monitoring-references/ (2019). Accessed 13 Jan 2020

Oh, S., Byon Y.J., Jang K., Yeo, H.: Short-term Travel-time Prediction on Highway: A Review on Model-based Approach, pp. 298–310 (2017)

Plötz, P., Gnann, T., Jochem, P., Yilmaz, H.Ü., Kaschub T.: Impact of Electric Trucks Powered By Overhead Lines on the European Electricity System and CO2 Emissions, pp. 32–40 (2019)

Plötz, P. et al. : Fraunhofer Institute for Systems and Innovation Research ISI, Oeko-Institut, Insitute for Energy and Environmental Research, Alternative drive trains and fuels in road freight transport – recommendations for action in Germany. https://www.oeko.de/fileadmin/oekodoc/Climate-friendly-road-freight-transport.pdf (2018). Accessed 20 Jan 2020

Polat, G., Buyuksaracoglu, Y., Damci, A.: Scheduling asphalt highway construction operations using the combination of line-of-balance and discrete event simulation. Techniques, 1126–1130 (2009)

Rajendran, S., Harper, A.: A simulation-based approach to provide insights on Hyperloop network operations. Tranp. Res. Interdisc. Perspect. **4**, 100092 (2020)

Rajendran, S., Srinivas, S., Saha, C.: Analysis of operations of port using mathematical and simulation modelling. Int. J. Log. Syst. Manag. **20**(3), 325–347 (2015)

Ramadhas, A.S.: Alternative Fuels for Transportation, pp. 8–9. CRC Press, Boca Raton, FL (2011)

Ramaswamy, V., Schwarzkopf, M.D., Randel, W.J., Santer, B.D., Soden, B.J., Stenchikov, G.L.: Anthropogenic and natural influences in the evolution of lower stratospheric cooling. Science. **31**, 1138–1141 (2006)

Riegelhuth, G.: Das eHighway-Projekt ELISA - Schwerlastverkehr per Oberleitung., Kongress: Elektromobilität im gewerblichen Umfeld. https://www.strom-bewegt.de/_aktion/emob2017/pdf/Riegelhuth.pdf (2017). Accessed 24 Feb 2020

Rodrigue, J.P., Comtois, C., Slack, B.: The Geography of Transport Systems. https://transportgeography.org/wp-content/uploads/GTS_Third_Edition.pdf (2013). Accessed 7 Jan 2020

Santer, B.D. et al.: Contributions of anthropogenic and natural forcing to recent tropopause height changes. Science. **301**, 479–483 (2003)

Scania: Presseinformation. https://www.scania.com/group/en/wp-content/uploads/sites/2/2018/09/p18083101de-vielseitige-hybrid-lkw-fur-den-stadtverkehr-1.pdf (2018). Accessed 18 Feb 2020

Siemens: eHighway - elektrifizierter Straßengüterverkehr. https://assets.new.siemens.com/siemens/assets/api/uuid:11c243e4061ed65b0a84a24d068d547b0a5a8afc/version:1500536895/broschuere-ehighway-de.pdf (2017). Accessed 27 Jan 2020

Six Sigma Daily. https://www.sixsigmadaily.com/what-is-dmadv/ (2012). Accessed 18 Jan 2020

Smith, D., Srinivas, S.: A simulation-based evaluation of warehouse check-in strategies for improving inbound logistics operations. Simul. Model. Pract. Theory. **94**, 303–320 (2019)

Trucking Info: Trucking statistics. https://www.truckinfo.net/trucking/stats.html (2018). Accessed 15 Jan 2020

United Nations: Climate Change. https://unfccc.int/process-and-meetings/the-paris-agreement/what-is-the-paris-agreement (2015). Accessed 22 Dec 2019

Xia, T., Nitschke, M., Zhang, Y., Shah, P., Crabb, S., Hansen, A.: Traffic-related air pollution and health co-benefits of alternative transport in Adelaide, South Australia. Environ. Int. **74**, 281–290 (2015)

Zhao, H., Wang, D.Q., Fulton, L., Jaller, M.: A comparison of zero-emission highway trucking technologies, A research report from the university of california institute of transportation studies. https://escholarship.org/content/qt1584b5z9/qt1584b5z9.pdf?t=pgskah (2018). Accessed 12 Jan 2020

Software Engineering Research and Practice

Visualizing Process and Memory Management Mechanisms

Hanh Pham(✉), Henry Becker, Kyle Wendholt, Manuel Reyes, Alec Lehmphul, Owen Pacchiana, Jack Lin, Michael Scotto, William Rubin, Mary Seelmann, Aaron Traver, Aleks Pilmanis, Christopher Brady, Charles Agudelo, Daniel Amoruso, Shelby Hinton, Marco Lemus, Jaedan Smith, Christian Collado, Tyler Wendover, Tevin Skeete, Joshua Morris, Michael Manzi, and Aram Agajanian

State University of New York at New Paltz, New Paltz, NY, USA
phamh@newpaltz.edu

Abstract. Visualization has emerged as a powerful approach for analyzing and understanding complex systems. Process and memory management mechanisms in computer systems can be difficult to be monitored and analyzed due to their dynamic nature. To address this challenge, this paper presents a set of software tools for visualizing key process and memory management mechanisms in computer systems. The given software tools provide visualizations of several fundamental CPU scheduling algorithms such as Shortest Job First (SJF), Shortest-Remaining Time First (SRTF), Priority Scheduling, and Round-Robin, as well as fundamental memory allocation mechanisms such as First-Fit (FF), Best-Fit (BF), and Worst-Fit (WF). In addition, they can demonstrate and monitor page replacement mechanisms such as Least-Recently-Used (LRU), Optimal, Most-Frequently-Used (MFU), and Least-Frequently-Used (LFU). These visualizations enable users not only to observe the behavior of individual mechanism but also to compare alternative strategies within the same category. The current system has three operational modes: view, advanced, and research, which allow users to analyze and investigate process and memory management mechanisms at varying access and functional levels. This software also supports experimental and comparative analysis.

Keywords: Software tools · Operating Systems · Visualization · CPU Scheduling · Memory Allocation · Page Replacement · Simulation

1 Introduction

Understanding and investigating complex computer mechanisms can present significant challenges. Visualization has been widely acknowledged as an effective approach for analyzing and monitoring complex systems [1, 7, 10, 19, 23, 30, 32] and mechanisms [9, 14, 21, 24, 29]. Process and memory management mechanisms are complex because they involve numerous parameters about processes and resources which dynamically affect system outcomes. These mechanisms, such as process scheduling and page replacement are often difficult to monitor, understand, and analyze in details.

H. R. Arabnia et al. (Eds.): CSCI 2025, CCIS 2974, pp. 93–106, 2027.
https://doi.org/10.1007/978-3-032-28475-4_6

To address these challenges, software simulators have been built for process and memory management mechanisms [3, 11, 15, 18, 28]. These simulators allow users to explore system behaviors in a controlled environment, thereby making otherwise abstract concepts more tangible. With visualization tools they add an interactive dimension that helps illustrate runtime behaviors, making the learning process more engaging and comprehensible. Over the past few decades, researchers have developed various visualization and simulation software for process and memory management.

Specifically, for CPU scheduling, Killen et al. [16] created Marble MLFQ, a visualization tool specifically focused on the multilevel feedback queue algorithm. Koya [17] developed Sched-ITS, an interactive tutoring system that provides guided instructions for CPU scheduling concepts through visualization. Sailah and Fadil [28] created an early simulator to demonstrate the behavior of different algorithms such as FCFS and Round Robin. Meanwhile, Leung and Zhang [18] animated the Linux processor scheduling algorithm, enabling learners to explore real-world CPU scheduling logic through visual demonstrations. Lowe and Liebrich [21] contributed VizzScheduler, a framework designed to visualize scheduling algorithms. Putchala and Bryant [27] focused on process synchronization with Synchron-ITS, which visualized the interactions of threads and shared memory for better conceptual understanding.

For memory management, Lopes et al. [20] developed SIME, a simulator for memory management, including allocation and paging techniques. Bhat et al. [4] extended this concept with a system for the visualization of virtual memory and memory allocation. Blanco et al. [5] introduced Vismep, a tool for visualizing memory consumption, enabling users to identify and understand patterns of resource usage. Similarly, Souza et al. [31] developed an interactive visualization platform for investigating memory consumption on Android devices, contributing to memory optimization and energy efficiency research. Azhari et al. [1] tackled memory leak detection problem with an efficient heap monitoring tool, demonstrating how visualization aids in root-cause analysis and performance tuning.

The visualization of memory performance in complex systems leads to analytical and diagnostic applications. Gimenez et al. [10] presented MemAxes, a framework for visualizing and analyzing complex memory performance behaviors, enabling users to identify performance bottlenecks and optimize workloads in high-performance computing (HPC) environments. Moreta and Telea [24] contributed to the visualization of dynamic memory allocations, allowing developers to monitor how applications interact with memory over time. Zellmann et al. [32] advanced visualization research by developing techniques for sparse volume data analysis using multi-core and GPU systems, reflecting the growing importance of visualization in high-performance environments. Similarly, Mateevitsi et al. [23] explored CPU visualization for optimizing HPC workflows, demonstrating that visualization not only aids in education but also in operational efficiency and scientific computing performance tuning.

Combination of visualization for both process and memory management in the same software include Hartley's work [13] which proposed the use of XTANGO, a system that allowed for the animation of process and memory management mechanisms to enhance conceptual understanding. This line of work was expanded by Jones and Newman [15], who introduced RCOS.java, a simulated CPU scheduling and memory allocation. Similarly, Maia et al. [22] proposed SOsim, to support hands-on learning and analysis of process and memory management mechanisms. To further enhance teaching effectiveness, Dall and Nieh [8] incorporated visualization into code review for OS education, emphasizing the benefits of examining the behavior of real code alongside visual process representations. Buendia and Cano [6] introduced Webgene OS, a generative and web-based architecture for teaching OS concepts. Alharbi et al. [2] and Hannaford et al. [12] developed an integrated standard environment enabling an interactive learning experience that unified scheduling, memory management, and synchronization modules. Similarly, Mustafa [25] introduced a modern OS visualization system to support enhanced learning through interactive experiments.

Beyond individual visualization tools, several studies have emphasized the analytical benefits of visual frameworks for understanding system-level performance. Chassin de Kergommeaux et al. [7] introduced Paje, an interactive visualization tool for tuning multi-threaded parallel applications, illustrating how visualization aids in performance analysis and optimization. Li et al. [19] developed PartFlow, a visualization system for mobile edge computing applications, supporting partitioning and workload offloading analysis. Soares and da Silva Oliveira [30] proposed an efficient architecture for real-time data visualization for IoT devices, demonstrating visualization's growing role in distributed and real-time systems.

Overall, while these examples demonstrate that visualization can bridge the gap between conceptual understanding and practical comprehension, enhances the analysis of complex computer systems and mechanisms, they lack openness where users can import their own mechanism design and implementation.

This paper describes an open, interactive, and web-based software for visualizing key process and memory management mechanisms in computer systems where users have full control of the visualizations, can easily modify input parameters and import their own design and implementation of mechanisms, visualize the results, compare them with the existing ones, and interact with the system in real-time. The rest of this paper is organized as follows. The software for simulating and visualizing process and memory management is presented in Sect. 2. Analysis of experimental results for process management mechanisms is explained in Sect. 3. Experimental results for memory allocation are provided in Sect. 4. Experimental results for page replacement are shown in Sect. 5. Conclusions and future work are provided in Sect. 6.

2 Software for Process and Memory Management Visualization

The design and development of these software tools are guided by the need to visualize process and memory management mechanisms in a manner that is open and interactive. To fulfill this objective, a client-server and web-based system was built which allows users to visualize various process and memory management mechanisms, enabling users to modify input parameters to observe their effects and enter user own output data or results so that users can visualize and check their accuracy easier. Users can also compare the performance of alternative mechanisms within the same category and submit their own implementations for analysis and comparison.

2.1 Software Architecture

Initially, the system which hosts the given software tools was implemented based on the design proposed in [26]. However, to make it more effective, initial modules were reorganized. The architecture of this web-based system is illustrated in Fig. 1. Mechanism Unit lets users select a specific process or memory mechanism to work with. Mode Unit lets users choose a mode from View, Advanced, or Research. Access Unit, through the Login entry point defines which modes are available to a given user. Submit Unit lets users work in the Advance mode via a Submission or in the Research mode via an Experiment. Admin Unit provides a dashboard of activities for the given user. Depending on the user's role: regular, manager, or admin, the dashboard can let users manage user's own record of submissions and experiments or records of the whole system.

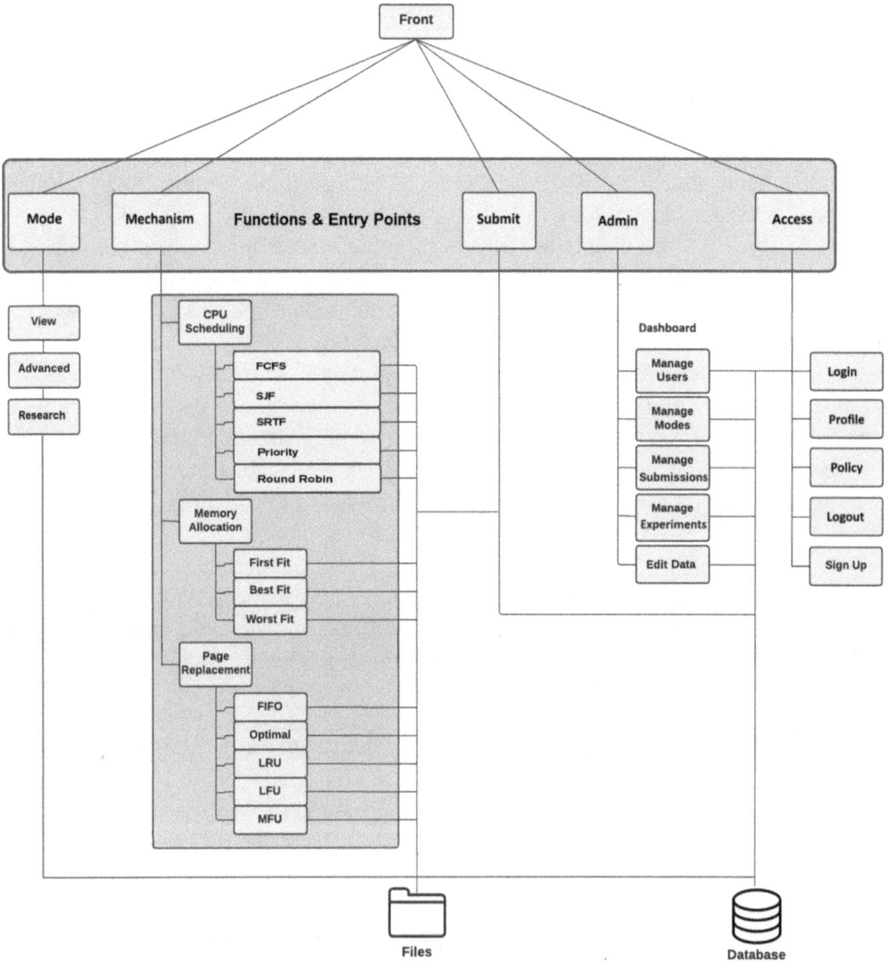

Fig. 1. Overall Software Architecture

2.2 Mechanism Visualization

This part of the software visualizes a process or memory mechanism based on its parameters or input data and its results or output data. In the case of full visualization, when the users at the client side select a mechanism, for example, Least-Recently-Used for page replacement, the mechanism ID or XXX, in this case is 023, will be defined and sent to the PHP/JS main thread at the server side as shown in Fig. 2.

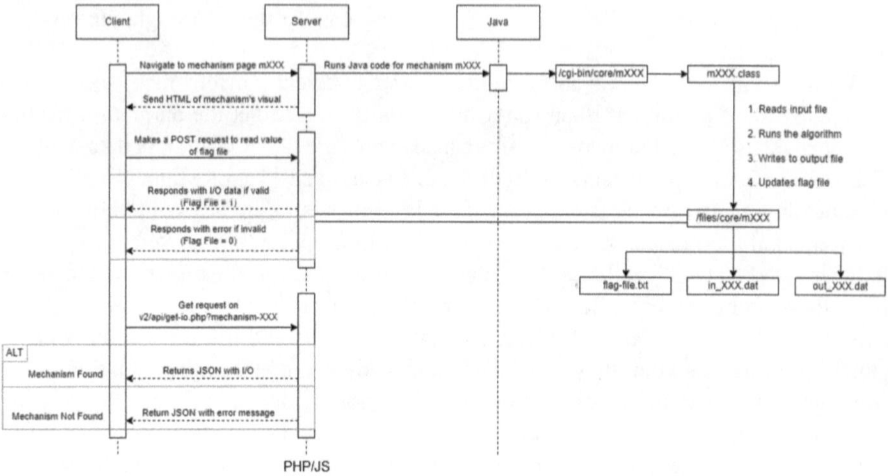

Fig. 2. Visualization of Mechanisms

The system will use the given mechanism ID to allocate and call the corresponding precompiled Java codes mXXX.class in a repository of codes in /cgi-bin/core/mXXX/ which implements the core functionalities of the given process or memory mechanism. This program should go to the data repository /files/core/mXXX/ and allocate and read the corresponding input data stored in the "in-XXX.dat" file, where XXX is the ID of the given mechanism. The data input, in this case a reference string or queue for page requests, is the default system data, stored in the "in-XXX.dat" file. The codes or program reads the input data and writes the results, in this case the contents of all memory frames for each step, to "out-XXX.dat" file, and switches the status in flag-file.txt to ready, as shown in Fig. 2. In parallel, the main thread, from where the mechanism program mXXX.class is forked out, keeps checking the status in flag-file.txt. Once the status turns to ready, that means the output data is produced and stored in the "out-XXX.dat" file, the main thread will use both input and output data to generate HTML/CSS/JS codes for animation. These codes will be sent to the client side and display the visualization of the given process or memory mechanism.

Besides the full visualization case, a lighter version is built for the case when the system output data for a process or memory mechanism is already well tested. In this case, the system will use the preproduced and prestored output data without running the Java codes to generate HTML/CSS/JS codes for animation. This helps to substantially reduce the system and server workload. This light visualization version can be used in

View mode as well as in the Advanced mode where users choose to run their codes on their own and local computers and submit only the output data to visualize the results for easy validations.

2.3 Submission Management

While users can see the visualization of process and memory mechanisms in the View mode using default input data, users can bring in different input data, output data, and codes in the Advanced and Research modes. They can do these through submissions and experiments.

With a submission in Advanced mode, users can select a mechanism, specify and modify the input data for it, then run a full visualization to get the animation for this specific and customized input data. Users can also provide their own output data, without the codes, for a given input data and get the results visualized through a light visualization, without running any codes or program. This kind of submissions would help users to easily check if their results are correct or not. The third option for the users is to provide their own codes, which reflexes their own design or solution for a given mechanism. Then, the system will compile and run user's codes or program which should read the given input data and produce fresh output data or results and use them to provide an animation to users so that they can see if their own design and implementation of the given process or memory mechanism is correct or not, as shown in Fig. 3.

With a submission in Research mode, users can compare several alternative mechanisms in the same category by having the codes for those mechanisms read the same input data. The system will collect different output data and visualize those results in a comparative manner.

2.4 Execution of User's Codes

As the proposed software is a web-based system which can have many restrictions for security purpose, letting users upload their own codes and run on the server can be risky. One solution is to restrict to Java codes only and give them the privilege to run on the local server, relying on the security of the JVM. This leads to fast execution. Another and a safer solution is to send users' codes to a remote server which provides a separate container for each submission to run, through a Docker engine. This leads to slower execution but can reduce the workload of the main server.

3 Experiments for Process Management

For process management, this software provides visualizations for fundamental CPU scheduling algorithms such as Shortest-Job-First (SJF), Shortest-Remaining-Time-First (SRTF), Priority, and Round-Robin. Users can also analyze the preemptive and non-preemptive versions of SJF and Priority. Within Priority methods two variations are available: Priority Low (lowest number means highest priority) and Priority High (highest number means highest priority).

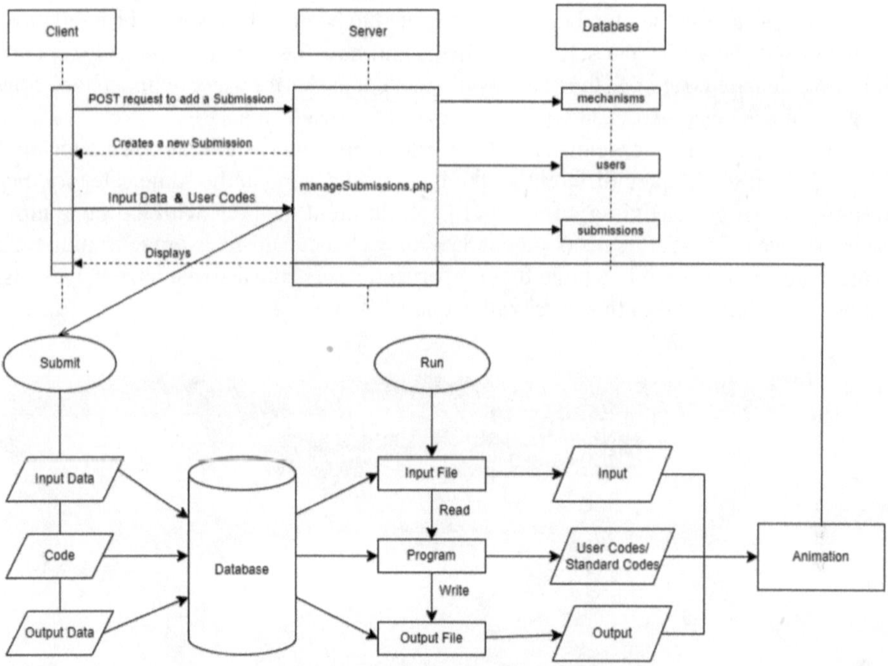

Fig. 3. Submission Management Flow

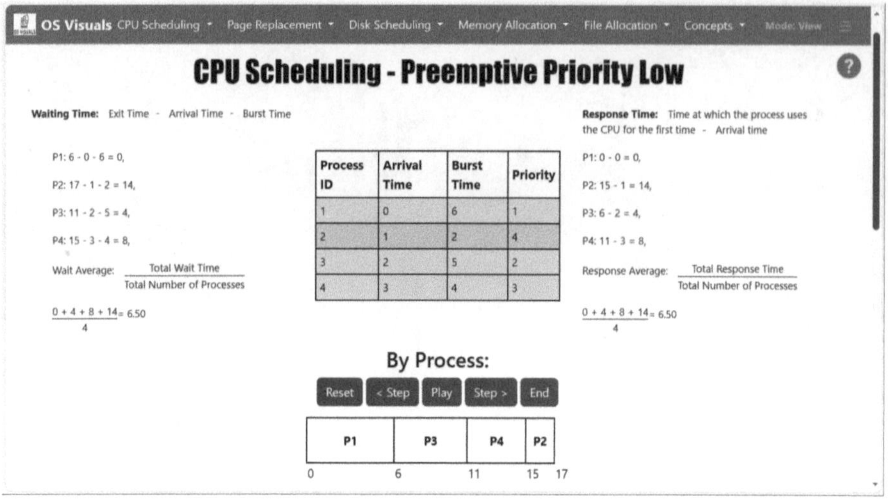

Fig. 4. Visualization of a Process Management Mechanism

In the View mode, any user can select a process management mechanism and analyze it through its visualization. For example, when the Preemptive Priority Low is selected, the system will use a default input data as shown in the central table in Fig. 4.

The users can use the blue buttons on the panel to control the pace of the animation, so that it will show how the schedule is made and how the waiting time and response time are calculated, step by step. This would make it easier for users to understand how the given mechanism works and able to analyze its impacts in details.

In Research mode, users can generate a random input data set or enter their own input data and use it to compare different methods or mechanisms in the same category. For example, for CPU scheduling, users can run all different methods with the same input data as shown in Fig. 5. Results or output data for each mechanism, average waiting time in this case, are recorded in a table for comparison. Users can also visualize the results, ranging from 5.0 to 9.0 in this case, with a chart.

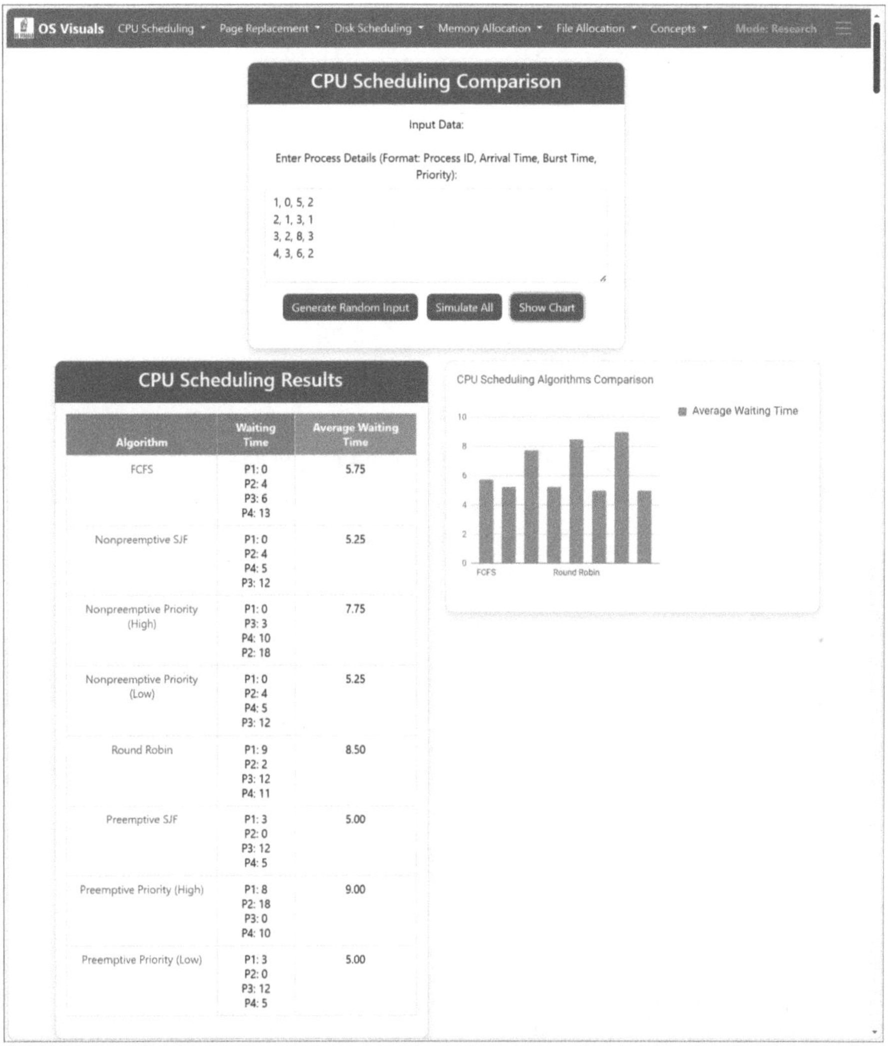

Fig. 5. An experiment with a Process Management mechanism in Research mode

In Advanced mode, users can edit input data, output data, as well as enter their own codes and design for a mechanism as shown in Fig. 6, this and some other figures are modified screenshots to be suitable for publication. By clicking on "Proceed to View" they can use the light visualization, without running any codes, to get an animation for a result or output of a mechanism. This serves well when users find this result or output by hand or testing their codes in their own computer without uploading into the system but need to visualize their results so that they can easily check if the results are correct.

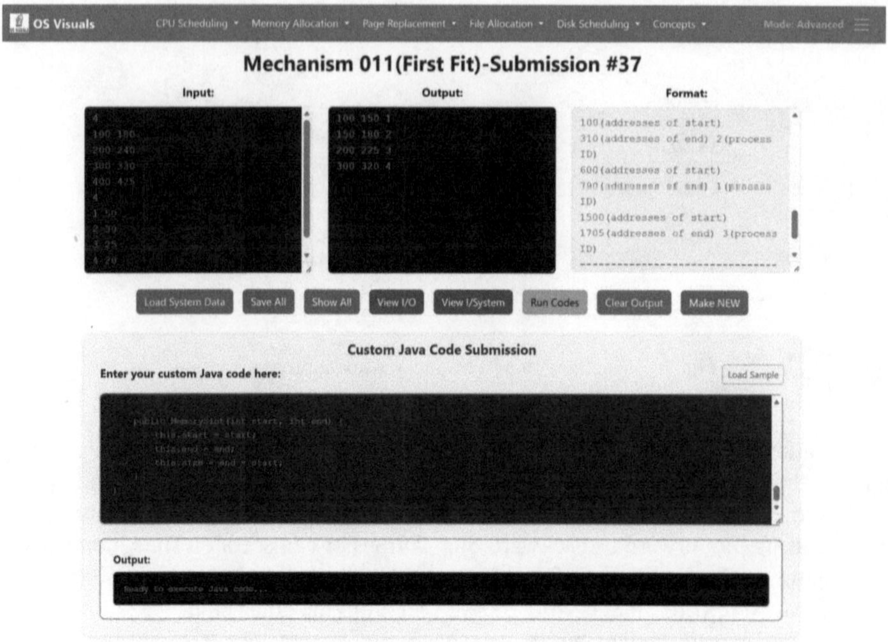

Fig. 6. Executing User's codes in Advanced mode

If users want to test their codes, they can drop them in the Java codes textbox and then click on "Compile and Run Java" and then "Proceed to View" to get a full visualization. In this case the user's codes may run within the Java Virtual Machine on the same web server or may be sent over to another remote server and run there within a container provided by a Docker engine. In first case, user's codes will be connected to the given input data, the produced output data will be stored in the same data repository. In the case of running on a remote server, the given input data together with user's codes will be sent and run there while output data will return.

4 Experiments for Memory Allocation

The built-in tools can provide visualizations for fundamental memory allocation mechanisms such as First-Fit (FF), Best-Fit (BF), and Worst-Fit (WF). In the View mode, users can select a memory allocation mechanism and analyze it through its visualization. For example, when the Best-Fit method is selected, the system will use a default input data as shown in Fig. 7.

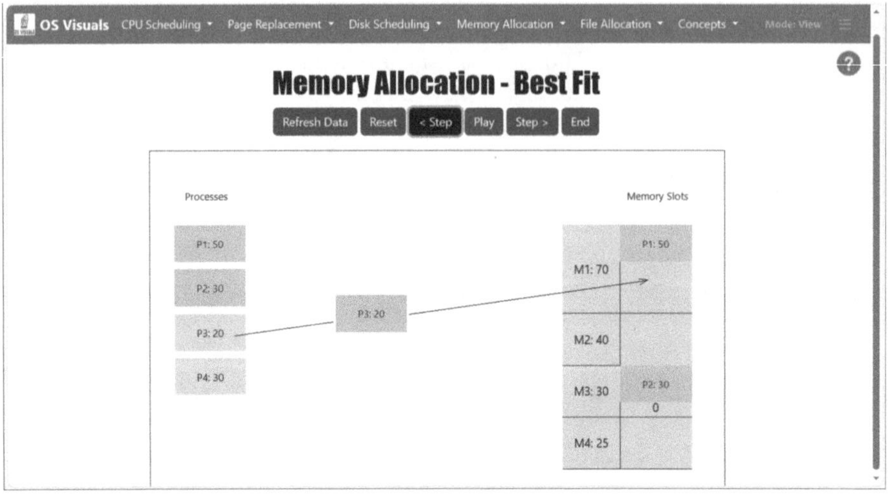

Fig. 7. Visualization of a Memory Allocation Mechanism

An animation is created to show a list of processes on the left and available memory slots on the right. The Play button on the control panel will automatically move each process to the right memory slot until all processes are examined. Users can control the pace by using the forward or backward Step buttons or can see all results immediately by clicking on the End button. Reset button starts the animation from the beginning. Refresh button reloads all data from the web server to the client side.

In the Advanced mode, similarly to what is in a process management mechanism, users can select a specific memory allocation mechanism from First-Fit (FF), Best-Fit (BF), or Worst-Fit (WF) and one of the following options. In the first option, they can edit or enter their own input data and do a full visualization which will use the system default codes, which read the given input data and produce new output data, and create an animation for the users. In the second option, users can just enter their own pair of input and output data do a light visualization, where an animation is created purely based on the given input and output data, without running any codes. In the third option, besides specifying their own input data, users can provide their own codes which reflex their own design and implementation for the chosen memory allocation mechanism. Then, they can compile and run their codes to get a full visualization as shown in Fig. 6. Again, user's codes can run within the Java Virtual Machine on the same web server or may be sent over to another remote server to run within a Docker container.

In Research mode, users can generate a random input data set and use it to compare First-Fit (FF), Best-Fit (BF), and Worst-Fit (WF) mechanisms in a style similar to what are in Fig. 5. Results or output data for each mechanism, number of unallocated processes in this case, are recorded in a table for comparison. Users can also visualize the results, ranging from 1 to 3 in this case, with a chart, again similar to what are in Fig. 5.

5 Experiments for Page Replacement

This software can provide visualizations for fundamental page replacement mechanisms such as FIFO, Least-Recently-Used (LRU), Optimal, Most-Frequently-Used (MFU), and Least-Frequently-Used (LFU).

In the View mode, users can choose a page replacement mechanism and analyze it through its visualization. For example, when the Optimal method is selected, the system will use a default input data, which is a reference string or a queue of requests with page numbers, as shown in the second line in Fig. 8. An animation is created to show the contents of the main memory or RAM in each step. The Play button on the control panel will automatically serve the next page number in the queue and show the new contents of RAM for this step while retaining the contents of RAM for all previous steps. This continues until all requests in the queue are served. Users can control the pace by using the forward or backward Step buttons. Users can also see the final result immediately by clicking on the End button. Reset button clears out all allocations and starts the animation from the beginning. Refresh button will reload the input and output data for the given page replacement mechanism from the web server to the client side.

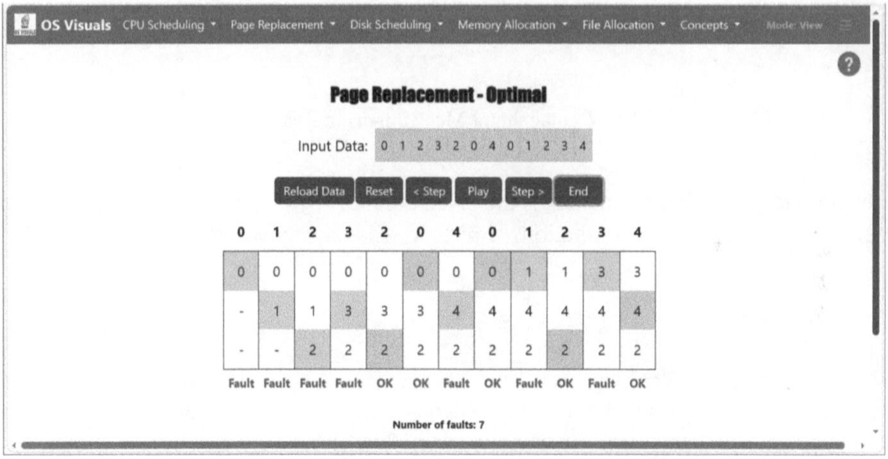

Fig. 8. Visualization of Page Replacement Mechanisms

In the Advanced mode, similarly to a memory allocation mechanism, users can edit input data, output data, as well as enter their own codes and design for a mechanism as shown in Fig. 6. Users have option to do either light (build animation using given input and output data, without running any codes) or full visualization to get an animation for a result or output of a mechanism.

In Research mode, users can specify their own input data or generate a random input data set and use it to compare FIFO, LRU, Optimal, MFU, and LFU mechanisms as shown in Fig. 9. Results or output data for each mechanism, number of page faults in this case, are recorded in a table for comparison. Users can also visualize the results, ranging from 12 to 17 in this case, with a chart. Then, users can visualize how exactly

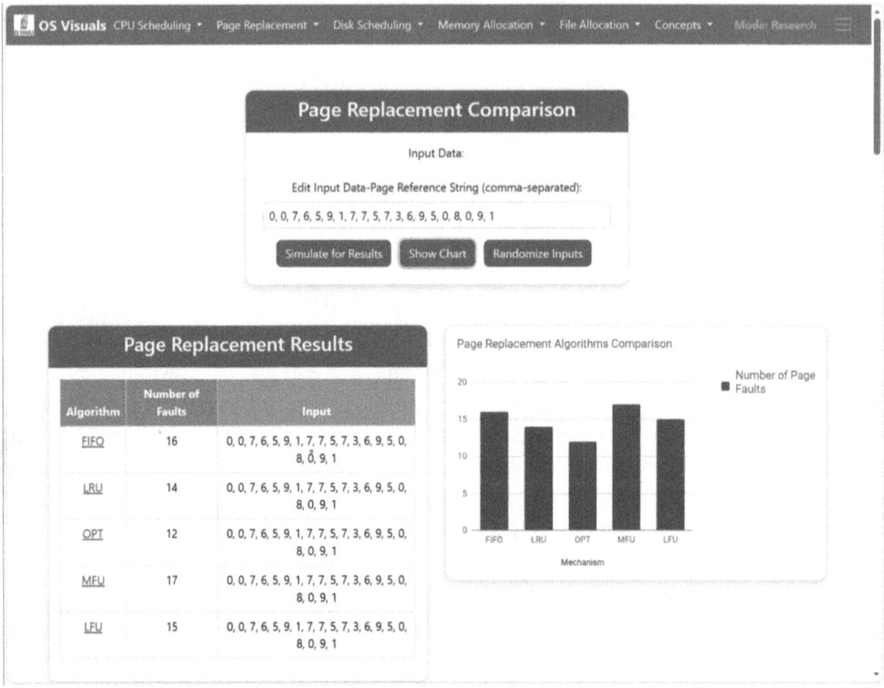

Fig. 9. Page Replacement Mechanisms in Research mode

each individual mechanism achieves the given outcome by clicking on the name of each method in the table to see animations similar to what are shown in Fig. 8.

6 Conclusions

We have developed a set of web-based software tools for visualizing process and memory management mechanisms in operating systems. Users can analyze various mechanisms through their visualizations in View mode where they can control the pace of animations. They can run various experiments by editing input data, output data, as well as entering their own codes which reflex their own design for a mechanism in Advanced mode. Users can also compare different methods or mechanisms in the same category by running them with a random input data set or their own input data in Research mode. These visualization and simulation tools can be very helpful in understanding and analyzing in depth various process and memory management mechanisms in operating systems. By enabling users to both observe and experiment with these mechanisms with visualization and simulation capabilities, these software tools can be a very effective means to ultimately mastering the inner workings of operating systems. Current limitations of this software include issues in dealing with multiple and concurrent access under the same user account, to the same submission from different client machines at the same time, and problems with transitioning from executing user codes to the visualization of its results for some mechanisms in advanced mode. To further advance these software tools, future work will involve refinement of all modules resolving the mentioned issues, along with the integration of AI models to guide and explain complex mechanisms to users.

References

1. Azhari, V., Bhamra, S., Ezzati-Jivan, N., Tetreault, F.: Efficient heap monitoring tool for memory leak detection and root-cause analysis. In: Proc. IEEE Int. Conf. Big Data, pp. 3020–3030. IEEE (2021)
2. Alharbi, A., Henskens, F., Hannaford, M.: Integrated standard environment for the teaching and learning of operating systems algorithms using visualizations. In: Int. Multi-Conf. Comput. Global Inf. Technol., pp. 205–208 (2010)
3. Barladian, B.K.H., et al.: Visualization component for the aircraft real-time operating system JetOS. Program. Comput. Softw. **46**, 167–175 (2020)
4. Bhat, W.A., et al.: Virtualization and visualization of virtual memory system for effective teaching–learning. Comput. Appl. Eng. Educ. **27**, 1286–1294 (2019)
5. Blanco, A.F., Bergel, A., Alcocer, J.P.S., Córdova, A.Q.: Visualizing memory consumption with Vismep. In: IEEE Working Conf. Softw. Vis. (VISSOFT), pp. 108–118 (2022)
6. Buendia, F., Cano, J.-C.: Webgene OS: a generative and web-based learning architecture to teach operating systems in undergraduate courses. IEEE Trans. Educ. **49**(4), 464–473 (2006)
7. Chassin de Kergommeaux, J., de Oliveira Stein, B., Bernard: Pajé: an interactive visualization tool for tuning multi-threaded parallel applications. Parallel Comput. **26**(10), 1253–1274 (2000)
8. Dall, C., Nieh, J.: Teaching operating systems using code review. In: Proc. 45th ACM Tech. Symp. Comput. Sci. Educ., pp. 549–554 (2014)
9. Farghally, M.F., et al.: Evaluating the effectiveness of algorithm analysis visualizations. In: Proc. 2017 ACM SIGCSE Tech. Symp. Comput. Sci. Educ., pp. 201–206 (2017)
10. Gimenez, A., et al.: MemAxes: visualization and analytics for characterizing complex memory performance behaviors. IEEE Trans. Vis. Comput. Graph. **24**, 2180–2193 (2018)
11. Goldweber, M., Davoli, R., Morsiani, M.: The Kaya OS project and the μMPS hardware emulator. ACM SIGCSE Bull. **37**(3), 49–53 (2005)
12. Hannaford, M., Henskens, F., Alharbi, A.: Integrated standard environment for the teaching and learning of operating systems algorithms using visualizations. In: Int. Multi-Conf. Comput. Global Inf. Technol., pp. 205–208 (2010)
13. Hartley, S.: Animating operating systems algorithms with XTANGO. ACM SIGCSE Bull. **26**, 344–348 (1995)
14. Hundhausen, C.D., et al.: A meta-study of algorithm visualization effectiveness. J. Vis. Lang. Comput. **13**, 259–290 (2002)
15. Jones, D., Newman, A.: RCOS.java: a simulated operating system with animations. In: Proc. Comput.-Based Learn. Sci. Conf. (2001)
16. Killen, S., et al.: Marble MLFQ: an educational visualization tool for the multilevel feedback queue algorithm. In: 8th IEEE Annu. Inf. Technol., Electron. Mobile Commun. Conf. (IEMCON), pp. 663–669 (2017)
17. Koya, B.: Sched-ITS: an interactive tutoring system to teach CPU scheduling concepts in an operating systems course (2017)
18. Leung, K.K., Zhang, D.: Animation of Linux processor scheduling algorithm. In: Proc. 7th IEEE Int. Symp. Multimedia (ISM'05), pp. 353–360 (2005)
19. Li, B., Li, M., Zhao, J., Cai, W.: PartFlow: a visualization tool for application partitioning and workload offloading in mobile edge computing. In: IEEE 18th Pacific Vis. Conf. (PacificVis), pp. 108–117 (2025)

20. Lopes, A., de Souza, D., de Carvalho, J., Silva, W., de Sousa, V.: SIME: memory simulator for the teaching of operating systems. In: Int. Symp. Comput. Educ. (SIIE), pp. 1–5 (2012)
21. Lowe, W., Liebrich, A.: VizzScheduler—a framework for the visualization of scheduling algorithms. In: Euro-Par 2001 Parallel Process., pp. 62–66 (2001)
22. Maia, L.P., Machado, F.B., Pacheco, A.C.: A constructivist framework for operating systems education: a pedagogic proposal using the SOsim. ACM SIGCSE Bull. **37**(3), 218–222 (2005)
23. Mateevitsi, V.A., et al.: Visuals on the house: optimizing HPC workflows with no-cost CPU visualization. In: IEEE 14th Symp. Large Data Anal. Vis. (LDAV), pp. 69–70 (2024)
24. Moreta, S., Telea, A.C.: Visualizing dynamic memory allocations. 4th IEEE Int. Workshop Vis. Softw. Understand. Anal., pp. 31–38 (2007)
25. Mustafa, B.: Visualizing the modern operating system simulation experiments supporting enhanced learning. In: Proc. 2011 Conf. Inf. Technol. Educ. (SIGITE'11), pp. 209–214 (2011)
26. Pham, H.: A web-based interactive and visualized approach to simulations of operating systems. In: Proc. Int. Conf. Inf. Comput. Technol. (ICICT), pp. 148–152. IEEE CPS (2022)
27. Putchala, M.K., Bryant, A.R.: Synchron-ITS: an interactive tutoring system to teach process synchronization and shared memory concepts in an operating systems course. In: Int. Conf. Collaboration Technol. Syst., pp. 180–187 (2016)
28. Sailah, H.M., Fadil, Y.A.: CPU Scheduling Simulation. J. Eng. Sci. (2009)
29. Shaffer, C.A. et al.: Algorithm visualization: the state of the field. ACM Trans. Comput. Educ. **10**(3), Art. 9, 1–22 (2010)
30. Soares, A.G., da Silva Oliveira, P.: An efficient architecture for real-time data processing and visualization for IoT devices. In: Int. Conf. Artif. Intell., Comput., Data Sci. Appl. (ACDSA), pp. 1–6 (2025)
31. Souza, G., Matias, P., Filho, R.M., Monteiro, E., Barreto, R., de Freitas, R.: Where did my memory go? An interactive visualization approach to investigate memory consumption on Android devices. In: IEEE Working Conf. Softw. Vis. (VISSOFT), pp. 106–110 (2024)
32. Zellmann, S., Schulze, J.P., Lang, U.: Binned k-d tree construction for sparse volume data on multi-core and GPU systems. IEEE Trans. Vis. Comput. Graph. **27**(3), 1904–1915 (2021)

Low-Cost Autonomous Indoor Navigation in Unstructured Environments

John Tadrous(✉)

Gonzaga University, Spokane, WA 99258, USA
tadrous@gonzaga.edu

Abstract. Critical indoor applications such as rescue operations in collapsed structures and close-quarters battle (CQB) tactics require a "send-and-forget" robot functionality. The user places the robot at a starting point and tasks it to autonomously navigate to a desired location. In this work, we develop and test a low cost indoor navigation framework that autonomously guides two-wheeled robots to arbitrary target points without pre-training or assistance from deployed infrastructure. Our framework utilizes LiDAR sensors, an inertial moment unit (IMU), and wheel encoders. Experimentation with Texas Instruments' TIRSLK has shown that the robot reaches its target location with an average distance accuracy of 98.3%.

Keywords: Embedded Systems · Indoor navigation · Sensor fusion and integration · CSCI-RTSE

1 Introduction

Autonomous indoor navigation has emerged as a major area of research, driven by the increasing demand for intelligent mobile systems in applications such as service robotics, logistics, surveillance, and assistive technologies.

Simultaneous Localization and Mapping (SLAM) has proven to be an effective approach for enabling mobile robots to build and navigate through maps of unknown environments [4,6]. SLAM inherently assumes a closed navigation environment where mapping requires the robot to return to a previous location several times.

In some time-critical applications, the closed navigation environment may not be feasible. For instance, emergency rescue operations and CQB tactics require robots to travel to inaccessible spots and execute tasks. In this work, we consider indoor autonomous navigation whereby prior knowledge regarding the navigation environment is not available, and infrastructure guiding the robot's trajectory is not present.

Dead reckoning techniques typically rely on local sensor measurements to estimate the robot's pose by integrating motion over time. Approaches such as sensor fusion, Kalman filtering, complementary filtering, and deep learning-based error correction have been explored to enhance dead reckoning reliability in dynamic and unstructured environments [5,7].

H. R. Arabnia et al. (Eds.): CSCI 2025, CCIS 2974, pp. 107–117, 2027.
https://doi.org/10.1007/978-3-032-28475-4_7

In [3], an obstacle avoidance mechanism has been implemented using only LiDAR data without vision-assisted modules. The system has shown performance merits in low luminosity regimes. However, positioning is attained using pre-deployed infrastructure. In [8], the authors utilize LiDAR and IMU data for autonomous electric wheelchair navigation. The proposed obstacle avoidance system has performed well; however, navigation to preset points requires the wheelchair to scan the environment and create a map and path plan, which is time and power-consuming for some time-critical applications.

In [6], mapping and path planning are required to guide the navigation process, but these steps come after the robot runs several scans (loops) of the indoor environment to create a reliable map, which requires additional navigation and exploration time that may not be suitable for some applications. In this work, we integrate IMU's gyroscope readings with wheel encoders and LiDAR sensor inputs to maintain a sufficiently accurate position of the robot, and allow the robot to navigate to a desired point without the need to construct a map or receive guidance from pre-deployed infrastructure.

In [9], a dead reckoning approach to robot positioning has been developed by specially mounting wheel encoders on the non-active wheels. While the achieved error performance has been promising, testing has been done only in a straight-line, unobstructed segment and not in an arbitrary indoor environment with possible blocking objects and walls. Our main contributions in this work are as follows.

- We develop a low-cost robot positioning system that relies only on encoders and IMU to track the instantaneous location and heading of the robot using a sensor fusion approach. Our system requires a simple mathematical analysis that reduces the time and power consumption of the robot.
- We develop an autonomous navigation framework featuring interrupt-based and foreground functions for object avoidance and navigation. No depth cameras, pre-training, or deployed infrastructure is required.
- We implement and validate our system on the TIRSLK [1] robotic system and test it through real-world experimentation. It attains a minimum average accuracy of 98.3% to the specified target points.

The rest of this paper is organized as follows. In Sect. 2, we lay out the robot positioning model. In Sect. 3, we present our proposed autonomous indoor navigation approach. In Sect. 4, we discuss the experimentation setup. Finally, we conclude the paper in Sect. 5.

2 Navigation System Model

2.1 Hardware Model

We consider a two-wheeled robot with two identical wheels that share the same axis of rotation, as shown in Fig. 1. Each wheel has a diameter d and is independently controlled by a gear motor with a quadrature encoder that generates n equally spaced pulses per wheel rotation. The robot's width is measured through the outer edges of the wheels to be w. The robot's center of gravity (CoG) is located at the midpoint of the wheels' axis and is used to capture the robot's position at any point in time.

Each motor is independently controlled by a direction signal and a PWM signal. At least one LiDAR sensor is mounted on the robot to determine the distance to nearby objects and enable object avoidance.

A gyroscope is mounted on the robot's surface to provide an independent estimate of the robot's heading. Gyroscope readings are sampled independently by the main microcontroller. The microcontroller runs at a sufficiently fast processing speed and is capable of interfacing with various sensors through different communication interfaces (e.g., I2C, UART, etc.).

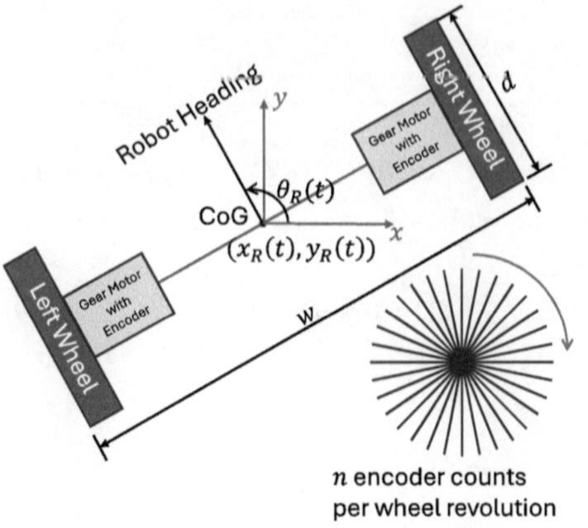

Fig. 1. Top view of a symmetric robot with the center of gravity (CoG) located at the midpoint along the axis of two identical active wheels. Each wheel is independently controlled by a gear motor with a quadrature encoder. The quadrature encoder generates n pulses per wheel rotation.

2.2 Positioning Analysis

The robot's position at time t is captured by the coordinates of its CoG, namely $(x_R(t), y_R(t))$, relative to the absolute x and y axes. The robot's heading angle $\theta_R(t)$ measures the angle between the x axis and the heading direction at time t.

Each of the n equally spaced encoder pulses signals a step movement by the respective wheel. The direction of each step is also captured by the encoder, allowing the robot to locally track its movement distance and direction.

Let $t + \delta t$ be the time instant at which one wheel has completed a movement step[1], then, based on that specific wheel position (left or right), the new robot

[1] Even when the two wheels are running, we assume the robot's CPU runs fast enough to detect and respond to encoders' pulses without missing a step by either wheel.

location and heading can be determined using the robot width w, wheel diameter d, and encoder counts n.

Define the angle $\alpha := \frac{d\times 360^\circ}{2nw}$, the chord $C := w \sin\frac{\alpha}{2}$, and the change in position and heading as $\Delta x := x_R(t+\delta t) - x_R(t)$, $\Delta y := y_R(t+\delta t) - y_R(t)$, $\Delta\theta_R := \theta_R(t+\delta t) - \theta_R(t)$, respectively.

Table 1 lists the robot's position and heading changes upon detecting a single step move by either the left or right wheel. Through these equations, the robot can track its current movement state as it navigates the surrounding environment.

Table 1. Changes in robot position and heading upon detecting a step movement by either robot wheel.

Motion Step	Left Wheel	Right Wheel
Forward	$\Delta\theta = -\frac{\alpha}{2}$ $\Delta x = C\cos(\theta_R(t) - \frac{\alpha}{2})$ $\Delta y = C\sin(\theta_R(t) - \frac{\alpha}{2})$	$\Delta\theta = \frac{\alpha}{2}$ $\Delta x = C\cos(\theta_R(t) + \frac{\alpha}{2})$ $\Delta y = C\sin(\theta_R(t) + \frac{\alpha}{2})$
Backward	$\Delta\theta = \frac{\alpha}{2}$ $\Delta x = -C\cos(\theta_R(t) + \frac{\alpha}{2})$ $\Delta y = -C\sin(\theta_R(t) + \frac{\alpha}{2})$	$\Delta\theta = -\frac{\alpha}{2}$ $\Delta x = -C\cos(\theta_R(t) - \frac{\alpha}{2})$ $\Delta y = -C\sin(\theta_R(t) - \frac{\alpha}{2})$

2.3 Autonomous Navigation Problem

Assuming the robot is initially positioned at location $x_R(0), y_R(0)$ with heading $\theta_R(0)$, it is tasked to autonomously navigate to a desired target location (x^*, y^*). The robot is required to reach its target location while consistently sensing for obstacles and performing necessary collision avoidance and repositioning functions.

The robot is agnostic to the surrounding environment and is not pre-trained on the plane layout, nor is it assisted with any beacon signals from deployed infrastructure.

3 Proposed Navigation Approach

Our approach aims at decomposing the robot's movement into straight line segments and turns about its CoG. In theory, straight line movement will only affect the robot's location, whereas turns about CoG will only change the heading while conserving the location.

In practice, realizing exact straight line movement and turning about CoG is not realistic because both motors are not perfectly identical, and their time constants and response to changes in PWM signals are not the same. Additionally, external irregularities on the navigation surface will affect the robot's trajectory. Accordingly, we realize PI controllers to stabilize the RPM speeds of the two wheels separately during motion.

At any time t, we maintain an estimate location $(\hat{x}_R(t), \hat{y}_R(t))$ and estimate heading $\hat{\theta}_R(t)$ of the robot's actual location and heading. Those estimates are updated by event-based interrupts that are triggered upon step completion.

Our high-priority interrupt-based tasks and main foreground functions for navigating the robot are explained in the following.

3.1 Interrupt-Based Tasks

Gyroscope Periodic Task: The gyroscope is periodically sampled through a timer-based interrupt that returns an estimate heading $\hat{\theta}_G(iT_G)$ at times $t = T_G, 2T_G, 3T_G, \cdots$, where T_G is the sampling period, and i is a positive integer. The raw gyroscope reading measures the rate of change of angular speed about the z-axis, which is processed and translated into the robot's estimated heading.

Let $Z_{Gyr}(iT_G)$ be the gyroscope reading at the ith sampling period. The estimated rate of change in heading between the $i-1^{st}$ and i^{th} sampling periods is obtained by a function $\phi(Z_{Gyr})$, determined based on the gyroscope manufacturing parameters and operational sensitivity. The estimated new heading based on the gyroscope data is

$$\hat{\theta}_G(iT_G) = \hat{\theta}_G((i-1)T_G) + \phi(Z_{Gyr}(iT_G))T_G.$$

Motor Encoder Tasks: We associate an interrupt with each wheel-generated pulse at every completed step. Upon detecting a wheel step at time $t + \delta t$, an encoder-based heading estimate $\hat{\theta}_E(t + \delta t)$ is updated as

$$\hat{\theta}_E(t + \delta t) = \hat{\theta}_E(t) + \Delta\theta,$$

where the quantity $\Delta\theta$ is determined according to Table 1, based on the direction and position of the wheel completing the step.

The robot heading estimate $\hat{\theta}_R(t+\delta t)$ is updated using a sensor fusion equation

$$\hat{\theta}_R(t + \delta t) := \gamma\hat{\theta}_E(t + \delta t) + (1 - \gamma)\hat{\theta}_G(t + \delta t), \quad (1)$$

where $\gamma \in [0, 1]$ adjusts the weight of each heading estimation component. The gyroscope estimate $\hat{\theta}_G(t+\delta t)$ is obtained from the most recent gyroscope sample.

We update the estimated location $\hat{x}(t + \delta t), \hat{y}(t + \delta t)$ based on the equations of Table 1, with $x_R(t), y_R(t)$ and $\theta_R(t)$ replaced with their estimates $\hat{x}_R(t), \hat{y}(t)$ and $\hat{\theta}_R(t)$ from the previous update.

The time spent by each wheel to complete every movement step is tracked through an input capture module. This is used to enable the efficient realization of the PI controllers to stabilize the RPM speed.

3.2 Robot Navigation Functions

Three main functions are used to manage the robot's navigation system as follows.

Align to point: Based on the robot's current location and heading, this function steers the robot about its CoG so that its heading aligns with a desired point, which is passed to this function. It also returns the number of wheel steps needed to reach the desired input point in a straight line. Given the current location and heading estimates $\hat{x}_R(t), \hat{y}_R(t)$, $\hat{\theta}_R(t)$, the function calculates the angle between the current robot heading and the desired location, as well as the direction of rotation to align the robot towards the target point. The function returns the remaining number of steps for the robot to reach the desired point in a straight line.

Robot Forward: This function moves the robot forward for a specific number of wheel steps, while consistently checking the surroundings for obstacles and controlling the wheels to maintain motion in a straight line trajectory. As the robot moves, the function samples the LiDAR sensor(s) and motor encoders every T_s time units. If the LiDAR sensor shows a potential collision, the function stops the robot and returns an *incomplete* output status. Otherwise, motor encoder data is used to assess the current RPM speed of each wheel, and a control PWM output is adjusted based on a PI controller for each wheel to maintain equal speeds on both wheels.

The function returns a *complete* status output when it moves the wheels for the desired number of steps without encountering an obstacle. If the distance to be traveled is above a certain threshold of η_A steps, the function realigns the robot to the desired target point after passing η_B steps.

Object Avoidance: This function steers the robot about its CoG away from a detected obstacle until there is enough clearance to move forward. The function further moves the robot forward in a straight line for a fixed number of steps by calling the Robot Forward function discussed above. While moving forward after steering from an object, the function returns an output status *incomplete* if the forward movement is obstructed, or *complete* otherwise.

To turn away from an object, the function steers the robot left or right, opposite to the side on the collision track. If the obstacle faces the front of the robot, the function randomizes its steering direction with equal probability, as illustrated in Fig. 2. This randomization ensures the robot will ultimately escape from any object by the recurrence of random walks.

4 Experimentation

4.1 Hardware Setup

We implement our framework on the Texas Instruments' Robotic System Learning Kit (TiRSLK) [1], as shown in Fig. 3. It encompasses two active wheels, each powered by a separate DC motor attached to a quadrature encoder, and a

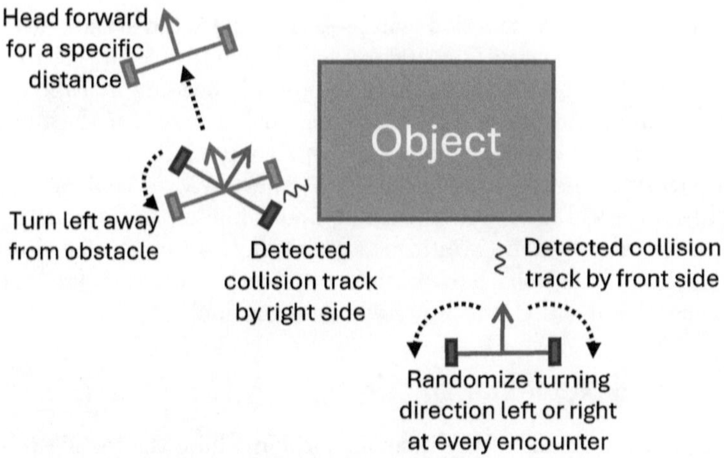

Fig. 2. Approach to object avoidance based on whether front or side collision track is detected.

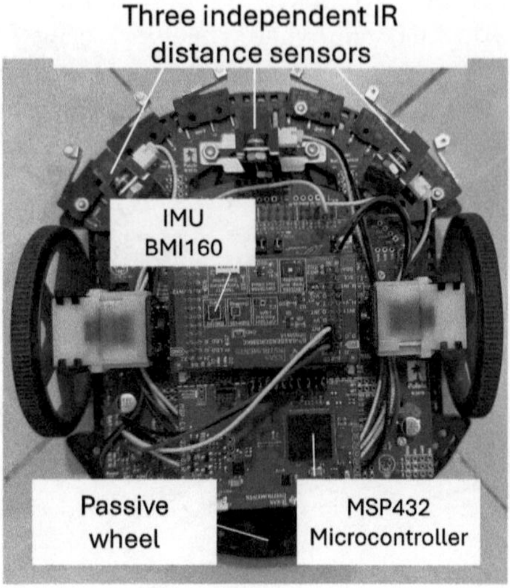

Fig. 3. Texas Instruments' Robot System Learning Kit (TiRSLK).

passive wheel at the rear of the robot to balance its weight. Three independent analog Sharp IR distance sensors are mounted at the front of the robot to assess distance to objects within a 10 to 80 cm range, and 120° combined field of view.

The robotic kit is controlled by an MSP432 microcontroller with a Cortex-M4 CPU running at 48 MHz. The gyroscope sensor is realized by the BMI160 IMU on Texas Instruments' Sensors BoosterPack^{TM} [2]. The gyroscope is sampled every $T_G = 30$ ms with a sensitivity of 131.2 LSB/°/s.

The Robot Forward function samples the LiDAR distance sensors and updates its PI controls every $T_s = 100$ ms. The PI controllers are set to maintain wheels at 60 RPM. The function refines the robot's heading by calling Align to Point, after completing $\eta_B = 75\%$ of its forward segment, if the full-segment length is greater than $\eta_A = 1440$ steps.

When steering the robot about its CoG, the PWM is kept at a low duty cycle of 12% to maintain accurate turning and minimize the impact of inertia. The sensor fusion equation (1) utilizes a weighting factor $\gamma = 0.3$. Finally, the measured robot dimensions have been $d = 69$ mm, and $w = 146$ mm. Each encoder generates $n = 360$ pulses per wheel revolution.

4.2 Experimentation Site

Figure 4 presents a layout of the indoor floorplan where the testing took place. The site features a corridor, two rooms, furniture, and an obstacle placed to study the impact of Object Avoidance on navigation. The start point is set at reference location $(x_R(0), y_R(0)) = (0, 0)$ cm, and the initial heading is $\theta_R(0) = 90°$. The target point is set at $(x^*, y^*) = (243.84, 792.48)$ cm, and the intermediate point is set at $(x^I, y^I) = (0, 548.64)$ cm, where the superscript I refers to that point being intermediate.

The purpose of the intermediate point is to save the robot's time and energy instead of bouncing several times to avoid collision with the right wall.

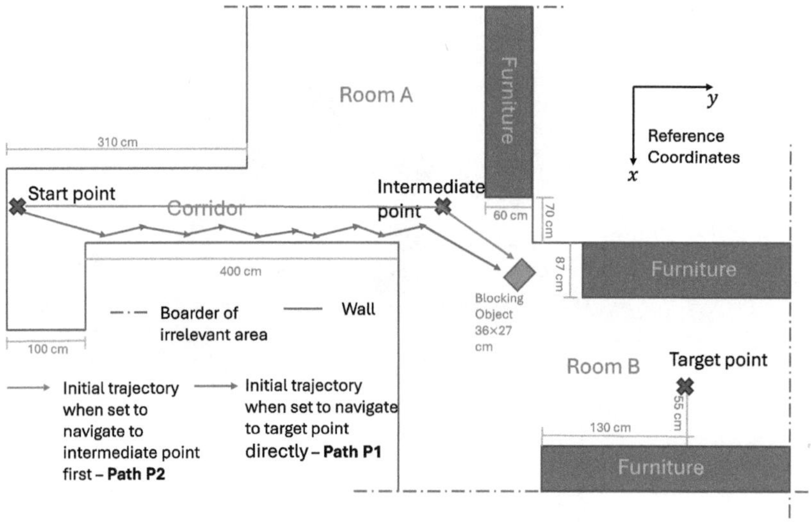

Fig. 4. Experimentation floorplan.

4.3 Experiment Results

We consider the distance error to the intermediate and target points shown in Fig. 4. The framework is tested through two different paths. The first, $P1$, is when the robot is tasked to navigate to (x^*, y^*) directly. The second, $P2$, requires the user to set the robot to navigate first to the intermediate point, before heading to the target point.

Let D^* and D^I be the distances between the start point and target point, and between the start point and intermediate point, respectively. The distance between the intermediate point and the target point is D^{I*}. Let (X_R^*, Y_R^*), (X_R^I, Y_R^I) be the robot's measured coordinates after completing its trip to the target and intermediate points, respectively. The measurements are collected using a scaled ruler and a square set.

The percentage distance error at the target point under path P_i is measured as

$$Er^*(i) := \begin{cases} \frac{\sqrt{(X_R^*-x^*)^2+(Y_R^*-y^*)^2}}{D^*} \times 100\%, & i = 1, \\ \frac{\sqrt{(X_R^*-x^*)^2+(Y_R^*-y^*)^2}}{D^I+D^{I*}} \times 100\%, & i = 2. \end{cases} \tag{2}$$

We normalize the error by the total distance that is *preset* to travel. Similarly, the percentage distance error at the intermediate point is calculated and denoted by Er^I.

For each path, $P1$ and $P2$, we let the robot autonomously navigate from the starting point for 30 times. Figure 5 shows pictures of the robot maneuvering around the obstacle that is detected by the front sensor.

Fig. 5. Pictures of the robot maneuvering around an obstacle.

The distribution of percentage distance error is depicted in Figs. 6, 7 for target and intermediate points. The error at the target point is shown under the two paths $P1$, $P2$. The average percentage distance error at the target point has been 1.66% and 1.47% under path $P1$ and $P2$, respectively. At the intermediate point, under path $P2$, the distribution of percentage distance error is shown in Fig. 7, with an average value of 1.44%, which is close to the average distance

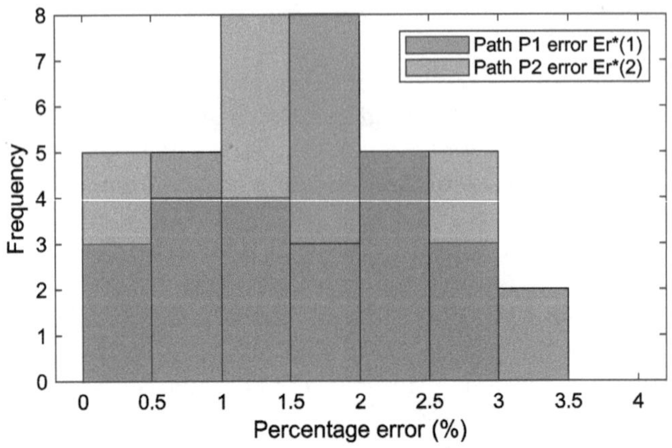

Fig. 6. Distribution of percentage distance error $Er^*(i)$ at target point.

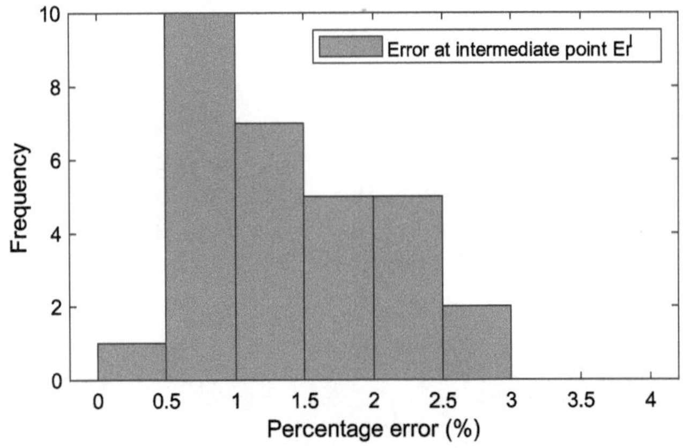

Fig. 7. Distribution of percentage distance error Er^I at intermediate point.

error at the target point under path $P2$. This finding may suggest that a constant portion of the error comes from the manual positioning of the robot at the start point, which is only guided by markings on the ground. However, the averaged error values of 1.66% or 1.47% at the target point are still reasonable for various applications with no prior training or guidance from beacon signals.

5 Conclusion

We have developed a low-cost autonomous indoor navigation framework for two-wheeled robots. The framework does not require a depth camera or a complex vision system. It assists in critical applications where normal SLAM and prior training of the robot are not available or infeasible to attain, and when there is no infrastructure to help localize the robot. These applications include rescue operations in collapsed structures and military CQB tactics. A sensor fusion approach that harnesses IMU gyroscope data together with wheel encoders has been considered and integrated with object avoidance and localization through LiDAR sensors. A simple motion tracking model has been adopted for low power consumption and faster processing. Implementation on the low-cost TIRSLK kits and testing in an indoor environment have shown the accuracy merits.

References

1. https://www.ti.com/lit/ml/sekp166/sekp166.pdf
2. https://www.ti.com/tool/BP-BASSENSORSMKII
3. Baras, N., Nantzios, G., Ziouzios, D., Dasygenis, M.: Autonomous obstacle avoidance vehicle using lidar and an embedded system. In: 2019 8th International Conference on Modern Circuits and Systems Technologies (MOCAST), pp. 1–4 (2019). https://doi.org/10.1109/MOCAST.2019.8742065
4. Cadena, C., et al.: Past, present, and future of simultaneous localization and mapping: toward the robust-perception age. IEEE Trans. Rob. **32**(6), 1309–1332 (2016). https://doi.org/10.1109/TRO.2016.2624754
5. Foxlin, E.: Pedestrian tracking with shoe-mounted inertial sensors. IEEE Comput. Graphics Appl. **25**(6), 38–46 (2005). https://doi.org/10.1109/MCG.2005.140
6. Noh, S., Park, J., Park, J.: Autonomous mobile robot navigation in indoor environments: mapping, localization, and planning. In: 2020 International Conference on Information and Communication Technology Convergence (ICTC), pp. 908–913 (2020). https://doi.org/10.1109/ICTC49870.2020.9289333
7. Qin, T., Li, P., Shen, S.: Vins-mono: a robust and versatile monocular visual-inertial state estimator. IEEE Trans. Rob. **34**(4), 1004–1020 (2018). https://doi.org/10.1109/TRO.2018.2853729
8. Sun, J., Yu, X., Cao, X., Kong, X., Gao, P., Luo, H.: Slam based indoor autonomous navigation system for electric wheelchair. In: 2022 7th International Conference on Automation, Control and Robotics Engineering (CACRE), pp. 269–274 (2022). https://doi.org/10.1109/CACRE54574.2022.9834195
9. Zheng, W., Zhou, X., Zhang, X., Wei, M., Fu, P.: An improved indoor mobile robot positioning method based on dead reckoning. In: 2022 International Conference on Sensing, Measurement & Data Analytics in the era of Artificial Intelligence (ICSMD), pp. 1–6 (2022). https://doi.org/10.1109/ICSMD57530.2022.10058446

iHELP: An iOS-Based mHealth System for Care Partner Activation and Preventive Healthcare for AD/ADRD Caregivers

Md. Mehedi Hasan[1], Soarov Chakra Borty[1(✉)], Mary Dioise Ramos[2], and Nazmus Sakib[1]

[1] Department of Information Technology, Kennesaw State University, Kennesaw, USA
soarovchakraborty@gmail.com, nsakib1@kennesaw.edu
[2] School of Nursing, Louisiana State University, New Orleans, USA
mramo1@lsuhsc.edu

Abstract. Alzheimer's Disease and Related Dementias (AD/ADRD) are advanced and irreversible neurological conditions that create a significant impact on the brain cells and decrease a person's thinking and functional capabilities. To deal with this dangerous and devastating disease, the mHealth application iHELP is proposed. This paper presents the development of the iOS prototype to help individuals cope with Alzheimer's disease. Moreover, this study examines the unique challenges associated with the platform, its structure, development techniques, and impact on the community. iHELP features include personalized user profiles, training modules, community-based assistance, and emergency asset resources. Additionally, it offers multilingual support and key functionalities, including 24/7 helpline support, emergency visit requests, and a communication network that connects care partners with patients. Moreover, the medium offers options for engagement through interactive blogs and provides access to local events organized by various communities. This research aims to enhance the quality of life for patients, caregivers, and individuals living with ADRD.

Keywords: Alzheimer's Disease · Caregiver Support · Dementia Management · mHealth Application · iOS application

1 Introduction

Alzheimer's disease is one of the medical challenges of this modern era and also the primary reason for dementia [2,16]. It is characterized by a progressive deterioration in cognitive abilities and problems in recalling recent occurrences. By 2050, nearly 14 million Americans will be affected by Alzheimer's disease [1]. The aging baby boomer generation born between 1946 and 1964 after World War II

H. R. Arabnia et al. (Eds.): CSCI 2025, CCIS 2974, pp. 118–132, 2027.
https://doi.org/10.1007/978-3-032-28475-4_8

has significantly increased the number of Alzheimer's cases, but the caregiving support structure has not improved as much [8]. This rising void has burdened families, caregivers, and healthcare workers. Therefore, there is a necessity for greater caregiver efficiency. An adequately developed Caregiver Activation Program (CAP) can be a vital step in empowering caregivers with the necessary skills, resources, and support systems to manage Alzheimer's patients effectively. The program would focus on educating caregivers in dementia supervision methods, providing emotional and psychological support, and utilizing technology to enhance productivity and efficiency. Furthermore, current cure techniques alleviate symptoms, and no practical treatment for AD exists. Patients affected with Alzheimer's disease are often stuck in the past with their old memories, and it also affects caregivers financially and mentally when an individual loses a loved one in front of their eyes.

In America, Asians are the fastest-growing racial group [18]. With the increase in population, older Asian people face several diseases, and age is one of the crucial factors for Alzheimer's disease. For example, among Asian American communities in the United States, a large number of Filipino Americans (Fil-Am) experience ADRD, and they are the second-largest group [17,19]. Most Dementia cases have been experienced by Filipino people based on the past 14-year record among Asian subgroups. However, culturally and linguistically limited material is available to educate the Fil-Am group about ADRD. They often misinterpret the disease, delay diagnosis, and do not seek help at a US medical center. Traditional Filipino diets, sedentary lifestyles, and specific genetic characteristics, like the APOE $\epsilon 4$ allele, are leading reasons for AD [7]. Again, Filipinos in the US are not usually covered by health insurance and often experience delayed medical care, and at the time, they realize the disease has spread too far [14,25]. It also has a significant impact on family members, both economically and emotionally. As major cities have established networks, the Fil-Am group living outside of developed areas suffers the most from a lack of appropriate help groups, respite care, and essential resources, which leaves many families isolated. Scaling essential initiatives also requires significant investment and partnership with multiple sectors, and sometimes, it is quite challenging to achieve in reality. Moreover, research engagement between the Fil-Am is limited due to doubts about the system and a lack of genuine efforts, and their experiences are usually excluded from analyses. Moreover, many data-driven analytical approaches have been effectively applied in previous studies [4,6,21] to identify early signs of health deterioration and optimize digital health interventions.

This project, iHELP (Innovative Health, Engagement, and Literacy Program), is designed to assist patients, caregivers, and respective groups involving Fil-Am community members, stakeholders, and organizations. iHELP provides training materials like videos and resources to enlighten people and support by empowering care partners and making culturally congruent care choices aligned with their values and preferences. Moreover, it helps users to track dementia care centers by location, cost, and type. Furthermore, this app can provide notifications about upcoming events and activities and bring caregivers' networks together. Additionally, an interactive blog enables users to share their thoughts, opinions, and experiences by asking questions and connecting with researchers, thereby assisting in enhancing peer support networks. Furthermore, users can request emergency visits, share their profiles with nearby caregivers, and receive immediate help during critical situations. Moreover, iHELP provides multilingual support, allowing groups like Fil-Am to easily access these resources. Features like surveys and a user-friendly interface make it suitable for people of all age groups. Additionally, acquiring less data, storage, and battery makes this app resource-efficient, and frequent updates and customer feedback ensure that the app will stay up to date over time.

This paper contains the following sections. Analysis of existing research is described in Sect. 2. The system architecture, including its methodology and various features, is presented in Sect. 3. Section 4 discusses the user interface design. Eventually, Sect. 5 illustrates a summary of our work.

2 Related Work

Existing studies have demonstrated how caregivers utilize technology for caregiving chores, including organizing responsibilities and observing caregiving activities. These technological studies encompass computers, smartphones, tablets, and mobile health (mHealth) applications specifically developed for caregiving activities [10,15,28]. Moreover, scientists have evaluated caregivers' requirements and selected technologies that assist caregiving duties and deliver educational content [10]. Recently, attention has turned towards Alzheimer's disease and related dementia caregivers' technology usage for health-related objectives [13,15,22]. Studies display that caregivers utilize mHealth applications to achieve medical decision-making (31%), obtain health knowledge (73%), and health-related goals (48%) [12,23,26]. Smartwatches used by ADRD caregivers account for around 7% of the total, but there is no explanation of how caregivers use these devices, as reported by Rathnayake et al. [23]. Moreover, research shows mobile app utilization has increased for ADRD caregivers, who often use social media apps. Cross-sectional studies have shown increased mobile device privilege levels among ADRD caregivers at 98%, 89%, and 99%, respectively, by Xie et al., Lee et al., and Rathnayake et al. [13,23,29]. Moreover, these studies highlighted frequent caregiver engagement with mHealth and social media apps, demonstrating caregivers' vulnerability to receiving caregiving-related knowledge and assets through these digital platforms [13,23]. Additionally, the study by Wójcik et al. provides insight into how caregivers utilize smartphones, including seeking support, performing smartphone actions, contacting families and health specialists, accessing available information, and engaging in recreational activities [28].

Many clinical analyses have been utilized in different phases of dementia, including preliminary, transitional, and concluding stages [20]. The study of caregiving trajectory focuses on caregivers' duties, starting with initially admiring the caregiver's position and evolving into personal care, then seeking outside help or resources for assistance, and finally, the caregiving ends [24]. Boots et al. observed that caregivers in the late phase of dementia, when reflecting on their views, don't match those of caregivers currently in the initial stage [3]. This conflict indicates that caregivers' perceptions alter over time as dementia advances, caregivers acquire new perspectives, and caregiving becomes more severe and complicated.

This iHELP system delivers an extensive, community-based environment personalized for ADRD patients and caregivers. iHELP integrates education through multimedia training materials, providing caregivers with hands-on knowledge specific to ADRD. By enlightening caregivers and enabling early intervention through perception and network support, iHELP facilitates the implementation of preventive techniques. Its functionality of locating dementia care facilities by location, cost, and type relieves caregivers during emergencies. The app also connects caregivers beyond information boundaries through participatory blogs, networks, and direct contact with researchers. Additionally, emergency support ensures patient safety in critical situations. Furthermore, the multilingual feature enhances accessibility, adaptability, and inclusivity by accommodating diverse linguistic backgrounds.

3 Methodology

This section provides an overview of the system's key functionalities and architecture. It emphasizes the primary features and unique aspects that have been developed to fulfill user requirements. This section is divided into several subsections to provide a thorough understanding of the system. Particularly, Subsect. 3.1 provides details of the development process of the iHELP app. Subsection 3.2 provides an overview of the system's key features, Subsect. 3.3 describes the underlying system architecture, and Subsect. 3.4 outlines the experimental environment required to operate the system.

3.1 Development Process of the iHELP App

The development of the iHELP app followed the CeHRes Roadmap [11] to ensure a user-centered and research-driven approach. In the Contextual Inquiry phase, the requirements and cultural significances of individuals with ADRD were explored through interviews, surveys, and focus groups. Additionally, in the Value Specification phase, the key priorities are reducing caregiver burden, enhancing health literacy, and promoting socioemotional support. During the Design phase, these needs were translated into features such as personalized user profiles, culturally tailored training modules, and multilingual support. The Operationalization phase involved deploying the iHELP prototype in real-world caregiving contexts, integrating it with community resources, healthcare support, and emergency helpline services. Finally, the Summative Evaluation phase will assess the app's long-term impact on caregiver burnout, quality of life, and healthcare utilization through longitudinal studies and feedback analysis. Supporting processes, such as User Requirements gathering, Prototyping, Value Drivers, and the Business Model, guide each step, ensuring the iHELP app is scalable, sustainable, and capable of delivering a meaningful impact to caregivers and patients affected by ADRD. Figure 1 illustrates the entire development process, and Fig. 2 outlines the overall architecture.

3.2 Feature Overview

i) User Profile and Home Dashboard: Users use this part as a gateway for interacting with the platform. This module contains several features. It includes a sign-in option, where a secure backend API is used to authenticate users through a user-friendly UI. By utilizing the sign-up option, the backend databases complete the registration process. Users can update their personal information by

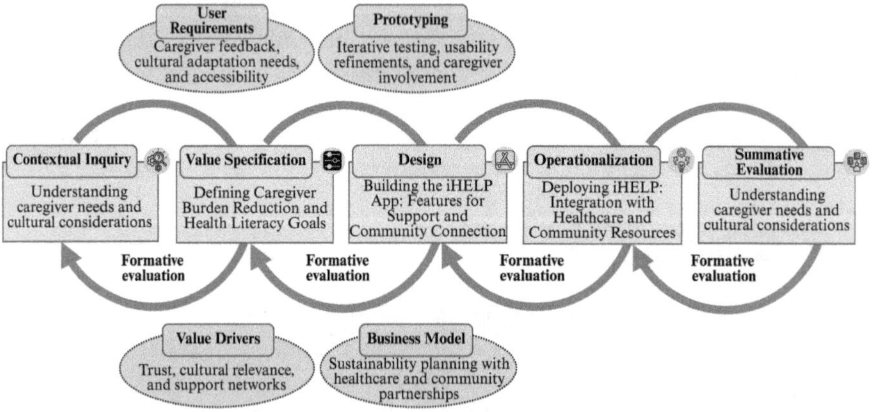

Fig. 1. Development Process of iHELP App using CeHRes Roadmap.

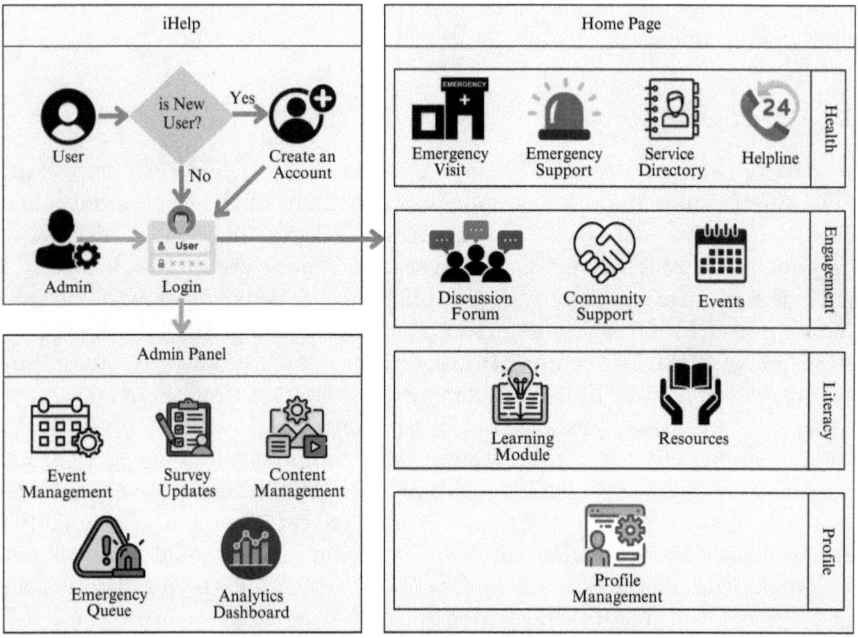

Fig. 2. Flowchart illustrating the system architecture and functional modules of the iHELP application.

using the edit personal details option. By using the forgot password feature, users can update their password if they forget it by following a few simple steps. The home page layout provides a smooth transition, prioritizes main feature elements, and is easy to navigate through key sections. The user dashboard will be updated automatically in real-time whenever a new event, content, notification, or latest update becomes available.

ii) Community Features: This app aims to empower caregivers of individuals with ADRD. Educational training modules are dynamic resources that can be easily accessed on the web or mobile devices for training purposes. By utilizing an event-driven backend system, users can access the most up-to-date content. Moreover, a community-based services directory is available, allowing users to search and filter nearby dementia care facilities. Further, they can get notifications of newly added services or updates. Recognizing that community engagement can directly benefit patients, the support network helps organize various workshops and motivates patients to participate. This can help them overcome loneliness and boost their confidence. A beneficial feature for patients is the interactive blog, where users share their daily activities, post questions, and respond to others' blogs. Future app versions will feature individual tabs to distinguish between caregiver-specific resources and patient-directed tools, ensuring transparency and relevance.

iii) Emergency Support: This section includes critical aid for patients. A 24/7 helpline feature is added for real-time emergency support at the most vital moments. Moreover, patients can expedite their visit requests and share their situation with nearby care partners for immediate assistance. Additionally, map-based assistance has been developed for patients, allowing users to locate emergency responders on interactive maps.

iv) Additional Features: The iHELP app has multilingual support, with languages dynamically updated based on location and user preference for broader accessibility. A dynamic text adaptation feature is added to transform into Filipino, English, and other languages. Moreover, daily feedback queries are updated through APIs to gather user input. Also, the About Us & Contact Us pages contain details regarding this app, and users can provide suggestions for improvement based on their experience.

3.3 System Architecture

i) Client: This mHealth system iOS prototype is compatible with both mobile and web platforms, offering a responsive and feature-rich experience for end-users. We utilized the Model-View-ViewModel (MVVM) design pattern for development purposes. The view layer illustrates the dynamic user interface, and the ViewModel manages all the data processing and user interactions. This architecture secures a clean codebase and supports essential updates and key features, including profile management, where users can create accounts, log in using email and password, and manage their profiles with a data synchronization service via APIs. Moreover, the app can generate automated emergency visit requests and allow users to locate nearby aid by viewing the map. Users can also get push notifications for real-time alerts such as surveys, emergency requests, and nearby events. The mobile client is designed to support all users on various devices. Authentication and secure session management are executed to protect user data and ensure compliance with HIPAA constraints [9].

ii) Backend System: The backend is designed using scalable technologies and serves as the core engine of the app's functionality. Moreover, it helps process client requests, handle databases, and execute complicated logic. MySQL is used as a database to store hospital details, doctor profiles, and user data in a structured format. Its scalability allows the iHELP app to handle extensive data as the user base expands [5]. Moreover, MySQL protects sensitive information and features several security measures, including encryption and user authentication and the combination of scalability and security is ideal for the app data. Additionally, the backend includes HIPAA-compliant security measures, and APIs are designed to prevent unauthorized access. Moreover, we collect the user's geographic coordinates through the mobile application to identify hospitals within a specified search radius. We apply the modified Haversine formula [27] for each hospital to calculate the great-circle distance from the user's location, as shown in Eqs. 1–4.

First, compute the change in coordinates:

$$\Delta\varphi = \varphi_2 - \varphi_1 \tag{1}$$

$$\Delta\lambda = \lambda_2 - \lambda_1 \tag{2}$$

Then compute the intermediate value:

$$a = \sin^2\left(\frac{\Delta\varphi}{2}\right) + \cos(\varphi_1)\cos(\varphi_2)\sin^2\left(\frac{\Delta\lambda}{2}\right) \tag{3}$$

Finally, calculate the distance:

$$d = 2r \arctan 2\left(\sqrt{a}, \sqrt{1-a}\right) \tag{4}$$

Where:

- φ_1 and φ_2 are the latitudes (in radians).
- λ_1 and λ_2 are the longitudes (in radians).
- $\Delta\varphi$ is the difference in latitude.
- $\Delta\lambda$ is the difference in longitude.
- r is the mean radius.
- a is an intermediate value.
- d is the great-circle distance.
- $\arctan 2(y, x)$ is the two-argument arctangent function.

iii) Admin Panel: The admin panel is developed to provide administrators with a user-friendly interface for efficient management and oversight. Moreover, Admins can create, schedule, edit, and update various activities and events. Furthermore, the admin can add or modify dynamic survey queries, which affect the client-side in real-time. Further, management can modernize educational materials, videos, blogs, and resources via a centralized interface. Additionally, admins can prioritize emergency requests based on patients' critical needs and ensure timely responses. Again, an analytics dashboard is used to track user engagement and event participation, as well as to determine the features most frequently used by users. HIPAA-compliant APIs are used to establish secure communication between the admin panel and the backend. This also ensures that all modifications are synchronized with the database, and clients can see those changes in real-time. This HIPAA-compliant ecosystem assures patient privacy and delivers a secure, robust, and user-friendly platform.

Moreover, the platform's elements are seamlessly integrated. The client module communicates with the backend APIs to get real-time updates, and the admin panel ensures that the latest resources are available. This architecture creates an efficient ecosystem for both users and administrators.

3.4 Environment Description

Xcode 14 is a primary tool for our project that runs on a Mac computer with the latest macOS version. The Swift 5.3 programming language, along with its modern syntax, is used to build the application, thereby increasing development efficiency. The testing process was conducted on both the iOS Simulator and various versions of iOS devices to ensure optimal performance and bug-free functionality across multiple iPhone models. This testing process ensures high benchmarks of application quality and reliability. Table 1 displays the metadata of the code, featuring details such as the current repository, version, and software used.

Table 1. The code's metadata includes version, dependencies, and system requirements.

No.	Description of the code's metadata	Metadata
1	Present iteration of the code	v1
2	Fixed reference to the code version in the repository	https://github.com/Dream-kid/iHelp
3	Permanent link to Capsule	https://github.com/Dream-kid/iHelp
4	Applicable legal license for the code	Apache License 2.0
5	Tool used for code version	Git
6	Programming languages used	Swift 5.3
7	Required tools, environments, and linked packages for compilation and execution	Mac with the latest macOS, Xcode 14.0 or later, Physical iOS Devices & iOS Simulator, Apple Developer Account
8	User manual	https://github.com/Dream-kid/iHelp/blob/main/README.md

MySQL delivers robust support for managing structured data and is appropriate for the app's requirements. It ensures a stable, speedy, scalable, and user-friendly platform. We used Bitbucket to manage the version control and code. A user-friendly git client source tree manages branches, tracks changes, and maintains a smooth workflow. These tools ensure the productivity and alignment of the contributors. The development process was conducted using Agile methodologies and Scrum approaches. The entire project was divided into separate sprints to achieve rapid outcomes. Both automated and manual testing were conducted to fulfill requirements. This development is designed to address the needs of caregivers and patients, while also prioritizing user-centric features

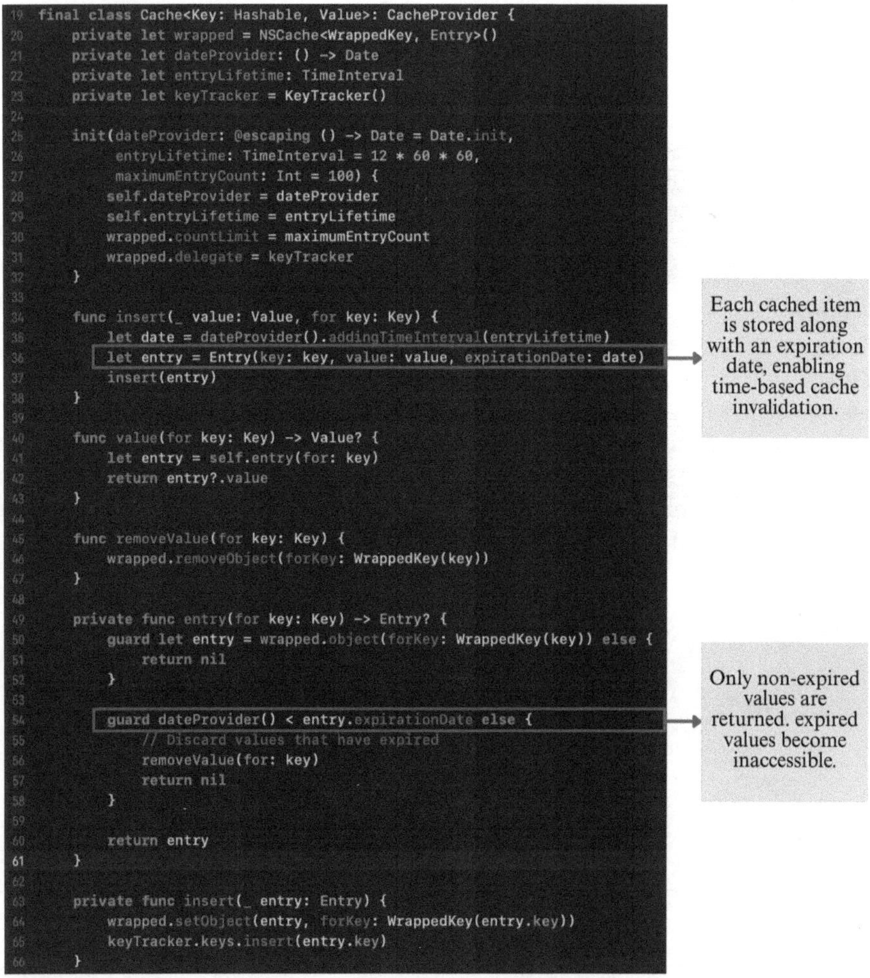

Fig. 3. Cache implementation with expiration-based invalidation and retrieval of non-expired values.

and secure functionalities. It is designed with an eye to prospective technological advancements, further ensuring long-term scalability and applicability. Figure 3, demonstrates the implementation of a cache system using Swift. The Cache class stores key-value pairs in memory using NSCache, ensuring values are discarded based on a given lifetime when they expire. This implementation is practical for efficient memory management, particularly for caching data that is frequently accessed but has a limited validity period.

4 System Demonstration

This section discusses the user interface and the main features of our iOS prototype. With a user-friendly interface and different features, iHELP is the digital solution to the gap between caregivers, patients, and necessary resources. Because of its organized sections and intuitive navigation, iHELP assists with daily caregiving tasks, helps users engage with the community, and enhances their knowledge about Alzheimer's disease. After opening the app, the first authentication page appears, offering users the option to log in if they have already created an account or sign up if they are new users. They can register themselves as patients or caregivers. By achieving successful authentication,

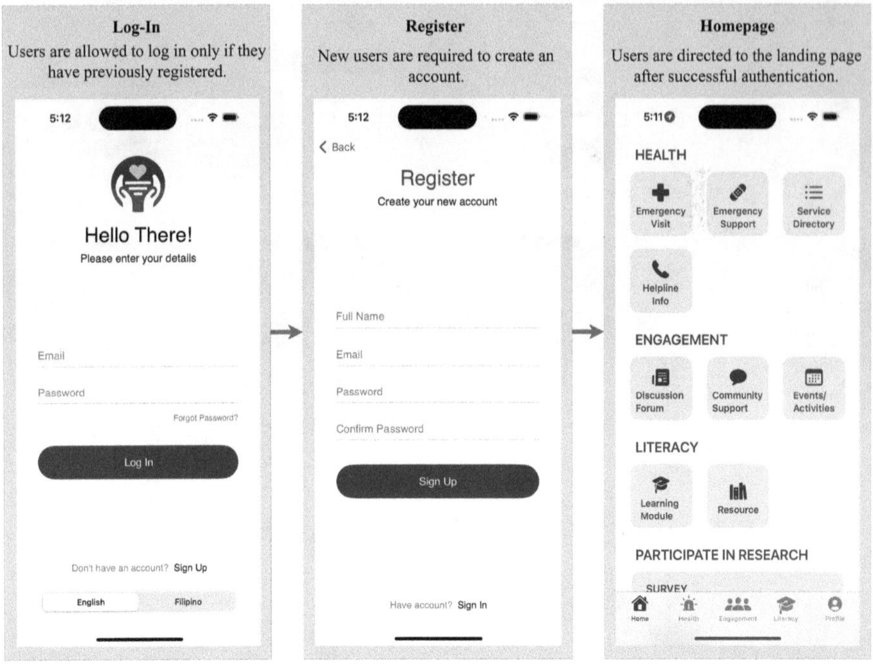

Fig. 4. User flow screenshots, showing the login, registration, and homepage interfaces following successful authentication.

users will be directed to a landing page with various features, serving as a significant hub for interaction. Figure 4 represents the application's flow. The home page contains several primary and secondary elements, enabling users to navigate through all the details easily. Due to its structured layout and clean design, the learning curve for first-time users is reduced when accessing critical tools and resources. Figure 5 provides a clear representation of the home screen within the iHELP application, featuring four main elements, and the landing page serves as a gateway to health, engagement, literacy, and the profile section.

Users can request emergency support in the health section in their native language. Moreover, they can identify the type of emergency and share their current location in the event of a crisis. Additionally, indications of user type, such as patient or caregiver, can be viewed in this UI. Furthermore, there is an additional tab called helpline info. The user can find details such as contact

Fig. 5. Screenshots of the iHELP mobile application showing emergency support, service directories, user profiles, literacy resources, community engagement, events, and multilingual options.

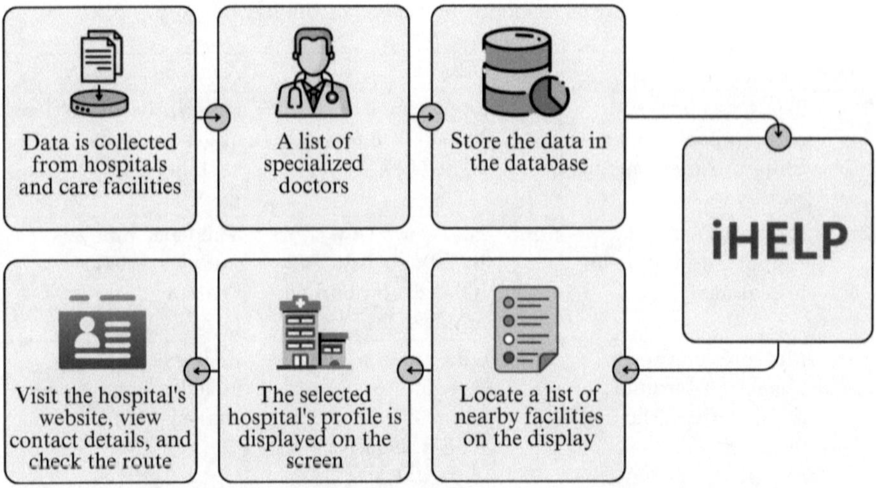

Fig. 6. iHELP workflow: collecting hospital data, storing it, and providing users with nearby facility details and contact information.

numbers, email addresses, and locations on several groups related to Alzheimer's patient care. Moreover, users can utilize the built-in map to locate nearby facilities and caregiver providers near their current locations. Figure 6 represents this workflow. The service directory includes various health-related services, such as hospitals, clinics, doctors, and medical equipment. The directory format can be valuable for finding typical users who need assistance.

In the engagement section, there are fields such as discussion forums, community support, and events. The discussion forum has interactive blogs where users can share their thoughts, discuss challenges, comment on each other's blogs, and create a mutual support network. In the community support section, several programs and workshops are available for patients with Alzheimer's disease, organized by various community groups. Moreover, in the event tab, users can view the latest content, webinars, and events in the event calendar.

In the literacy section describes in Fig. 5, users can see learning modules and resources. By tapping into learning modules, users can access a range of tutorial videos, blogs, and links related to dementia, including information about Alzheimer's disease, its effects, and strategies for managing it. Users can discover articles, publications, and research associated with this disease from various resources. Further, the user can find the caregiver's contact number and availability in this section. Families who are new to Alzheimer's care can consider this section as a beginner's guideline for understanding the struggles and solutions related to the disease. This illustrates how the iHELP application converts caregiving into a manageable and supportable system. Users can update their passwords and choose their preferred language in the profile section. Further, they can participate in the survey to improve the application and understand

Table 2. Post-Development Phases of the iHELP App.

Phase	Aim	Time	Data	Analysis
1	Test feasibility and early impact among Filipino American caregivers	6 mo	Engagement logs (n = 200–300), interviews (n = 50–60)	Descriptive statistics, usage patterns, usability survey
2	Pilot test usability, support, and app effect in real use	6 mo	Interviews (n = 30–40), surveys (n = 100–120) on burden, support, literacy	Thematic analysis, survey statistics, Pearson's r
3	Measure long-term impact on burnout, QoL, and healthcare use	12 mo	Longitudinal surveys (n = 50–70), follow-ups (n = 40–60), health records (n = 200–300)	Regression, mixed-effects, survival analysis
4	Add AI features for personalized, adaptive support	18 mo	User logs (n = 500–1000), AI rec. data	ML models, precision–recall, PR–AUC

the terms & conditions and privacy policy, and learn about the frequently asked questions and explanations.

The iHELP application is a digital tool and a lifeline for caregiver, making everything available at their fingertips. Its competent design and commitment make it a valuable resource for caregivers and patients, ensuring no one has to fight with Alzheimer's disease alone. As the app grows, it shows how technology can enhance lives and support communities. Table 2 outlines the post-development phases of the iHELP app. Future work will concentrate on continuous app improvement based on user feedback, scaling the app for broader user groups, and exploring its integration with healthcare systems. Additionally, further studies will evaluate the app's long-term effectiveness in diverse populations, and the development of advanced AI capabilities will offer even more tailored support to caregivers.

5 Conclusions

iHELP is a transformative tool for enhancing the patient experience and providing caregiving support for individuals affected by ADRD. This paper proposes an iOS prototype of a mHealth system since most Americans currently use iPhones over Android devices. Aside from the iOS platform, we also created a fully operational web-based approach to enable accessibility beyond devices. The web-based system enables caregivers, healthcare providers, and family members to view and manage care-related data efficiently and effectively. We will develop an Android-based version to promote inclusivity and engage more users, particularly in areas where Android devices are prevalent. Partnerships with healthcare experts will ensure that the system aligns with clinical benchmarks and best practices in dementia care. By pioneering groundbreaking advancements in the medical field for patients affected with Alzheimer's disease, this initiative will revolutionize how sufferers acquire care, and they will live their remaining lives with dignity and freedom.

Acknowledgments. This research was supported in part by the 2022 Alzheimer's Association Research Grant to Promote Diversity (AARG-D), titled "Developing Care Partner Activation Program for Filipino Americans". The authors gratefully acknowledge this support, which contributed to the advancement of culturally tailored approaches for caregiver activation and Alzheimer's care among diverse communities.

References

1. Association, A.: 2015 alzheimer's disease facts and figures. Alzheimer's & Dementia **11**(3), 332–384 (2015)
2. Blennow, K., de Leon, M.J., Zetterberg, H.: Alzheimer's disease. The Lancet **368**(9533), 387–403 (2006)
3. Boots, L.M., Wolfs, C.A., Verhey, F.R., Kempen, G.I., de Vugt, M.E.: Qualitative study on needs and wishes of early-stage dementia caregivers: the paradox between needing and accepting help. Int. Psychogeriatr. **27**(6), 927–936 (2015)
4. Chakraborty, S., Hasan, K.A., Paul, S.: A reinforcement learning framework for lung segmentation of covid-19 and pneumonia affected chest x-ray image. In: International Conference on Pattern Recognition and Machine Intelligence. pp. 224–232. Springer (2021)
5. Chakraborty, S., Paul, S., Hasan, K.A.: Performance comparison for data retrieval from nosql and sql databases: a case study for covid-19 genome sequence dataset. In: 2021 2nd International Conference on Robotics, electrical and signal processing techniques (ICREST). pp. 324–328. IEEE (2021)
6. Chakraborty, S., Paul, S., Hasan, K.A.: A transfer learning-based approach with deep cnn for covid-19-and pneumonia-affected chest x-ray image classification. SN computer science **3**, 1–10 (2022)
7. Cosentino, S., Scarmeas, N., Helzner, E., Glymour, M.M., Brandt, J., Albert, M., Blacker, D., Stern, Y.: Apoe $\varepsilon 4$ allele predicts faster cognitive decline in mild alzheimer disease. Neurology **70**(19_part_2), 1842–1849 (2008)
8. Delli Carpini, M.X.: Baby boomers. In: The Forum. vol. 12, pp. 417–445. De Gruyter (2014)
9. English, A., Ford, C.A.: The hipaa privacy rule and adolescents: legal questions and clinical challenges. Perspect. Sex. Reprod. Health **36**(2), 80–86 (2004)
10. Frisbee, K.L.: Variations in the use of mhealth tools: the va mobile health study. JMIR Mhealth Uhealth **4**(3), e3726 (2016)
11. van Gemert-Pijnen, L., Span, M.: Cehres roadmap to improve dementia care. In: Handbook of smart homes, health care and well-being, pp. 1–11. Springer (2016)
12. Kim, H., Paige Powell, M., Bhuyan, S.S.: Seeking medical information using mobile apps and the internet: are family caregivers different from the general public? J. Med. Syst. **41**, 1–8 (2017)
13. Lee, J.A., Nguyen, H., Park, J., Tran, L., Nguyen, T., Huynh, Y.: Usages of computers and smartphones to develop dementia care education program for asian american family caregivers. Healthcare Informatics Research **23**(4), 338–342 (2017)

14. Lee, S., Martinez, G., Ma, G.X., Hsu, C.E., Robinson, E.S., Bawa, J., Juon, H.S.: Barriers to health care access in 13 asian american communities. Am. J. Health Behav. **34**(1), 21–30 (2010)
15. Makowka, J., Lau, T., Kachnowski, S., Pugliese, L., Woodriff, M., Griffin, M., Crowley, O., Lam, V., Lee, G., Harward, N.: Caregivers & technology: What they want and need. AARP Real Possibilities Project CATALYST the Power of We pp. 1–42 (2016)
16. Masters, C.L., Bateman, R., Blennow, K., Rowe, C.C., Sperling, R.A., Cummings, J.L.: Alzheimer's disease. Nature reviews disease primers **1**(1), 1–18 (2015)
17. Ocampo, A.C.: The Latinos of Asia: How Filipino Americans break the rules of race. Stanford University Press (2016)
18. Ong, J., Ong, P., Ong, E.: The future of asian america in 2040. AAPI Nexus: Policy, Practice and Community **14**(1), 14–29 (2016)
19. Paik, S.J., Choe, S.M.M., Witenstein, M.A.: Filipinos in the us: Historical, social, and educational experiences. Social and Education History **5**(2), 133–160 (2016)
20. Perneczky, R., Wagenpfeil, S., Komossa, K., Grimmer, T., Diehl, J., Kurz, A.: Mapping scores onto stages: mini-mental state examination and clinical dementia rating. Am. J. Geriatr. Psychiatry **14**(2), 139–144 (2006)
21. Prapty, A.S., Shitu, T.T.: An efficient decision tree establishment and performance analysis with different machine learning approaches on polycystic ovary syndrome. In: 2020 23rd International conference on computer and information technology (ICCIT). pp. 1–5. IEEE (2020)
22. Rathnayake, S., Jones, C., Calleja, P., Moyle, W.: Family carers' perspectives of managing activities of daily living and use of mhealth applications in dementia care: A qualitative study. J. Clin. Nurs. **28**(23–24), 4460–4470 (2019)
23. Rathnayake, S., Moyle, W., Jones, C., Calleja, P.: Family carers' needs related to management of functional disability in dementia care and use of mhealth applications in health information seeking: An online survey. Collegian **27**(3), 288–297 (2020)
24. RJV, M.: Family caregiving. change, continuity and diversity. Interventions in dementia care: Toward improving quality of life (2000)
25. Sanchez, F., Gaw, A.: Mental health care of filipino americans. Psychiatr. Serv. **58**(6), 810–815 (2007)
26. Shaffer, K.M., Chow, P.I., Cohn, W.F., Ingersoll, K.S., Ritterband, L.M.: Informal caregivers' use of internet-based health resources: an analysis of the health information national trends survey. JMIR aging **1**(2), e11051 (2018)
27. Sinnott, R.W.: Virtues of the haversine. Sky and telescope **68**(2), 158 (1984)
28. Wójcik, D., Szczechowiak, K., Konopka, P., Owczarek, M., Kuzia, A., Rydlewska-Liszkowska, I., Pikala, M.: Informal dementia caregivers: current technology use and acceptance of technology in care. Int. J. Environ. Res. Public Health **18**(6), 3167 (2021)
29. Xie, B., Champion, J.D., Kwak, J., Fleischmann, K.R.: Mobile health, information preferences, and surrogate decision-making preferences of family caregivers of people with dementia in rural hispanic communities: cross-sectional questionnaire study. J. Med. Internet Res. **20**(12), e11682 (2018)

Approach for Management Antipattern Detection on Scrum Software Projects

Pamela Flores(✉), Marcela Mosquera, Jorge Segovia, and Carlos E. Anchundia

Escuela Politécnica Nacional, Quito, Ecuador
{pamela.flores,evelyn.mosquerae,jorge.segovia, carlos.anchundia}@epn.edu.ec

Abstract. The project management phase of the software development lifecycle (SDLC) is crucial for planning, controlling, and coordinating scope, schedule, cost, and quality for the successful development of a software product. This phase involves managing risks, changes, and communication between stakeholders. Bad practices can lead to recurring, inefficient, and routine solutions that threaten project success and generate technical debt. These bad practices, which have adverse effects, are known as antipatterns. Their identification and mitigation could be especially relevant under the Scrum framework, as it is the most widely used framework for managing software projects. It is an agile and flexible approach to developing complex products that seeks to maximize the value delivered to the customer through short, continuous work cycles.

This article presents an approach for detecting and mapping antipatterns (DMAP) and a tool designed to apply this approach to management antipatterns in Scrum projects. This approach has been developed using the Design Science Research (DSR) methodology, which includes a case study selection protocol, a dichotomous questionnaire, and the NPLF scale adapted from the ISO/IEC 33020:2019 standard. By analyzing available documentation, the tool transforms static evidence into reproducible and objective diagnoses, reducing subjectivity in the assessment process. The resulting artifact generates a structured report highlighting the identified Scrum antipatterns and their relationship to management issues, facilitating decision-making and contributing to academic and professional practice. This proposal represents a first step toward semi-automated solutions for the early detection of antipatterns in software development.

Keywords: Management antipatterns · Scrum · Process assessment · Artifact

1 Introduction

Software Engineering applies systematic principles to the development, operation, and maintenance of high-quality systems, structuring the process into the main phases of the software development lifecycle (SDLC) [1]. The SDLC phases

H. R. Arabnia et al. (Eds.): CSCI 2025, CCIS 2974, pp. 133–145, 2027.
https://doi.org/10.1007/978-3-032-28475-4_9

provide an orderly framework to ensure that the software meets requirements, ensuring robustness and scalability throughout its lifecycle. Within the SDLC, the project management phase is essential, as it defines the planning, control, and coordination of resources, laying the foundation for the success of the project [2]. In the project management phase, effective management involves setting clear objectives, realistic deadlines, and defined responsibilities. This approach is particularly reflected in the use of agile methodologies and frameworks such as Scrum , which allow for clearer and more structured evidence of project management practices, prioritizing incremental delivery and adaptability [3]. However, when project management practices in agile methodologies or frameworks are not applied properly, they can lead to dysfunctions that generate structural problems, compromise subsequent phases, and affect the final quality of the software [4,5].

The literature extensively documents the issues that affect the success of software projects. Among these, management deficiencies such as inadequate planning, unclear requirements, and insufficient risk management stand out, leading to significant project failures [6,7]. In addition, management antipatterns — such as Fire Drill, Glass Case Plan, and Inflexible Plan—have been identified that are often perceived as justified decisions but actually mask structural problems. Early detection of these antipatterns remains a challenge due to informal documentation and the lack of standardized tools for their identification [8,9].

Antipatterns, defined as recurring solutions that appear valid but are counterproductive, represent structural defects in management processes that degrade quality and increase technical debt [6]. In the context of Scrum, examples such as large Product Backlogs or failure to deliver the Sprint Goal amplify these deficiencies, reinforcing the need for specific tools for their early detection [10]. Within this framework, the objective of this research is to identify and map Scrum antipatterns and project management antipatterns, proposing a lightweight artifact that allows for their detection in a structured, reproducible, and useful way for decision-making.

To achieve this purpose, we present the Detection & Mapping of Antipatterns (DMAP) procedure, a lightweight artifact that uses a dichotomous questionnaire and the NPLF scale, adapted via a conceptual reformulation of the original model based on ISO/IEC 33020:2019 [11], which provides a structured score to identify dysfunctional Scrum practices and map them to management antipatterns, generating structured reports that facilitate decision-making [10]. The DMAP procedure offers a reproducible method that transforms static traces such as meeting minutes and Product Backlogs into actionable diagnostics. The research developed in this article lays the foundation for the development of phase-specific detection tools, aligning with comprehensive approaches to SDLC [12].

The remainder of this paper is organized as follows: Sect. 2 reviews previous work on the detection of antipatterns in project management; Sect. 3 describes the design and operation of DMAP; Sect. 4 discusses the findings, implications, and limitations; and Sect. 5 presents the conclusions and future research directions.

2 Literature Review

We formulated two research questions to guide our review: **RQ1:** What are the most relevant reported software management antipatterns in the existing literature?, **RQ2:** How can software management antipatterns be systematically identified during project execution?

The literature on project management has identified key antipatterns that affect planning, control, and communication, contributing significantly to project failure [4]. Standish Group reports consistently show that more than half of software projects fail to meet their objectives, and that deficiencies in project management are identified as one of the main causes of failure [6]. These defects, detailed in Table 1, are mainly related to management issues, such as inadequate planning, unclear requirements, and poor risk identification, which contribute to 53% of project failures, compared to only 3% of technical problems [7,13]. These practices, often accepted as justified decisions, can mask structural problems that negatively impact project success [14].

Table 1. Antipatterns in Project Management

Antipattern	Description
Fire Drill	Long periods of planning or inactivity followed by an intense and chaotic phase just before key deliverables [5,14].
Glass Case Plan	Initial plan that is not updated, assuming it will be sufficient for the entire project [5,15].
Inflexible Plan	Plans that do not adapt to changes in requirements or priorities [5].
Road to Nowhere	Lack of clear planning in the project [5,14].
Project Mismanagement	Lack of adequate control and monitoring during project execution [14].
Process Disintegration	Breakdown of management processes, resulting in a lack of consistency and coordination between project activities [14].
Myopic Delivery	Excessive focus on short-term deliverables, ignoring the long-term strategic objectives of the project [14].
Thrown Over the Wall	Transfer of responsibilities or products without adequate communication between teams or phases, causing misunderstandings and delays [14].
Mushroom Management	Keeping the team in the dark about key decisions or relevant information, limiting their ability to contribute effectively [14].

The current state of antipattern detection in project management has been developed from various methodological approaches. Initially, template-based mechanisms were proposed in order to standardize the descriptions of antipatterns, emphasizing their symptoms, contexts, and consequences [5,15]. Subsequently, ontology-based approaches, particularly using Web Ontology Language (OWL), attempted to formalize the relationships between antipatterns and their manifestations [16]. Later, probabilistic models, such as Bayesian networks, were introduced with the aim of estimating the probability of occurrence of an antipattern; however, these approaches relied heavily on expert knowledge, which limited their scalability and generalizability in different contexts [17].

Based on these theoretical foundations, tools were developed to facilitate the identification of antipatterns in real management environments. Among them is SPARSE, a desktop application that used semantic technologies and allowed managers to select observed symptoms to obtain associated antipatterns and their possible relationships. However, it is not designed to analyze static artifacts such as project plans, which limits its ability to adapt to new antipatterns or contexts [8]. This proposal subsequently evolved into DENSE, a web application that expanded its functionality by allowing users to enter not only symptoms, but also causes and consequences. In addition, DENSE incorporated detection mechanisms based on semantic inferences and a knowledge base that is constantly updated through user contributions and online sources. However, semantic inferences are not completely automatic, as input data must be validated by experts, introducing subjectivity [9].

Other work has focused on leveraging existing Application Lifecycle Management (ALM) tools. A representative case is Software Process Antipatterns Detector (SPADe), whose detection mechanism combines a manual transcription of the symptoms observed in the project in the form of formal parameters and indicators with a probabilistic approach based on metrics and weights to confirm the presence of an antipattern. This can introduce biases or ambiguities and limits automation in projects without ALM or with incomplete data [18]. Along the same lines, proposals have been explored that use the Software & Systems Process Engineering Metamodel (SPEM) standard, implemented in the EPF Composer tool. In this case, symptoms and consequences are modeled as tasks, roles, or artifacts expected in the process, allowing quantitative formulas to be established, evaluation thresholds to be defined, and SQL queries to be generated that can be applied to project data to verify the presence of antipatterns, which requires human intervention, preventing complete automation [19]. These efforts highlight the challenge of detecting antipatterns due to their informal documentation, variable representations, and dependence on manual or context-specific methods [2,4,20].

Despite these advances, there is a critical gap in the systematic connection between Scrum antipatterns and project management antipatterns, particularly in the context of static analysis. Current tools such as DENSE—the successor to SPARSE—and SPADe require manual symptom inputs or dynamic data, limiting their applicability to first-phase static artifacts such as project plans or backlogs. Furthermore, human-centered antipatterns (e.g., communication failures in planning) are difficult to formalize, and there are no standardized mechanisms for classifying and detecting project management-specific antipatterns using static analysis. This gap, identified in the literature, highlights the need for a conceptual artifact that links Scrum antipatterns to management problems and proposes detection tailored to static planning documents, enabling early identification.

3 Self-assessment Questionnaire (DMAP)

The detection of antipatterns in the software project management phase was developed using the Design Science Research (DSR) methodology, following an iterative process of construction and validation of the artifact called Detection & Mapping of Antipatterns (DMAP). This artifact allows the identification of dysfunctional practices in projects that use Scrum, mapping them to management antipatterns documented in the literature and classifying their relationship using the standardized NPLF scale. DMAP integrates a selection protocol, a dichotomous questionnaire, and a mapping module, offering a practical and reproducible tool that reduces subjectivity by relying on documentary evidence, such as Scrum roles, events, or practices [10,21]. This approach not only detects management problems early on, but also generates actionable knowledge to improve processes in agile environments, where Scrum predominates as a framework [10].

3.1 Case Study Selection Protocol

The DMAP procedure begins with a protocol to identify projects that exclusively use Scrum, ensuring their suitability for the analysis of management antipatterns. The selection criteria evaluate the quality of the documentation of key Scrum activities, as detailed in Table 2.

Table 2. Key Scrum elements expected in the case studies

Scrum Activity	Expected general parameters
Daily Scrum	Description of the daily meeting, regularity of its occurrence, duration, and form of participation by team members.
Product Backlog	General management of the backlog, frequency of updates, and adaptations made.
Sprint Planning	Description of Sprint planning, definition of objectives and criteria for selecting tasks.
Sprint (execution)	Sprint development, goal achievement and obstacles encountered.
Sprint Review	Presentation of results, feedback received and stakeholder participation.
Sprint Retrospective	Process evaluation, improvement agreements, and follow-up on commitments.

Software development projects that use Scrum without combining it with other methodologies, such as iPlus, are prioritized, excluding hybrid approaches to ensure consistency in the analysis [10]. Documentation is evaluated on three levels:

- **Full compliance:** Activities are described in sufficient detail and clarity, allowing the questionnaire to be answered completely or almost completely.
- **Partial compliance:** The activities are mentioned, but without the necessary level of detail. This leaves some questions in the questionnaire unanswerable with certainty.
- **Not documented or non-compliant:** The activities are not described, or the information is so superficial that it prevents the questionnaire from being applied reliably.

The selection protocol ensures that each Scrum case study has the necessary documentation to be evaluated, thus minimizing the risk of incomplete or biased analysis. This guarantees that the dichotomous questionnaire, designed to systematically identify observable antipatterns in Scrum activities, can be applied effectively and reliably, providing solid results for the analysis of project practices.

3.2 Dichotomous Questionnaire

The dichotomous questionnaire is at the heart of the DMAP procedure, as it allows for the systematic, objective, and reproducible detection of Scrum antipatterns in the selected case studies. Its design is based on the catalog of antipatterns described in The Scrum Anti-Patterns Guide [10], from which recurring dysfunctional practices were extracted, defined in terms of observable symptoms and common consequences. To ensure consistency and reproducibility, the procedure follows a series of clearly defined steps:

1. **Document entry and standardization:**
   The process begins with the collection of Scrum activity documents, such as meeting minutes, progress reports, Sprint plans, and Product Backlogs. It is essential that these documents comply with a minimum structure that allows the symptoms and consequences of antipatterns to be clearly identified. This standardization ensures that the information can be evaluated without ambiguity and that the analysis is replicable across different projects.
2. **Application of the dichotomous questionnaire:**
   Each antipattern is evaluated using three dichotomous (yes/no) questions designed to reflect the documented symptoms and consequences. These questions are constructed from specialized literature on Scrum antipatterns, as shown in Table 3, ensuring academic validity and objectivity in the analysis. This stage allows evaluators to respond consistently and systematically, minimizing the influence of subjective judgments.

Table 3. Examples of dichotomous questions for the Daily Scrum activity

Antipattern	Question	Answer
Status report	Is a structured format (such as the 3 questions or similar [10]) used to guide the meeting?	Yes
No routine	Does the meeting always take place in the same place or platform and at a set time?	No
Not enforcing the time-box	Does the meeting usually last longer than 15 min?	Yes

3. **Evaluation and determination of antipatterns:** The responses obtained are compared with the criteria defined for each antipattern. An antipattern is considered present when the responses match the expected symptoms and consequences. Additionally, observations on the quality of the evidence and the completeness of the information are recorded, ensuring traceability and reliability in the results.
4. **Recording and reporting results:** Finally, a structured record of the responses is generated, organized by project, activity, and antipattern, indicating the presence or absence of dysfunctional practices.

In this way, the dichotomous questionnaire ensures the systematic and reproducible identification of Scrum antipatterns, based on clear and structured documentary evidence. This allows for an objective analysis of dysfunctional practices in the evaluated projects, laying the foundation for their mapping to management antipatterns using the NPLF scale, which facilitates a comprehensive and feasible diagnosis of the problems detected.

3.3 NPLF Scale and Antipattern Mapping

To evaluate the relationship between Scrum antipatterns and project management antipatterns, we use the NPLF scale, a methodological guide that systematically organizes the correspondence between both types of antipatterns, adapted via a conceptual reformulation of the original model from the international standard ISO/IEC 33020:2019 [11]. The scale classifies the degree of linkage into four levels, from "Not linked" to "Strong linkage," as detailed in Table 4. This NPLF scale reduces subjectivity and ensures transparency in identifying theoretical and practical relationships, considering symptoms, causes, and consequences shared between antipatterns.

Based on this scale, the Scrum antipatterns are mapped to the management antipatterns, considering the evidence from the cases analyzed. The procedure allows for the identification of conceptual and practical relationships, determining the relevance of each link and consolidating information based on observable symptoms, causes, and consequences. As a result, a structured report is obtained that integrates the presence, degree of linkage, and evidence of each antipattern,

Table 4. NPLF scale for evaluating the relationship between Scrum and management antipatterns

Level	Description
N (Not linked)	No evidence of a significant relationship between the Scrum antipattern and the management antipattern.
P (Weak link)	Theoretically coherent relationship, but with limited or generic evidence.
L (Moderate link)	Coherent and contextualized relationship, with indirect or partial evidence.
F (Strong link)	Clear and direct relationship, supported by compelling evidence in the document or annexes.

as shown in Table 5, which facilitates decision-making in project management and the implementation of process improvements, team training, and documentation strengthening [4].

Table 5. Examples of Mapping Scrum Antipatterns to Management Antipatterns

Scrum Antipattern	Management Antipattern	Degree of Linkage (NPLF)
Status Report (Daily Scrum)	Glass Case Plan	L
Large Product Backlog	Fire Drill, Inflexible Plan	F
No business objective (Sprint Planning)	Road to Nowhere	L
Failure to deliver the Sprint Goal	Fire Drill, Project Mismanagement	F

When multiple Scrum antipatterns are related to the same management antipattern, a single degree of linkage is assigned, corresponding to the highest level observed according to the hierarchy ($N < P < L < F$). For example, if Status Report and Obsolete Issues are linked to Glass Case Plan with a degree of L, and Very Little Planning with a degree of F, the final degree for Glass Case Plan will be F, as it represents the strongest evidence among all possible links. This procedure ensures that the analysis is consistent and prioritizes the relationships with the greatest impact.

The DMAP procedure generates a structured report that consolidates the presence, level of linkage, and evidence of each antipattern detected in the analyzed projects. This report provides project managers with clear, data-driven information to make improvement-oriented decisions, such as process refactoring, training in agile practices, or strengthening documentation [5]. Unlike reactive approaches, DMAP allows for early and proactive detection, aligning with the need for tools that integrate agile practices with management diagnostics [1].

4 Discussion

The DMAP procedure makes a significant contribution to the detection of management antipatterns within projects that use Scrum, by combining static analysis of artifacts with the NPLF scale. The findings demonstrate that this proposal provides tangible value in both agile and academic environments. In development teams, DMAP provides structured reports that allow Scrum Masters and Product Owners to identify frequent dysfunctional practices—such as oversized Product Backlogs or failure to deliver Sprint Goals—facilitating early intervention during planning and execution [4]. In academic contexts, meanwhile, the procedure functions as an objective assessment tool that promotes the adoption of sound agile practices, preventing the accumulation of technical debt in case studies [2] and encouraging critical reflection on the use of Scrum. This dual applicability highlights the relevance of DMAP in addressing the gap identified in the literature, where other tools such as SPADe and DENSE rely on manual inputs and fail to leverage static planning documents [8,9].

Beyond its immediate usefulness, the results suggest that DMAP provides a conceptual framework that systematically connects Scrum antipatterns with project management antipatterns. This chain reveals that early deficiencies in planning, such as Glass Case Plan or Inflexible Plan, can propagate to later phases of the software lifecycle, generating ambiguities in requirements, excessive dependencies in design, or construction problems such as Spaghetti Code. The evidence obtained reinforces the thesis that antipatterns are not isolated events, but interrelated phenomena whose accumulation increases the risk of failure [6,12]. In this sense, DMAP aligns with efforts in the literature that propose comprehensive detection architectures, but adds a more accessible approach by relying on documentary evidence and a dichotomous questionnaire that reduces subjectivity [10,21].

However, the implementation of DMAP also faces limitations that condition its scope. Its dependence on static documentation represents a challenge in agile frameworks or methodologies, where records are often informal, incomplete, or inconsistent [2,4]. This restricts the applicability of the procedure in projects with low levels of documentation and reduces its ability to provide a reliable diagnosis when evidence is partial. Furthermore, unlike probabilistic approaches based on Bayesian networks, DMAP lacks a predictive component that would allow it to anticipate the emergence of antipatterns in dynamic contexts [17]. In addition, although the NPLF scale provides standardization and transparency, it requires further validation in broader industrial scenarios to confirm its robustness and generalizability in the face of the variability of real projects [11].

Despite these limitations, the results suggest important practical and theoretical implications. In the professional sphere, DMAP reinforces the need for lightweight tools that integrate agile practices with managerial diagnostics and bridge Scrum and traditional project management [1]. At the academic level, its implementation promotes evidence-based training processes and offers a methodological resource that helps to bridge the gap identified between research and practice. Furthermore, by facilitating the early detection of dysfunctional patterns, DMAP enhances teams' ability to prevent the spread of structural failures, improving the sustainability and success of projects [5,14].

Finally, the results open up clear lines for future work. One promising direction is to complement DMAP with dynamic analysis techniques that process metrics in real time, which would extend its applicability to projects with minimal or constantly evolving records. Similarly, the incorporation of machine learning models would strengthen the predictive capacity of the procedure, anticipating the emergence of antipatterns rather than merely detecting them retrospectively. Likewise, validating the NPLF scale in industrial environments and with large-scale projects would lend greater solidity and legitimacy to the findings, consolidating DMAP as a practical and transferable tool in both research and professional software project management [11,20].

5 Conclusions

We answer the first research question by presenting a report on common antipatterns found in the literature on the project management phase and describing the basic architecture of the DMAP artifact (Fig. 1), developed from the identification of a central limitation in the state of the art: the absence of mechanisms that allow Scrum antipatterns to be systematically linked to management antipatterns, especially in scenarios where documentation is static, incomplete, or informal. To this end, DMAP was designed and specified as an artifact whose conceptual architecture offers a rigorous, reproducible, and applicable diagnosis in the early phases of the software lifecycle.

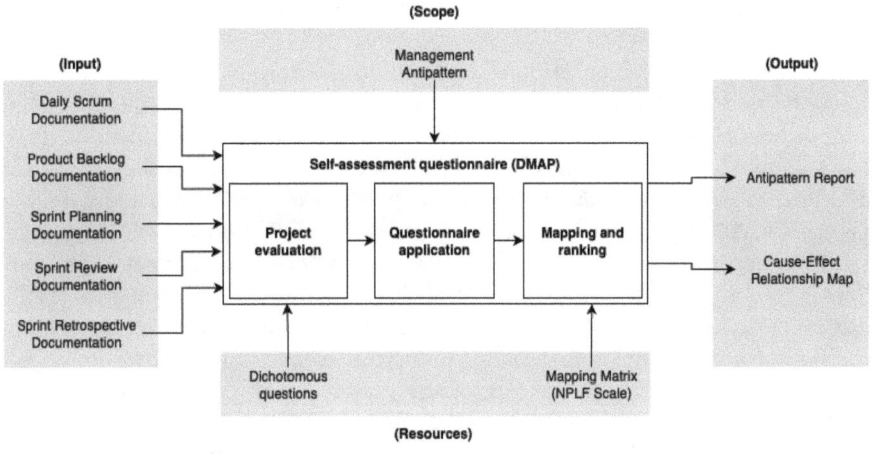

Fig. 1. IDEF-0 diagram of the DMAP artifact.

The main contributions of this research are embodied in the architecture of DMAP, which introduces several key innovations. First, it uses a dichotomous questionnaire, which standardizes the detection of Scrum antipatterns through closed questions that reduce the ambiguity of diagnoses. Second, it incorporates the NPLF scale, adapted from the ISO/IEC 33020:2019 standard, which allows dysfunctional practices to be mapped to management antipatterns with a standardized and traceable framework. Finally, it generates structured reports that transform documentary evidence, such as meeting minutes and Product Backlogs, into actionable diagnoses that strengthen decision-making. It is crucial to note that the artifact is not intended to replace the human judgment of the manager or Scrum Master; on the contrary, it is designed as a decision support system that provides objective and traceable evidence to identify risky practices. Its true value lies not in the automation of judgment, but in the automation of analysis, allowing those responsible to focus their attention on aspects of the project that require deeper and more strategic deliberation.

This research also answers the second research question, which concerns how software management antipatterns can be systematically identified during project execution. This study presents an artifact that does not replace the judgment of the manager or Scrum Master; it functions as a decision support system, automating analysis and providing traceable evidence to focus human deliberation.

Methodologically, the research confirms the relevance of the Design Science Research (DSR) approach for the construction and validation of artifacts intended for the early detection of antipatterns, showing how an incremental, evidence-oriented design can generate knowledge applicable in both academic and professional contexts. In the educational field, DMAP provides a reproducible procedure for evaluating case studies, strengthening the effective adoption of agile practices. In professional practice, it is emerging as an early detection mechanism that can reduce the accumulation of technical debt, improve planning, and increase the managerial maturity of teams. In the teaching field, DMAP is an effective pedagogical tool for raising students' awareness of the impact of their practices on management results. However, the main limitations of the study are related to the dependence on static documentation, which in agile environments is often incomplete or informal, and the need to validate the NPLF scale in larger industrial projects to ensure its robustness. In this regard, future work includes the integration of DMAP with ALM tools, the use of dynamic and predictive techniques based on artificial intelligence, and the extension of its validation in real scenarios. Overall, this work provides a conceptual and methodological framework that connects Scrum antipatterns with those of project management, laying the foundations for the development of comprehensive detection solutions throughout the SDLC and offering a solid path for the continuous improvement of software project management.

Based on the above, the projection of this work is clear. The next step is the empirical validation of the artifact in academic and industrial projects, which will allow its effectiveness to be measured in real-world scenarios. Subsequently, work will be done on integrating DMAP with dynamic ALM tools and expanding its catalog of antipatterns. The long-term vision is to consolidate a comprehensive ecosystem for detecting antipatterns, capable of covering different phases of the software lifecycle, providing real-time feedback, and thus strengthening a virtuous cycle of planning, execution, and diagnosis.

DMAP's main contribution lies in its innovative approach to systematizing the detection of antipatterns in the management phase, connecting agile practices with structural dysfunctions in a reproducible manner. Unlike previous methods that require complex infrastructure or are limited to retrospective analysis, DMAP leverages Scrum traceability to offer a practical tool that facilitates early intervention, such as optimizing planning or improving communication between the team and stakeholders. Its user guide and online accessibility make it especially valuable in academic settings and organizations with limited resources, where early detection of problems can reduce technical debt and improve project outcomes. These limitations highlight the need for adjustments to expand DMAP coverage. Future research could explore the integration of automated tools, such as text analysis of meeting minutes or backlogs using artificial intelligence, to overcome the reliance on manual documentation. Likewise, validating DMAP in industrial projects with diverse teams and combining it with quantitative metrics, such as sprint completion rate, would strengthen its robustness. Developing unified taxonomies that integrate technical and managerial antipatterns could also provide a more holistic view of project quality. In conclusion, DMAP establishes a solid foundation for improving agile project management, fostering a culture of continuous improvement and laying the groundwork for future advances in the detection of antipatterns in software engineering.

Acknowledgments. This research was supported by *Escuela Politécnica Nacional del Ecuador* through the research project PIS-24-07.

References

1. Sommerville, I.: Software Engineering. 10th ed. Pearson Educación (2015)
2. Stamelos, I.: Software project management anti-patterns. J. Syst. Softw. **83**(1), 52–59 (2010)
3. Schwaber, K., Sutherland, J.: The Definitive Guide to Scrum: The Rules of the Game (2020)
4. Brada, P., Picha, P.: Software process anti-patterns catalogue. In: Proceedings of the 24th European Conference on Pattern Languages of Programs (EuroPLoP 2019). ACM, pp. 1–10 (2019)
5. Brown, W., Malveau, R., McCormick, H., Mowbray, T.: AntiPatterns: Refactoring Software, Architectures, and Projects in Crisis, 1st edn. Wiley, USA (1998)
6. Standish Group, "Chaos report on software projects. In: Project Smart, Boston, MA, USA (2014), información técnica
7. Gulla, J.: Seven reasons it projects fail. In: IBM Systems Magazine (2012)
8. Settas, D., Cerone, A., Fenz, S.: Towards automatic generation of ontology-based antipattern bayesian network models. In: 2011 9th International Conference on Software Engineering Research, Management and Applications (SERA), pp. 46–53 (2011)

9. Fitsilis, P., Tilentzidis, K., Moustakas, I., Stamelos, I., Settas, D.: Enabling open software project management data with antipatterns. In: Proceedings of the 19th International Academic MindTrek Conference, pp. 194–201. ACM (2015)
10. Wolpers, S.: The Scrum Anti-Patterns Guide: A Hands-on Manual from the Trenches (2023)
11. ISO/IEC, Information Technology — Process Assessment — Process Measurement Framework for Assessment of Process Capability, Geneva, Switzerland (2019), international Standard ISO/IEC 33020:2019(E)
12. Picha, P., et al.: Process anti-pattern detection - a case study. In: EuroPLop 2022, vol. 27, pp. 1–18 (2022)
13. Project Management Solutions, "Strategies for project recovery: a pm solutions research report. In: PM Solutions (2011)
14. Laplante, P.A., Neill, C.J.: Antipatterns: Identification, Refactoring, and Management. Auerbach Publications, New York, NY, USA (2005)
15. Buschmann, F., Meunier, R., Rohnert, H., Sommerlad, P., Stal, M.: A System of Patterns: Pattern Oriented Software Architecture. Wiley, New York, NY, USA (1996)
16. Settas, D., Stamelos, I.: Resolving complexity and interdependence in software project management antipatterns using the dependency structure matrix. In: Lee, R. (ed.) Software Engineering Research, Management & Applications, vol. 150, pp. 205–217. Springer-Verlag, Berlin Heidelberg (2008)
17. Settas, D., Bibi, S., Sfetsos, P., Stamelos, I., Gerogiannis, V.: Using bayesian belief networks to model software project management antipatterns. In: Proceedings of the 4th International Conference on Software Engineering Research, Management and Applications (SERA 2006), pp. 117–124 (2006)
18. Picha, P., Brada, P.: Software process anti-pattern detection in project data. In: Proceedings of the 24th European Conference on Pattern Languages of Programs (EuroPLoP 2019), pp. 1–12. ACM (2019)
19. Šimečková, L., Brada, P., Picha, P.: SPEM-based process anti-pattern models for detection in project data. In: 2020 46th Euromicro Conference on Software Engineering and Advanced Applications (SEAA), pp. 89–92 (2020)
20. Smiari, P., Bibi, S., Stamelos, I.: Knowledge acquisition during software development: modeling with anti-patterns. In: Nalepa, G.J., Baumeister, J. (eds.) Synergies Between Knowledge Engineering and Software Engineering, ser. Advances in Intelligent Systems and Computing, Springer, Cham, Switzerland, AG, vol. 626, pp. 75–92 (2018)
21. Hevner, A.R., March, S.T., Park, J., Ram, S.: Design science in information systems research. MIS Q. **28**(1), 75–105 (2004)

Agile Methodologies in Supply Chain Management: A Systematic Review of Practices and Enabling Technologies

Zeel Patel, Reece Raymond Tuscano, and Soma Datta(✉)

University of Houston-Clear Lake, Houston, TX 77058, USA
datta@uhcl.edu

Abstract. Volatility and frequent disruptions have renewed interest in supply chain agility and the enabling role of digital technologies. This paper synthesizes peer-reviewed evidence on agile supply chain practices and their outcomes across healthcare, automotive, and pharmaceutical sectors. Following a structured search of IEEE Xplore, Scopus, Web of Science, and Google Scholar, supplemented by seminal prior works, we analyze how cross-functional teaming, modular processes, and iterative planning interact with AI-enabled analytics, IoT-based visibility, and blockchain traceability to enhance responsiveness, flexibility, and resilience. We present a cross-industry framework that maps practices to capabilities (visibility, predictability, adaptability) and performance outcomes (lead-time reduction, service level, risk mitigation). Implementation challenges—data quality, cultural resistance, integration costs, and governance—are discussed with pragmatic recommendations. We conclude with a research agenda that advances testable propositions on digital enablers, contextual moderators (regulatory intensity, market turbulence), and alignment with leagile strategies.

Keywords: Agile supply chain · Leagility · Digital transformation · Supply chain flexibility · Emerging technologies · Supply chain resilience

1 Introduction

Supply chains today are dynamic networks rather than linear systems and must respond swiftly and flexibly to uncertainty in an interconnected world. Conventional management styles often play catch-up with fast-changing customer patterns, world trade policy, and new technologies. This creates a need for approaches that deliver not only efficiency but also flexibility and resilience attributes that agile approaches can provide.

The COVID-19 pandemic exposed limitations of traditional, efficiency-oriented models—manifested in delays, stockouts, and escalating costs—thereby elevating agility and resilience as strategic imperatives.

H. R. Arabnia et al. (Eds.): CSCI 2025, CCIS 2974, pp. 146–152, 2027.
https://doi.org/10.1007/978-3-032-28475-4_10

1.1 Background

Historical analyses of Supply Chain Management (SCM) show its evolution from logistical support to a strategic discipline that shapes organizational performance. Disruptions driven by technology shifts, changing customer requirements, and shocks such as natural disasters and pandemics highlighted weaknesses in conventional frameworks. Many organizations adopted lean systems to minimize cost by eliminating waste [1, 2]. Agile SCM has emerged as a complementary strategic approach emphasizing flexibility, speed, and responsiveness, particularly in high-volatility contexts such as healthcare and automotive [3, 4].

1.2 Research Questions

- What can agile methodologies tell us about improving flexibility and resilience within the supply chain?
- What problems and opportunities do digital transformation technologies present in agile supply chains?

1.3 Motivation

Growing disruption intensity makes adaptability a priority. Recent research indicates significant potential for agile SCM to improve responsiveness, customer satisfaction, and lead-time reduction [5, 6]. Applying agile principles prepares organizations not only to endure but also to thrive in high-volatility environments.

1.4 Structure of the Paper

Section 2 reviews the literature on agile SCM based on a structured search across IEEE Xplore, Scopus, Web of Science, and Google Scholar. Section 3 synthesizes empirical evidence across industries. Section 4 examines the role of emerging technologies. Section 5 discusses challenges, opportunities, and implications for practice. Section 6 concludes and outlines future research directions.

2 Methods

We conducted a structured literature search across IEEE Xplore, Scopus, Web of Science, and Google Scholar for 2004–2025 using combinations of 'agile', 'leagility', 'supply chain', 'AI', 'IoT', and 'blockchain'. Seminal pre-2015 works were retained where they underpin theory.

Inclusion criteria: peer-reviewed, English-language studies that empirically or conceptually address agility-related practices in SCM. Exclusions: non-SCM domains, non-scholarly sources. After deduplication and two-stage screening (titles/abstracts, then full texts), we extracted data on context, practices, digital enablers, and outcomes.

3 Literature Review

There is broad agreement that agile SCM is a strategic requirement in today's volatile environment. Agility enables rapid response to market changes and disruptions by instilling flexibility, modularity, and partnering across the value chain [8, 10]. The hybrid notion of leagility combines lean efficiency with agile adaptability [11]. Digital transformation—AI, IoT, and blockchain—facilitates predictive insights, transparency, and end-to-end visibility [2, 3, 5]. Organizations practicing digital-agile report reduced lead times, improved customer satisfaction, and enhanced risk resilience, with case evidence across manufacturing and healthcare [2, 4]. The agile features and benefits in SCM are shown in Table 1.

Table 1. Agile features and benefits in SCM.

Feature	Description	Benefit to SCM
Flexibility	Ability to adjust production and supply rapidly	Quick response to demand fluctuations
Real-time Data	Continuous data acquisition and monitoring	Improved decision-making and responsiveness
Modularization	Division of processes into adaptable sub-processes	Easier customization and scalability
Cross-Functional Teams	Diverse teams collaborating across functions	Faster innovation and problem-solving
Iterative Processes	Cyclical, feedback-driven development	Continuous improvement and adaptability

4 Case Studies and Empirical Evidence

This section discusses cross-industry agile practices. Table 2 lists those industries that compare agile practices practiced by them.

4.1 Healthcare

Healthcare organizations that adopted leagile practices improved emergency preparedness by balancing cost efficiency with responsiveness [1]. Agile supplier collaboration shortened reaction time for critical items (e.g., PPE, ventilators) during COVID-19 [9]. Some national health authorities and provider networks have deployed collaborative procurement and ordering platforms, enabling near real-time coordination—illustrative of agile information-sharing mechanisms.

4.2 Automotive

Automotive supply chains are complex and demand-sensitive. Agile practices have improved coordination, reduced lead times, and enhanced customer satisfaction [6]. The transition to EVs necessitates rapid supply chain reconfiguration, where agile approaches help integrate new components and partners quickly [11]. Just-in-Time (JIT) inventory, when combined with agile planning, reduces holding costs while maintaining responsiveness.

4.3 Pharmaceutical

Pharmaceutical supply chains operate under strict regulation and demand uncertainty. Agile strategies, supported by digital technologies, enhance visibility, compliance, and responsiveness [3, 14].

4.4 Cross-Industry Comparison

Common patterns emerge across sectors: agile information sharing in healthcare; demand-driven planning and JIT in automotive; and traceability and compliance emphasis in pharmaceuticals.

Table 2. Comparison of agile practices across industries.

Industry	Agile Practices	Impact on SCM
Healthcare	Real-time collaboration with suppliers	Timely access to critical supplies (e.g., PPE, ventilators)
Automotive	JIT inventory with agile planning	Reduced inventory costs; faster response to demand changes
Pharmaceutical	AI & Blockchain for traceability	Improved visibility; reduced counterfeiting risk; timely access to medications
Cross-Industry	Customer-driven adaptability	Higher satisfaction and market responsiveness

5 The Role of Emerging Technologies in Agile SCM

The impacts of digital technologies and agility are explained here. A summary of these is shown in Table 3.

5.1 Artificial Intelligence and Machine Learning

AI/ML enables predictive analytics for demand and inventory, supporting faster, data-driven decisions [2, 3, 11]. Machine learning identifies patterns in historical and streaming data, improving sourcing, production, and distribution planning.

5.2 Internet of Things (IoT)

IoT sensors provide end-to-end visibility into the location and condition of goods, improving disruption management and responsiveness [8]. The integration of IoT with AI enables autonomic adjustments and continuous optimization.

5.3 Blockchain

Blockchain enhances transparency, trust, and traceability through secure, tamper-evident records [7, 15]. It reduces fraud and supports regulatory compliance, especially in sectors such as pharmaceuticals [5].

Table 3. Digital technologies and agility impacts

Technology	Function in SCM	Impact on Agility
AI/ML	Predictive analytics; demand forecasting	Enhanced responsiveness; inventory optimization
IoT	Real-time tracking of goods and conditions	Improved visibility; faster disruption response
Blockchain	Secure, transparent transaction records	Increased trust, traceability, and compliance

5.4 Discussion

AI, IoT, and blockchain collectively strengthen core agility capabilities—predictability, visibility, and traceability—but performance depends on data quality, technical expertise, security, interoperability, and governance [3, 6, 14]. Technology is necessary but insufficient; organizational readiness, culture, and strategy alignment remain decisive. The concept of leagility highlights industry-specific balances between lean efficiency and agile responsiveness.

6 Challenges and Opportunities in Agile SCM

6.1 Challenges

Key challenges include multi-partner coordination (especially in healthcare and automotive), investment requirements for digital infrastructure (including IoT, AI, and blockchain) [16], and cultural resistance to agile operating models.

6.2 Opportunities

Opportunities center on resilience, responsiveness, and competitive differentiation enabled by data-driven collaboration and visibility [4].

The challenges and opportunities that were found in agile SCM are listed in Table 4.

Table 4. Challenges and opportunities

Type	Description	Implication
Challenge	Coordination among multiple partners	Requires robust communication and integration strategies
Challenge	High implementation costs for digital technologies	Demands strategic investment and long-term planning
Challenge	Cultural resistance to agile transformation	Necessitates change management and training
Opportunity	Resilience through technology	Better preparedness for disruptions
Opportunity	Customer-centric agility	Increased market share and profitability

6.3 Practical Responses to Challenges

Leaders should champion cross-functional teaming, invest in scalable data platforms, and align vendor partnerships to accelerate digital enablement. Treat technological investments as strategic, not purely operational costs; train teams in agile ways of working.

7 Conclusion and Future Research Directions

Agile methodologies help supply chains cope with volatility by improving responsiveness, reducing lead times, and enhancing customer satisfaction. Digital technologies amplify these effects by increasing transparency and prediction. Organizations that overcome integration and cultural hurdles are better positioned to build sustainable competitive advantage.

Future research should quantify the impact of analytics maturity on agility outcomes, particularly in the context of AI and predictive analytics. Additionally, evaluating the causal links between blockchain-enabled transparency and trust, compliance, and risk mitigation will provide deeper insights into its role in supply chain governance. Developing leagile frameworks that integrate lean and agile principles for effective disruption response remains essential for enhancing supply chain resilience. Cross-industry applications should be explored further by identifying sector-specific best practices and boundary conditions to ensure adaptability and relevance. Finally, aligning agility with environmental and social performance is critical for advancing sustainability in agile supply chain management.

Prioritize training and development, strengthen relationships with technology partners, and adopt iterative deployment roadmaps to scale agile SCM capabilities.

References

1. Li, X., Martins, A.L.: Leagility in the healthcare research: a systematic review. BMC Health Serv. Res. **24**(1) (2024). https://doi.org/10.1186/s12913-024-10771-0

2. He, J., Fan, M., Fan, Y.: Digital transformation and supply chain efficiency improvement: an empirical study from A-share listed companies in China. PLoS One. **19**(4) (2024). https://doi.org/10.1371/journal.pone.0302133
3. Azeez, R.: Effects of Emerging Technologies on the Wellbeing of Procurement and Supply Chain Professionals: Evidence from Nigeria. https://www.jare-sh.com/downloads/mar_2024/azeez.pdf (2024)
4. AlKahtani, M., Rehman, A.U., Al-Zabidi, A., Choudhary, A.: Agile supply chain assessment: an empirical study on concepts, research and issues. Arab. J. Sci. Eng. **44**(3), 2551–2565 (2018). https://doi.org/10.1007/s13369-018-3299-7
5. Mu, X., Liu, J., Pu, Y.: Differential games in the supply chain of innovative products with consumer purchase regret induced returns. PLoS One. **19**(8) (2024). https://doi.org/10.1371/journal.pone.0302154
6. Wu, C., Barnes, D.: Design of agile supply chains including the trade-off between number of partners and reliability. Int. J. Adv. Manuf. Technol. **97**(9–12), 3683–3700 (2018). https://doi.org/10.1007/s00170-018-2205-5
7. Marin, I., Marin, M.-P., Nicolae, C.: Farm Supply Chain Management Improvement Through the Use of Lean, Agile and DevOps Methodologies. https://animalsciencejournal.usamv.ro/pdf/2020/issue_2/Art50.pdf (2020)
8. Oliveira-Dias, D., Moyano-Fuentes, J., Maqueira-Marín, J.M.: Understanding the relationships between information technology and lean and agile supply chain strategies: a systematic literature review. Ann. Oper. Res. **312**, 973–1005 (2022). https://doi.org/10.1007/s10479-022-04520-x
9. Kuupiel, D., Bawontuo, V., Mashamba-Thompson, T.: Improving the accessibility and efficiency of point-of-care diagnostics services in low- and middle-income countries: lean and agile supply chain management. Diagnostics. **7**(4), 58 (2017). https://doi.org/10.3390/diagnostics7040058
10. Alimardani, M., Rafiei, H., Rabbani, M.: A novel approach toward coordinated inventory management of an agile multi-echelon multi-product supply chain. Cogent Eng. **2**(1) (2015). https://doi.org/10.1080/23311916.2015.1025596
11. Abbasi, M., Hosnavi, R., Babazadeh, R.: Agile and flexible supply chain network design under uncertainty. Int. J. Ind. Eng. Theory Appl. Pract. https://www.semanticscholar.org/paper/Agile-and-flexible-supply-chain-network-design-Abbasi-Hosnavi/3845c1d2702d277c95208ce0d80ca6e1d5e406d3. (2014)
12. Lou, P., Zhou, Z., Chen, Y.-P., Ai, W.: Study on multi-agent-based agile supply chain management. Int. J. Adv. Manuf. Technol. **23**(3–4), 197–203 (2004). https://doi.org/10.1007/s00170-003-1626-x
13. Lalmazloumian, M., Wong, K.Y., Kannan Govindan, M.E., Kannan, D.: A robust optimization model for agile and build-to-order supply chain planning under uncertainties. Ann. Oper. Res. **240**(2), 435–470 (2016). https://doi.org/10.1007/s10479-013-1421-5
14. Ghatari, A.R., Mehralian, G., Zarenezhad, F., Rasekh, H.R.: Developing a model for agile supply: an empirical study from Iranian pharmaceutical supply chain. Iran. J. Pharm. Res. **12**(Suppl), 193–205 (2013). https://pubmed.ncbi.nlm.nih.gov/24250689/
15. Xu, N.-R., Liu, J.-B., Li, D.-X., Wang, J.: Research on evolutionary mechanism of agile supply chain network via complex network theory. Math. Probl. Eng. **2016**, 1–9 (2016). https://doi.org/10.1155/2016/4346580
16. Dolatabad, M.J., Azhdarifard, M., Dwijendra, N.K.A., Al-Sudani, A.Q.A.S.: Evaluating agile practices in green supply chain management using a fuzzy multicriteria approach. Discrete Dyn. Nat. Soc. **2022**, 1–12 (2022). https://doi.org/10.1155/2022/4290848

Education - Mathematics

Adaptation and Language Challenges in Tutorial Formats for Discrete Mathematics

Alan Bowen-James[1] and Mary Ruth Freislich[2(✉)]

[1] Universal Business School Sydney, Group Colleges Australia, Sydney, Australia
alan.bowen-james@ubss.edu.au
[2] School of Mathematics and Statistics, University of New South Wales, Sydney, Australia
m.freislich@unsw.edu.au

Abstract. The study extends previous investigation of the achievement of domestic and international students in a first-year Discrete Mathematics subject in an Australian university. Two different programs were studied, with no difference in syllabus or lectures, but different approaches to problem solving classes. The first was traditionally organized, with lectures supplemented by two small-group tutorials each week, led by a tutor giving solutions to problems, either answering students' questions, or chosen to illustrate typical tasks. The second program replaced one tutorial by a tutor-led class given to a larger group, and the second replaced by a board tutorial, in which students collaborated in small groups, writing their work on a board, with a tutor available to answer questions. Achievement scores came from examination scripts. In both programs, achievement in Term 2 was lower than that in Term 1, and that of international students lower than that of domestic students. Comparison between programs significantly favored the program with traditional tutorials. The present study extended the investigation to the two corresponding Term 3 groups. Within both programs the purpose of the extension was to examine the effect of a longer previous period of adaptation to university study, considering separately the two groups with different national backgrounds. In the case of the international students, the adaptation period also gave more time to accustom them to using spoken English. The comparison between programs differed from that in the previous work, because the teaching material for the board tutorial program in Term 3 had been extensively revised, to make it easier for students to handle typical tasks. In the traditional program, achievement of both subgroups did not differ significantly from the highest level in earlier terms, indicating benefit from the longer adaptation period. In the board tutorial program, the achievement of Term 3 domestic and international students was significantly higher than that in earlier terms, but international students still scored significantly lower than domestic students. The comparison between programs in term 3 still significantly favored traditional tutorials, but only because of the lower scores for international students in board tutorials. The conclusion was that the adaptation period and the revised teaching material had both been beneficial in the board tutorials, but that collaborative problem-solving tutorials were still a disadvantage for the international group, which was likely to be so because language difficulties made classroom discussion difficult for them. No significant gender effects were found within national groups.

H. R. Arabnia et al. (Eds.): CSCI 2025, CCIS 2974, pp. 155–172, 2027.
https://doi.org/10.1007/978-3-032-28475-4_11

Keywords: Discrete Mathematics · Flipped classrooms · Collaboration · Board tutorials

1 Introduction

The present study follows up on issues raised by the results of two previous studies by the authors [4, 5]. These dealt with students in a first-year Discrete Mathematics subject taught in a large Australian university. The subject is taken mainly by students in electrical engineering and computer science. The mathematics involved is very different from the core material encountered in students' previous mathematical studies, and requires a high level of explanation and proof, which means that many of the students find it challenging.

The first study investigated students in a traditional teaching program, consisting of lectures and tutorials, in which the small tutorial classes were led by a tutor solving problems on the blackboard and answering questions. Achievement comparisons were between groups defined by enrolment in successive terms and by national background, that is, domestic versus international students. International students at the university come mainly from China and India and make up about 40% of the Discrete Mathematics enrolment group. Achievement in Term 2 was lower than in Term 1, indicating no advantage in slightly later enrolment. International students had lower achievement scores than domestic students, interpreted as reflecting possible greater past focus on core mathematics and possibly a need for longer adaptation to international study. This motivated extending investigation to Term 3, which afforded a longer time for both groups to adapt to unfamiliar mathematics and to overcome transitional issues.

The second study dealt with the same Discrete Mathematics subject and examined the effect of a change in the organization of tutorial classes, in which one of two tutorial hours used tutor-led problem-solving, similar to the traditional tutorials, but in larger groups. The second was replaced by classes called board tutorials, in which students worked in small collaborative groups, writing their attempts at solutions to problems on a board as they emerged from discussion. This organization of class time is like that used in flipped classrooms, but involved flipping only one hour, instead of the whole time. In flipped programs instruction is delivered by video or written resources, out of class time. That is, the board tutorial program differed from flipped programs, because lecture classes for presentation of material were the same in the board tutorial program as in the traditional one. Within the board tutorial group, results of comparisons between terms and national groups were the same as in the traditional program, with achievement significantly lower in Term 2 than in Term 1, and also significantly lower for international than domestic students, in both terms. Comparison of programs found significantly lower achievement among students in board tutorials, for both the groups defined by national background. The conclusion would seem to be that reasons for extension to Term 3 would be the same. But there was one feature of the implementation of the board tutorial program that gave a new and different reason for extension of investigation. Starting from Term 2 of the first year of board tutorials, the teaching staff worked on revision of the teaching material used in the collaborative classes, with the intention of making solutions more accessible to students. There was much more revised material in Term 3 than in Term 2, and examination of this material indicated that the new material did make it easier for students to get started on the given tasks. More details are in Sect. 2. Thus, the new design of the material means that the Term 3 board tutorial program had an important design difference from earlier implementations, thus reopening the original comparative question, with a major focus on design.

2 Background

2.1 Previous Work by the Authors

The rationale for extending the present investigation to Term 3 students rests on two connected developments outlined above. First, earlier work [4, 5] showed that international students consistently underperformed relative to domestic students in both traditional and board tutorial formats, with the gap persisting across Terms 1 and 2. In traditional tutorials, this gap was interpreted as partly reflecting the challenges of adjusting to the very unfamiliar mathematical material in Discrete Mathematics, which is very different from the core material taught in secondary programs, with the challenge likely to be stronger for the international students, because their need for high entrance scores could cause a stronger focus on core material. For all students, a further challenge came from the methods required in Discrete Mathematics, whose level of abstraction requires extensive proof-writing and focus on logic, beyond levels encountered in previous studies. In board tutorials, the previous work, which showed lower achievement for students of both types of national background, indicated that an additional disadvantage was linked to the collaborative, discussion-driven classwork in the program.

Second, two changes in the teaching context made a Term 3 investigation particularly informative. One was the longer adaptation period available to Term 3 entrants, providing additional time for students to adjust to both university study and the linguistic demands of English-medium mathematics instruction. The other was the revision of board tutorial materials, which was undertaken after staff observed that early versions of the collaborative tasks were too difficult for many students to approach confidently. The most extensive revisions occurred in Term 3, making this the first full implementation of the redesigned materials.

Together, these factors created a unique opportunity to examine whether (a) additional time in the university environment helps close the performance gap for international students, and (b) whether carefully scaffolded collaborative tasks can reduce or eliminate the disadvantage previously observed for all students in board tutorials. The background to each of these factors (material revision, transition to university mathematics, collaborative learning formats, and the role of language in peer interaction) is outlined below.

2.2 Revision of Teaching Materials for Board Tutorials

Teaching staff reported that students in the first implementation of board tutorials (Term 1) found the tasks difficult to begin and complete, even when working in groups. The difficulty can be illustrated in description of the original Term 1 problem sheets. The first of these contained five questions requiring the application of set theory concepts, proof construction, the use of a novel binary operation on sets, and the identification of errors in an invalid proof. While these tasks aligned with intended learning outcomes, they were presented without scaffolding or worked examples, leaving many students unsure of how to initiate their solutions. As Lindner and colleagues show [18], tasks that are not understood pull attention into sense making rather than solution steps, while worked-example support outperforms problem-solving.

An analysis of students' examination responses to similar questions reinforced this impression: high error rates on proof-based and abstract set manipulation questions were common, even among otherwise capable students. This indicated that the cognitive load of interpreting and structuring solutions for unfamiliar problem types was impeding students' ability to demonstrate their mathematical understanding in the collaborative environment. Later problem sheets showed similar patterns.

Revisions to the task sheets were implemented in two stages over Terms 2 and 3. In Term 2, some of the most difficult proof tasks and very unfamiliar abstractions like that in the binary operation question, were removed or replaced with more accessible problem types. The revised materials included simpler examples illustrating basic set operations and relationships, along with a guided proof structured in discrete steps to model the expected reasoning process. These changes reflected an explicit attempt to lower extraneous cognitive load, that is, the mental effort associated with processing unfamiliar instructions, so that more attention could be devoted to the central mathematical ideas.

In Term 3, the revisions went further. The basic tasks were expanded to include explicit prompts for constructing objects such as a power set, which serve as entry points to more abstract reasoning. Hints were embedded within questions to direct students toward productive approaches and to help them avoid common reasoning pitfalls observed in earlier cohorts. The pace of syllabus coverage was also slowed, allowing more time for consolidation before introducing complex material. In addition, a tutorial sheet devoted entirely to graph theory was introduced. This was a deliberate response to staff concerns that students often encountered graph-theoretic concepts for the first time in the final examination without prior guided practice. By embedding graph theory within the collaborative environment, the Term 3 design aimed to familiarise students with the definitions, representations, and proof strategies unique to this topic.

From a pedagogical standpoint, these revisions were intended to replicate some of the advantages of traditional tutorials within the board tutorial structure. In traditional tutorials, the tutor's explanations, worked examples, and guided questioning naturally scaffold students' learning. Without these supports, board tutorial students relied heavily on peers and written materials. By embedding more explicit examples and structured pathways into the task sheets, the revisions sought to provide analogous scaffolding in a format compatible with independent group work.

2.3 Transition to University Mathematics

The literature shows that prior preparation is the strongest single predictor of achievement in first-year university mathematics [22] However, the shift from school mathematics to university-level learning also demands adaptation to new expectations, often requiring the development of more independent, self-regulated study strategies [23]. This is particularly true for Discrete Mathematics, a subject that is rarely included in secondary-level curricula [27] and introduces proof-oriented reasoning to many students for the first time [8]. For international students, the transition can be doubly challenging if their previous focus on examination results emphasized procedural fluency over abstract reasoning [4]. In such cases, unfamiliar content may interact with the need to adapt to new teaching formats, increasing the time required to reach full academic performance.

2.4 Collaborative Problem Solving in Class

There is no previous research that specifically investigates board tutorials, so the only closely related literature on collaborative problem solving in class time comes from studies of flipped classrooms. Flipped programs, however, embed classroom collaboration within a larger pedagogical shift: lectures are replaced or reduced, and students are expected to study mathematics content in written or video form independently before class. This important distinction limits the direct relevance of flipped classroom findings to board tutorials.

More importantly, the applicability of flipped classroom research is restricted for three further reasons. First, early studies often suffered from design defects. For example, the meta-analysis [13] identified faulty design in a significant minority of studies. Later work shows similar problems, such as lack of proper control groups [14] or the choice of unsuitable outcome measures [16]. Second, findings on achievement are mixed. The meta-analysis [13] reports moderate evidence in favour of flipped programs, but this is undermined by the methodological weaknesses already noted. They also observe that favourable results occur mostly when the mathematics content is relatively low in complexity, with benefits diminishing as the level of material increases. Third, and most crucially for assessing collaboration, most flipped classroom implementations confound the benefits of collaboration with the effects of frequent additional assessment. Quizzes in preparation and in-class work, plus graded assignments, are often part of the model. In some cases, research directly examines components of flipped programs; for example [15], found that a particular flipped program showed positive achievement effects only after frequent quizzes were introduced.

Although the flipped classroom literature therefore offers only limited direct guidance about achievement effects, more relevant insight comes from studies of students' experience of collaborative work in flipped mathematics classes. The report in [12] highlights well-known potential difficulties. Group composition is a recurring problem: uniformly high-achieving groups may see little need for genuine collaboration; mixed-ability groups can become dominated by the more able students; and uniformly low-achieving groups may produce few correct solutions.

Additional problems were found in [12] to arise from group members having different intentions: those seeking deeper understanding may be resisted by peers aiming only to finish assigned tasks quickly. Collaboration can also produce anxiety about contributing incorrect solutions and misleading the group. The report in [25] includes extensive examples of gender-associated difficulties in interaction. In the authors' previous work [5], no gender differences in achievement were found, but this does not exclude the possibility of gender-related interactional effects, which remain worth monitoring.

From a theoretical standpoint, the potential benefits of collaboration in mathematics are grounded in social-constructivist perspectives. While social-constructivist theory [7, 30] underpins much of the enthusiasm for collaborative learning in mathematics, its direct applicability to formats such as board tutorials is limited. Mathematics differs from many other disciplines in that correctness is established by formal proof and adherence to established definitions, rather than by negotiated consensus [10]. As a result, while peer dialogue can promote engagement and reveal misconceptions, it cannot substitute for individual mastery of precise methods. In practice, collaborative settings may also entrench shared errors when all group members hold similar misunderstandings, a risk well-documented in small-group mathematics work [31]. Furthermore, as argued in [16], minimally guided approaches, often inspired by social constructivism, can overload working memory, especially for novices grappling with the high intrinsic cognitive load of mathematical reasoning. In linguistically diverse classrooms, the reliance on rapid verbal interaction may further disadvantage second-language speakers, who can be less able to contribute before discussions move on [11]. These limitations suggest that while social-constructivist principles may offer insights into the role of interaction in learning, their practical implementation in mathematics education must be accompanied by strong discipline-specific scaffolding, explicit instructional support, and equitable participation structures: precisely the rationale for the extensive Term 3 revisions to the board tutorial materials.

Collaborative reasoning - explaining solutions, challenging proofs, and negotiating approaches - helps consolidate understanding and makes thinking visible, enabling misconceptions to be addressed [12]. Yet the benefits of classroom collaboration depend heavily on task design (the tasks must be accessible yet challenging, with appropriate scaffolding so that all group members can begin productively), group dynamics (participation must be balanced; groups dominated by one or two confident voices risk marginalising others), and instructional support (student engagement can be sustained by prompting participation and intervening when groups struggle [12]. In mathematics, scaffolding is often provided by the instructor through modelling, prompting, and feedback. Board tutorials, however, deliberately limit tutor intervention to promote independence, placing the burden of scaffolding on the written materials themselves. In early implementations, this burden was not met: Term 1 sheets contained challenging proof and set-operation tasks without worked examples, guiding prompts, or incremental progression. As Barron's [1] classic classroom study shows, groups stall early when confident members take over, leaving others as passive observers. This pattern restricts equity and learning potential, with international students disproportionately represented among those who hesitate to participate [26]

Term 3 materials were revised to address these shortcomings. Changes included simpler warm-up problems, guided proofs with step-by-step structure, explicit prompts for individual contributions, hints to avoid common errors, slower progression through difficult topics, and a dedicated graph theory sheet to provide earlier exposure before the exam. These changes aimed not just to improve mean performance but to reduce participation disparities, particularly between domestic and international students, in a format where no tutor ensures balanced interaction.

Second-language challenges in collaborative contexts. In the authors' previous work [5], no interaction was found between tutorial type and national background, leading to the conclusion that international students' lower achievement in board tutorials could not be further explained by greater difficulty collaborating in a second language. However, two factors justify revisiting this question in the present Term 3 study. First, Term 3 students have had more time to adapt to university life, which for international students includes additional time to adjust to using spoken English in academic discussion. This could lead to better board tutorial performance than in earlier terms. Second, the substantial Term 3 material revisions aimed to make the subject matter more accessible; if performance improves for all, any remaining domestic–international gap might more clearly reflect persistent language-related interactional challenges.

Anecdotal evidence from teaching staff supports this. International students have reported that they need longer to formulate ideas in English, and that group discussions often move on before they can comment or ask a question. One student in the study [11] expressed it plainly:

> Interjecting or jumping into conversation is hard. By the time I am ready to speak, it is over. I am so slow.

Research confirms that high formal English test scores do not necessarily equate to full command of spoken academic English. The work in [19] found substantial variation in oral proficiency among international students in the United States despite high test results. The review [3] review of 61 studies of international students in the United States. identifies language barriers as among their most common challenges. Studies [11, 17] highlight that classroom participation is often the area most affected by such barriers.

Closer to the present context, a recent study [29] found that East Asian students in New Zealand participated minimally in class discussions. A report [33] of interviews with Chinese students in Australia, in their first language, records frequent difficulty and anxiety in classroom interaction, especially in group discussions, where they struggled to convey their ideas in English. Students found informal interaction harder than prepared presentations. Similarly, work in the Philippines [28] includes cases where interruptions or questions could cause complete breakdowns in spoken English communication among English learners. Beyond performance issues, there is evidence from the United States [20] for instances of linguistically motivated discrimination against international students, which could heighten anxiety about speaking in groups.

These findings align with broader research on second-language use in collaborative learning, such as [21]. Students using a second language often face difficulty due to the following factors:

- Processing complexity: simultaneously interpreting mathematical and linguistic content.
- Formulation delay: slower response times can cause missed opportunities to contribute.
- Turn-taking challenges: informal norms favour quicker, more assertive speakers.
- Cognitive load from translation: mentally rehearsing in one's first language before speaking.
- Affective barriers: fear of error, misunderstanding, or slowing the group.

In traditional tutorials, structured turn-taking and tutor facilitation can mitigate these issues. In peer-led board tutorials, rapid exchanges and unstructured interaction can amplify them, leading to reduced participation and fewer opportunities to develop and articulate mathematical reasoning.

The present study therefore tests whether Term 3's improved scaffolding can reduce cognitive barriers enough to raise international students' performance, and whether any remaining gaps point to persistent interactional disadvantages tied to second-language use in collaborative settings.

3 Method

3.1 Sample

Simple random samples of approximately one in five students were drawn from Term 3 enrolments in the last year of the program with traditional tutorials, and the first year with board tutorials. Numbers are listed with the descriptive statistics in the Results section.

3.2 Data Collected

Program membership, gender and national background (domestic or international) were recorded. In the traditional program it was not expected that gender differences would be found, but the potential for gender affecting interaction in the collaborative program required its inclusion in analyses.

Achievement scores were obtained from students' responses in the final examination of the Discrete Mathematics subject. The examinations covered the same syllabus, and the questions were at a similar level of difficulty. Scores for each task were assigned using a common scale derived from the SOLO taxonomy [2], in the same way that was used in the previous two studies [4, 5]. The taxonomy ranks students' responses to a task at five levels depending on logical consistency and the use of evidence relevant to the task. A summary of the levels is in Table 1 below.

The derived scoring scale gives each task a score from 0 to 5, based on the first four SOLO levels, because the top level was considered inapplicable in examination responses. The first split was based on the consistency required by the SOLO Relational level, with 5 going to a logically consistent solution, and 4 to one containing a minor slip not affecting consistency. Any lack of consistency or completion scored at most 3, with 0 corresponding to the SOLO Prestructural level, 1 to the Unistructural level (one relevant and correct step), and 2 or 3 to the Multistructural level, that is, flawed by errors of logic or misunderstanding, but with at least 2 steps right, and 2 or 3 depending on how much of a valid solution is given.

Analyses. Gender differences were checked within programs and national backgrounds, because programs defined different contexts, and testing across national background would risk confusing gender comparisons by results for national background.

Three analyses involving achievement were carried out. Each of the first two involved achievement comparisons within each program, using factorial analysis of variance by

Table 1 The SOLO taxonomy

Level	Definition
Prestructural	Essentially no valid response.
Unistructural	One aspect of the problem correctly identified, but no diversity of aspects, so that questions of consistency cannot arise.
Multistructural	Multiple relevant information presented and used, but without considering relationships between different parts, so that inconsistency appears.
Relational	Multiple relevant information presented and used in a way that recognises relationships and achieves consistency within the given task.
Extended abstract	Multiplicity recognised and consistency achieved over a context beyond that of the given task.

terms (earlier terms versus Term 3) and national background. The third compared Term 3 results across programs, using factorial analysis of variance by program and national background.

4 Results

4.1 Gender Comparisons

Gender comparisons were made within tutorial types because the learning context could influence gender-related patterns of achievement and were made separately for domestic and international students to avoid conflating gender with national background differences. As in the case of previous terms [4, 5], no significant gender differences were found. This finding held across all combinations of tutorial type and national background, confirming that gender was not a confounding variable in subsequent analyses.

4.2 Descriptive Statistics

These are in Tables 2 and 3 below. The set of means in Term 3 shows a pattern that was very different from those in previous terms [4, 5], from which descriptive statistics, for comparison, are listed below in Table 3.

In the traditional program, the mean achievement scores of domestic and international students (23.72 and 23.16, respectively) were both close to the highest levels observed in earlier terms (Table 3). The difference between domestic and international students was small (0.56 points), suggesting that the achievement gap previously observed between these groups in earlier terms had largely closed.

In the board tutorial program, both domestic and international students showed higher mean scores in Term 3 compared with their counterparts in Terms 1 and 2. Domestic students' mean rose from 23.2 to 25.53 (+2.33), while international students' mean increased from 18.1 to 21.75 (+3.65). The larger relative gain for international students indicates that the revised Term 3 teaching materials may have been particularly helpful for

Table 2 Term 3 descriptive statistics

Tutorial method	Term 3		
		Domestic	International
Traditional	Mean	23.72	23.16
	St, dev.	9.75	8.96
	n	57	69
Board tutorials	Mean	25.53	21.75
	St. dev.	9.95	9.81
	n	57	89

Table 3 Terms 1 and 2 descriptive statistics

Tutorial method	Terms 1 and 2		
		Domestic	International
Traditional	Mean	24.9	21.6
	St, dev.	9.4	9.8
	n	234	145
Board tutorials	Mean	23.2	18.1
	St. dev.	10.1	10.5
	n	172	201

this group. However, the absolute difference between domestic and international means in board tutorials (3.78 points) remained substantial, and the statistical significance of the gap is examined below.

4.3 Comparisons within Traditional Tutorials

Results of a factorial analysis by terms and national background within the traditional tutorial program are in Table 4 immediately below.

The factorial ANOVA for traditional tutorials (Table 4) revealed: no significant effect of term ($p = 0.8654$), indicating stability in achievement across earlier terms and Term 3; a significant effect of national background ($p = 0.0029$), which reflected the lower achievement of international students overall; no significant interaction between term and national background. Least significant differences (Table 5) showed that the only significant pairwise difference was between domestic and international students in earlier terms ($p = 0.0213$). This confirms that, in Term 3, international students in traditional tutorials achieved scores statistically indistinguishable from those of domestic students, suggesting an equalisation effect over time. Details are in Appendix 1.

Table 4 Traditional tutorials: factorial analysis of variance

Source	Sum of squares	df	Mean squares	F	Probability
Effects	983.27	3	327.76	3.6428	0.0127*
Terms	2.59	1	3.59	0.0288	0.8654
National background	806.13	1	806.13	8.9596	0.0029**
Interaction	171.28	1	171.28	1.9036	0.1683
Residual	44986.89	500	89.97		
Total	45970.16	503			

Table 5 Traditional tutorials: least significant differences

Group	TT3 international	TT3 domestic	TT12 domestic
TT12 international	ns	ns	$p = 0.0213$*
TT3 international		ns	ns
TT3 domestic			ns

Note TT12 = Traditional, Terms 1 and 2, TT3 = Traditional, Term3

4.4 Comparisons within Board Tutorials

Results of a factorial analysis by terms and national background within the traditional tutorial program are in Table 6 immediately below.

Table 6 Board tutorials: factorial analysis of variance

Source	Sum of squares	df	Mean squares	F	Probability
Effects	3287.45	3	1095.82	10.5931	0.000***
Terms	2026.77	1	2026.77	19.5926	0.000***
National background	1460.22	1	1460.22	14.12	0.000***
Interaction	7.34	1	7.34	0.07	
Residual	54374.58	515	103.45		
Total	56562.03	518			

Differences between terms and between national groups are highly significant, and interaction very far from significance. The pattern for means indicates higher scores for Term 3, and, within terms, significantly lower scores for international students. These group differences are confirmed by the least significant differences, summarised below in Table 7, with details given in Appendix 2.

Table 7 Board tutorials: least significant differences

Group	BT12 domestic	BT3 international	BT3 domestic
BT12 international	$p = 0.0023$**	$p = 0.0014$**	$p = 0.0000$***
BT12 domestic		ns	$p = 0.0027$**
BT3 international			$p = 0.0302$*

Note BT12 = Board tutorial program, Terms 1 and 2, BT3 = Board tutorial program, Term 3

4.5 Comparisons Involving Tutorial Programs

Results for a factorial analysis of variance by tutorial program and nation are in Table 8 immediately below.

Table 8 Analysis of variance: tutorial type and national background

Source	Sum of squares	df	Mean squares	F	Probability
Effects	507.71	3	169.24	1.82880	0.1422
Tutorial type	0.09	1	o.09	0.0001	0.9753
Nation	335.92	1	335.92	3.6300	0.0578#
Interaction	169.50	1	169.50	1.8316	0.1771
Residual	24708.81	267	92.54		
Total	25216.52	270			

4.6 Summary

In summary the key quantitative trends show:

- Traditional tutorials: Stable achievement over time, with the national background gap closing in Term 3.
- Board tutorials: Significant improvement for all students in Term 3, but persistent national background gap.
- Between tutorial programs: No difference in overall Term 3 means, but domestic–international disparity confined to the board tutorial format.
- Effect of revised materials: Board tutorial students in Term 3 outperformed their predecessors, suggesting that redesigned tasks improved accessibility and engagement.
- Adaptation effects: International students in traditional tutorials appear to have benefitted from the extended adaptation period; in board tutorials, adaptation effects were present but insufficient to close the gap.

5 Discussion

Results for students in traditional tutorials serve to modify the conclusions of the first of the present sequence of studies [4], because Term 3 scores show neither the decrease in achievement by term nor the lower achievement results for international students found previously. The absence of a significant difference between terms for domestic students indicates that later enrolment in Term 3 neither conferred advantage nor disadvantage for this group. For international students, however, the Term 3 results are consistent with the view that an extended period of adaptation to international university study has had a favourable impact on achievement, effectively closing the gap with domestic peers observed in earlier terms.

The improvement for international students in the traditional tutorial program may be understood in terms of two overlapping forms of adaptation. First, there is **adaptation to university study of mathematics**, developing and implementing study strategies to handle mathematics delivered at greater speed and requiring greater skill and organization than earlier work, which also applies to domestic students. Second, important for international students, there is **interactional and linguistic adaptation**: the gradual development of fluency in using English for complex academic explanations, both orally and in writing. Over time, repeated engagement in tutor-led discussions of proofs and problem-solving strategies may help international students overcome some of the communicative delays and hesitations reported in anecdotal staff accounts, where the pace of conversation often moved beyond their intended contributions before they could be voiced. This interpretation is consistent with the body of research identifying extended exposure and participation as important factors in overcoming the language-based interactional barriers identified in the Background section.

Turning to the results for board tutorials, two aspects are particularly noteworthy. First, the significant increases in Term 3 achievement scores for both domestic and international students strongly suggest that the extensive revisions to the teaching materials were effective in making the collaborative problem-solving tasks more accessible. The revised set of problem sheets for Term 3 provided more explicit examples, scaffolded proofs, and guided hints, thus replicating some of the pedagogical functions of traditional tutorials. These changes probably reduced the initial cognitive load for students, enabling them to reach a point of engagement with the more challenging problems in a shorter time. The improvement for domestic students, in particular, suggests that the redesigned materials helped bridge the gap between lecture content and collaborative practice. Second, and more problematically, the achievement gap between domestic and international students persisted in board tutorials despite both groups benefiting from the revised materials and the longer adaptation period. This persistence points to a limitation in the reach of material-based interventions.

While improved scaffolding can support conceptual understanding and problem-solving skills, it does not directly address the **interactional demands** of fast-paced peer discussion in a second language. For international students, particularly those whose spoken English processing speed lags behind comprehension, the real-time nature of collaborative board work can remain a barrier even when the underlying mathematical content becomes more familiar. The additional cognitive effort required to formulate and articulate ideas in English, especially when group dynamics favour rapid exchanges, can result in reduced opportunities to contribute, diminished influence in the collaborative process, and, ultimately, a lower overall benefit from the session.

These findings reinforce the earlier conclusion that the board tutorial format, as currently implemented, places international students at a relative disadvantage. They also refine the interpretation: the disadvantage is not solely due to unfamiliarity with the content or lack of preparation but appears to be tied to the **mode of interaction** itself. In this context, 'adaptation', in the sense of acclimatisation to content and university culture, may proceed successfully, yet adaptation to the interactional norms of peer-based mathematical discussion may require specific, targeted support.

The persistence of the gap in board tutorials suggests that such support could involve structural modifications to collaborative sessions. Possible interventions include: introducing structured turn-taking protocols to slow the pace and ensure all group members have an opportunity to contribute; assigning specific discussion roles (for example, explainer, summariser, questioner) to distribute responsibility and create predictable entry points for participation; integrating a brief 'written response' phase before verbal discussion, allowing all students to formulate contributions without the pressure of immediate speech; using glossary-based support for key mathematical terminology to reduce cognitive load in real-time exchanges.

Future research should examine these strategies systematically, perhaps through pilot implementations in selected board tutorial groups, with both quantitative achievement measures and qualitative analysis of participation patterns. Given the evidence from other contexts that language barriers disproportionately affect collaborative discussion rather than solitary study tasks [11, 30, 33], such research could help determine whether the disadvantage observed here can be mitigated without abandoning the collaborative model entirely.

Finally, the present results highlight the value of mixed method approaches in mathematics education research. While the achievement data reveal clear patterns, they cannot fully capture the lived experience of students in collaborative settings. Observational studies, classroom recordings, and post-session interviews could clarify how group dynamics interact with linguistic factors to shape participation. Such work would not only strengthen the interpretation of the current findings but also inform the design of inclusive pedagogical strategies for linguistically diverse cohorts.

In summary, the Term 3 results show that international students in traditional tutorials can achieve parity with domestic students given sufficient adaptation time; while in board tutorials, substantial material revisions yield improvements for all but leave a persistent achievement gap for international students. This gap appears to stem from the inherent interactional demands of the collaborative format and addressing it will require pedagogical interventions that extend beyond the content of the tasks themselves.

Disclosure of Interests. The authors have no competing interests to declare that are relevant to the content of this article.

Appendix 1

Least significant differences: traditional tutorials by term and national background

Group		TT3 international	TT3 domestic	TT12 domestic
TT12 international	t	0.9471	1.2360	2.3220
	df	137	126	191
	Probability	0.3452	0.2188	0.0213*
TT3 international	t		0.3291	1.2283
	df		123	188
	Probability		0.7426	0.2208
TT2 domestic	t			0.7403
	df			177
	Probability			0.4304

Appendix 2

Least significant differences: board tutorials by term and national background

Group		BT3 international	BT3 domestic	BT12 domestic
BT12 international	t	3.0651	3.2304	5.2153
	df	372	288	256
	Probability	0.0023**	0.0013**	
BT3 international	t		0.7477	3.0378
	df		359	227
	Probability		0.4773	0.0022*
BT2 domestic	t			2.1890
	df			144
	Probability			0.0302*

Appendix 3

Least significant differences: Term 3. comparisons by tutorial type and national background

Group		TT3 international	TT3 domestic	BT3 domestic
BT3 international	t	0.9058	1.2036	2.3143
	df	155	144	144
	Probability	0.3664	0.2307	0.0221*
TT3 international	t		0.3245	1.3739
	df		123	123
	Probability		0.7461	0.1720
TT3 domestic	t			1.0770
	df			112
	Probability			0.3166

References

1. Barron, B.: When smart groups fail. J. Learn. Sci. **12**(3), 307–239 (2003)
2. Biggs, J.B., Collis, K.F.: Evaluating the Quality of Learning: the SOLO Taxonomy. Academic Press, New York, NY (1982)
3. Bokayev, B.: The socialization of international students in American society and its educational system: a comprehensive literature review. J. Soc. Stud. Educ. Res. **15**(3), 289–316 (2024)
4. Bowen-James, A., Freislich, M.R.: Transition challenges in discrete mathematics: comparing domestic and international student performance. In: Proceedings of the 2024 International Conference on Computer Science and Computational Intelligence (CSCI), December 11–13, Las Vegas, USA (2024)
5. Bowen-James, A., Freislich, M.R.: Collaboration, flipped classrooms and the challenge of board tutorials in mathematics education. In: Proceedings of the 2025 International Congress in Computer Science Computer Engineering and Applied Computing (CSCE), July 21–24, 2025, Las Vegas, USA (2025)
6. Chen, X.: Online collaboration tools and international students' sense of belonging in group work. J. Teach. Learn. Technol. **12**(1), 36325 (2023)
7. Cobb, P.: Where is the mind? Constructivist and sociocultural perspectives on mathematical development. Educ. Res. **23**(7), 13–20 (1994)
8. Di Martino, P., Gregorio, F.: The first-time phenomenon: successful students' mathematical crisis in secondary–tertiary transition. In: Proceedings of the 42nd Conference of the International Group for the Psychology of Mathematics Education, Umeå, Sweden (2018)
9. Duff, P.A.: Language socialization into academic discourse communities. Annu. Rev. Appl. Linguist. **30**, 169–192 (2010)
10. Ernest, P.: Social Constructivism as a Philosophy of Mathematics. SUNY Press, Albany NY (1998)

11. Findlay, M., Wan, G., Press, A., Jones, K.E., Maranto, M.: Hidden struggles: increasing international graduate students' sense of belonging. J. Comparat. Int. Higher Educ. **17**(1), 167–181 (2025)
12. Fredriksen, H., Hadjerrouit, S.: An activity-theory perspective on contradictions in flipped mathematics classes at the university level. Int. J. Math. Educ. Sci. Technol. **51**(4), 520–541 (2020)
13. Güler, M., Kokoç, M., Bütüner, S.O.: Does a flipped classroom work in mathematics education? A meta-analysis. Educ. Inform. Technol. **28**, 57–97 (2023)
14. Henriksen, M.: Reflections of a first-time flip in a differential equations classroom. Primus. **31**(8), 909–924 (2021)
15. Jensen-Vallin, J.: Teaching students to formulate questions. Primus. **27**(2), 189–201 (2017)
16. Kirschner, P.A., Sweller, J., Clark, R.E.: Why minimal guidance during instruction does not work. Educ. Psychol. **41**(2), 75–86 (2006)
17. Krouss, P., Lesseig, K.: Effects of a flipped classroom model in an introductory college mathematics course. Primus. **30**(5), 617–635 (2020)
18. Lin, S.Y., Scherz, S.D.: Challenges facing Asian international graduate students in the US: pedagogical considerations in higher education. J. Int. Stud. **4**(1), 16–33 (2014)
19. Lindner, M.A., Eitel, A., Dalehefte, I.M.: Learning to solve ill-defined statistics problems: does self-explanation matter as much as with worked examples? Instr. Sci. **50**, 1005–1031 (2022)
20. Martirosyan, N.M., Hwang, E., Wanjohi, R.: Impact of English proficiency on academic performance of international students. J. Int. Stud. **5**(1), 60–71 (2015)
21. Matsumoto, Y.: Multilingual international students' communicatory practices in US universities: rethinking appropriate Englishes through English as a lingua franca perspective. Harv. Educ. Rev. **92**(4), 486–507 (2025)
22. Morita, N.: Negotiating participation and identity in second language academic communities. TESOL Q. **38**(4), 573–603 (2004)
23. Nicholas, J., Poladian, L., Mack, J., Wilson, R.: Mathematics preparation for university: entry, pathways and impact on performance in first year science and mathematics subjects. Int. J. Innovat. Sci. Math. Educ. **23**(1), 37–51 (2015)
24. Rach, S., Heinze, A.: The transition from school to university in mathematics: which influence do school-related variables have? Int. J. Sci. Math. Educ. **15**, 1343–1363 (2017)
25. Reinholz, D.L., McMorris, H., Apkaian, N.: When active learning is inequitable: Women's and students of color's participation predicts their course grades more than participation of other students. J. Res. Math. Educ. **53**(5), 428–463 (2022)
26. Rensaa, R.J., Fredriksen, H.: Gender perspectives on a flipped classroom. Cogent Educ. **9**, 2115832 (2022)
27. Rios, J.: Language, comfort speaking, and collaboration: a QuantCrit analysis of multilingual students' experiences in introductory college mathematics courses. Int. J. Res. Undergrad. Math. Educ. **10**, 642–672 (2024)
28. Sandefur, J., Lockwood, E., Hart, E., Greefrath, G.: Teaching and learning discrete mathematics. ZDM Math. Educ. **54**, 753–775 (2022)
29. Separa, L.A.C., Generales, L.J., Medina, R.J.: Situational speaking difficulties of English second language learners in the Philippines. J. Southeast Asian Stud. **25**(1), 144–167 (2020)
30. Soltani, B., Zhang, L.J.: International language socialization in an English-medium university. Aust. Rev. Appl. Ling. **46**(3), 316–338 (2023)
31. Vygotsky, L.S.: Mind in Society: the Development of Higher Psychological Processes. Harvard University Press, Cambridge MA (1978)
32. Webb, N.M.: The teacher's role in promoting collaborative dialogue in the classroom. Br. J. Educ. Psychol. **79**(1), 1–28 (2009)

33. Zhang, H., Tan, H.: Foreign language anxiety and study abroad: Chinese students' voices of experiencing foreign language anxiety during their graduate study in Australia. Cogent Educ. **16**(1), 2367303 (2024)

Problem-Solving Using Logic and Reasoning, Mathematics, Algorithms, Python, and Generative AI: Part Four

Weizheng Gao[1(✉)] and Shanzhen Gao[2]

[1] Department of Mathematics, Computer Science, and Engineering Technology, Elizabeth City State University, Elizabeth City, NC 27909, USA
wegao@ecsu.edu

[2] Reginald F Lewis College of Business, Virginia State University, Petersburg, VA 23806, USA
sgao@vsu.edu

Abstract. Problem-solving is a multifaceted skill that plays a critical role in business, education, technology, and everyday life. Effective problem-solving requires the integration of logic and reasoning, mathematics, algorithms, programming, and increasingly, artificial intelligence (AI). Building on our previous studies—Parts One, Two, and Three—this fourth installment continues to explore the combined use of logic, mathematics, Python programming, and generative AI to enhance problem-solving competency. In this paper, we present a series of problem scenarios and analyze the responses generated by two leading generative AI models, ChatGPT and Google Gemini. Comparative evaluations demonstrate the potential of generative AI to support learning and reasoning, while highlighting important limitations, including occasional inaccuracies, logical inconsistencies, and a lack of mathematical rigor. These findings reinforce the importance of balancing AI-generated suggestions with human reasoning, mathematical knowledge, and algorithmic thinking. This work offers practical insights for educators, students, and researchers on responsibly leveraging AI as a complementary tool to enhance analytical thinking and problem-solving skills.

Keywords: Problem-Solving · Intelligent Decision-making · Generative AI · ChatGPT · Python Programming · Logic and Reasoning · Mathematics · Algorithms

1 Introduction

Problem-solving is a critical skill that enables individuals to navigate challenges, make informed decisions, and achieve goals across academic, professional, and real-life contexts. Modern problems, ranging from everyday personal situations to advanced business, mathematical, and technological issues, require more than intuition or isolated knowledge. Effective problem-solving draws upon logical reasoning, mathematical thinking, creativity, and structured analytical approaches. While the definition of problem-solving varies across disciplines, for example, being viewed as a cognitive process in psychology or a computational process in computer science, there is consensus that it involves

H. R. Arabnia et al. (Eds.): CSCI 2025, CCIS 2974, pp. 173–188, 2027.
https://doi.org/10.1007/978-3-032-28475-4_12

applying knowledge, reasoning, and strategy to overcome obstacles and reach desired outcomes [1–4, 7].

Equipping students with strong problem-solving skills has become increasingly important due to the rapid evolution of technology and the digital transformation in education, business, and society. Students today must master foundational competencies, including critical thinking, logic, mathematical reasoning, and creativity, as well as digital literacy, intelligent decision-making, time management, and organizational skills. These skills collectively enhance students' confidence and readiness to tackle complex real-world problems with resilience and adaptability.

As technological innovation accelerates, problem-solving has expanded beyond traditional methods to include computational and AI-driven approaches. Python programming has emerged as a valuable tool for developing algorithmic and analytical thinking, enabling learners to simulate scenarios, analyze data, and visualize solutions in dynamic ways. Generative AI, including ChatGPT, further enhances learning by providing immediate feedback, supporting exploratory thinking, and facilitating interactive engagement with problem-solving processes. However, AI-generated solutions also require careful evaluation for accuracy, logic, and validity, highlighting the importance of human oversight. This paper builds upon our previous publications—Parts One, Two, and Three—where we introduced a multidisciplinary approach that integrates logic, mathematics, algorithms, Python, and generative AI to support and enhance problem-solving ability [1–3]. Part Four continues this developmental progression by presenting new problem scenarios, comparing the responses of generative AI models, and examining both the instructional benefits and limitations of AI in problem-solving contexts. Through this installment, we aim to enhance the understanding of how human reasoning and generative AI-assisted problem-solving can complement each other, enhancing effectiveness, deepening comprehension, and preparing students to think critically in an AI-supported learning environment.

2 Literature Review

Problem-solving is a multidimensional skill that draws from logic, mathematics, technology, and cognitive development. Prior research highlights that effective problem-solving requires more than content knowledge—it demands strategic thinking, creativity, analytical reasoning, digital literacy, and the ability to integrate multiple domains when approaching complex challenges. The following review synthesizes literature across three thematic areas: (1) foundational perspectives on problem-solving, (2) the integration of AI, Python, and computational tools, and (3) knowledge management (KM) as a driver of problem-solving and decision-making in education and business contexts.

Foundations of Problem-Solving: Logic, Critical Thinking, and Creative Competence

Seminal research in mathematics education emphasizes the importance of explicit instruction in strategic problem-solving processes. Schoenfeld emphasized that successful mathematical problem-solving extends beyond procedural knowledge and requires heuristic strategies, metacognitive awareness, and reflective thinking, enabling learners to analyze and adapt different approaches to reach coherent solutions [10]. Complementing this, Snyder and Snyder highlighted that business education has historically underemphasized critical thinking, advocating for instructional practices such as collaborative learning, active inquiry, and applied problem scenarios to foster students' analytical capabilities and reasoning skills [11]. Together, these studies provide foundational guidance for strengthening logic-based competencies that are essential to effective problem-solving.

Beyond logic, creativity also plays a crucial role in shaping adaptable problem solvers. Khalid et al. argued for integrating Creative Problem-Solving (CPS) in mathematics classrooms to cultivate innovative thinking and flexible reasoning, particularly in environments shifting away from rote memorization toward higher-order thinking [13]. Their findings align with current educational goals of preparing learners to confront unpredictable, real-world challenges. Meanwhile, Araiza-Alba et al. explored the impact of immersive virtual reality (IVR) as a digital learning tool, demonstrating that interactive and multisensory environments enhance students' engagement, memory, and cognitive flexibility, leading to improved problem-solving performance and motivation [12]. Collectively, these foundational studies affirm that logical reasoning, critical thinking, and creativity form the core pillars of problem-solving across disciplines.

AI, Python, and Computational Approaches to Problem-Solving

The rise of computational thinking and artificial intelligence has transformed how problem-solving is taught, learned, and applied. A series of recent studies by Gao and colleagues illustrates the evolution of integrating logic, mathematics, algorithms, Python programming, and generative AI in problem-solving education. Their first publication in this series introduced a multidisciplinary model demonstrating how Python, algorithms, and AI can support logical and mathematical reasoning in real-world problem contexts [1]. The second installment expanded application scenarios and emphasized the value of algorithmic thinking, computational reasoning, and generative AI-assisted solutions to strengthen students' analytical confidence and decision-making skills [2]. Part Three advanced the instructional framework further by illustrating how generative AI tools, such as ChatGPT, can bridge human reasoning with automated problem-solving techniques, helping learners critically evaluate AI-generated responses while reinforcing domain knowledge [3]. Subsequent work by the same authors applied these concepts using Google Colab to help students practice collaborative and accessible computational problem-solving using AI, Python, and cloud-based programming environments [4].

Beyond general problem-solving, Python and AI have also been applied to specialized mathematical fields. Gao et al. demonstrated how Python and generative AI can support computational visualization and graph theory instruction, enabling more intuitive and interactive learning of complex structures through prompt-engineered simulations [9]. Su et al. contributed complementary insights by presenting mathematically rigorous findings on the dominion of grids in graph theory, showcasing how structured reasoning and proof techniques reinforce mathematical problem-solving skills [5]. Together, these computational studies show that pairing Python with AI not only enhances efficiency and accuracy but also expands students' capacity to explore, model, visualize, and verify mathematical and algorithmic solutions.

Knowledge Management as a Driver of Problem-Solving and Decision-Making
In addition to logic and computational tools, knowledge management (KM) frameworks make significant contributions to problem-solving in educational and business environments. Russ emphasized the importance of KM for sustainable development in an era of technological acceleration, arguing that organizations must leverage knowledge ecosystems, digital tools, and human capital to support continuous innovation and complex decision-making [8]. Building upon this concept, Gao and Gao proposed a KM mesosystem model to enhance business education by integrating KM strategies with AI-driven tools, fostering informed decision-making, reflective learning, and improved problem-solving performance across academic settings (in press) [6]. Furthermore, Gao et al. examined the interplay between AI and human logic, demonstrating that while generative AI accelerates analytical processes and supports solution development, human reasoning remains crucial for interpretation, validation, ethical judgment, and contextual relevance [7]. These perspectives collectively highlight that effective problem-solving today requires not only cognitive skill and technological proficiency but also the ability to manage, share, apply, and evaluate knowledge within evolving digital ecosystems.

Summary
Across the literature, a consistent theme emerges that problem-solving in the 21st century requires a holistic integration of logical reasoning, creativity, computational tools, and knowledge management. Foundational studies emphasize the importance of critical thinking, heuristic strategies, and creative cognitive development, while contemporary research demonstrates how AI and Python enhance computational efficiency and provide new learning pathways. Complementary KM-based frameworks further reinforce that problem-solving is not solely an individual cognitive process but a dynamic, collaborative, and technology-supported practice. This integrated perspective aligns with and informs the present study, which extends previous work by examining how logic, mathematics, algorithms, Python, and generative AI can be collectively leveraged to strengthen problem-solving competencies.

3 Research Methodology

This study adopted a qualitative, comparative research design to investigate the role of generative artificial intelligence in supporting problem-solving across multiple disciplinary contexts. Seven problem scenarios were intentionally developed to reflect a broad spectrum of cognitive and analytical domains, including logic and reasoning, mathematics, probability, critical thinking, real-world decision-making, graph theory, and Python programming. Each scenario was carefully structured to include elements requiring analytical reasoning, conceptual understanding, and, in some cases, computational execution.

The problem prompts were individually submitted to two state-of-the-art generative AI models—ChatGPT and Google Gemini—under standardized conditions. To preserve objectivity, no additional hints, clarifications, or iterative prompting were used, ensuring that the responses reflected each model's first-attempt reasoning capability. The generated responses were collected, documented, and evaluated using five criteria: (1)

accuracy of the final solution, (2) correctness of mathematical or computational steps, (3) clarity and coherence of explanation, (4) logical soundness, and (5) contextual appropriateness. Expert human review was performed to validate the results, identify reasoning errors, and assess the pedagogical value of each AI-generated solution. Comparative analysis enabled the identification of strengths, limitations, and recurring patterns in AI-assisted problem-solving, forming the basis for recommendations on integrating AI effectively in educational settings.

4 A Winning Lottery Pen

Problem 1
A small startup claims to have created a "Winning Lottery Pen" that can increase a person's chances of winning the lottery by 99%. The company says the pen contains "lucky magnetic ink" that aligns with "positive lottery energy" when filling out a ticket. Each pen costs $199, and the company claims that "many buyers have already won a moderate amount of money." However, the company has provided no evidence, and neither lottery officials nor customers have supported the product. Additionally, the customer reviews cannot be verified.

Questions (Short Answer)

(a) List two claims in the article that seem questionable or unrealistic.
(b) What evidence should the company provide before people believe the product works?
(c) Identify one red flag that suggests this product may not be trustworthy.

ChatGPT's Response (11:30 PM on 11/02/2025)**:**

(a) The pen can "increase a person's chances of winning the lottery by 99%."
The "lucky magnetic ink" aligns with "positive lottery energy" to improve winning odds.
(b) They should provide verifiable evidence, such as independent testing results, scientific proof of how the pen affects lottery outcomes, and confirmed winning records from official lottery authorities or third-party studies.
(c) A major red flag is that the company makes big claims without any proof, and customer reviews cannot be verified.

Google Gemini's Response (11:38 PM on 11/02/2025)**:**

(a) The explanation involving "lucky magnetic ink" and "positive lottery energy."
The claim of increasing a person's chances of winning the lottery by 99%.
(b) The company should provide verifiable, independently audited records of winners and statistically significant data (like results from a randomized controlled trial) confirmed by a reputable third party.
(c) The company has provided no evidence, and lottery officials or verifiable customers do not support the claims.

Remarks:
Both ChatGPT and Gemini provided accurate and informative answers.

This product appears to rely on superstition rather than science, and any credible evidence does not support the claims. Genuine lottery odds cannot be changed by a pen, ink, or "energy alignment." The high price, unverifiable testimonials, and lack of independent validation all suggest that the product is likely a scam or misleading marketing tactic. Consumers should exercise caution and avoid spending money on items that promise unrealistic results without substantiated evidence.

5 Computing Average Age

Problem 2
Alice has an annual income of one billion dollars and owns 77 cats, 88 donkeys, and 101 Dalmatians. When she was born, her mother was 100 times Alice's age. What is the average age of Alice and her mother? Assume ages are integers of years.

ChatGPT's Response (7:30 AM on 11/03/2025):

Indeterminate (any nonnegative integer).

Let Alice's current age be A and her mother's M (integers, so assume $A, M \geq 0$).
"At Alice's birth" (i.e., A years ago): Alice was 0 and her mother was $M - A$.
"Mother was 100 times Alice's age" ⇒ $M - A = 100 \cdot 0 = 0 \Rightarrow M = A$.

So they're the **same age now**, and the average is

$$\frac{A+M}{2} = \frac{A+A}{2} = A,$$

which can be **any** nonnegative integer. (E.g., 66 works, but so do 0, 1, 2, ...)

Google Gemini's Response (7:38 AM on 11/03/2025):

The information about the income and pets is a distraction. The statement that the mother was "100 times Alice's age" when Alice was born (0) is mathematically true but provides **no usable information** about their age difference (D).

In this type of riddle, when the age difference is rendered useless, the intended answer is usually the **arithmetic average of the two numbers given in the context of age (100 and 1, assuming Alice is 1 and the mother is 100)**:

$$\text{Average Age} = \frac{100+1}{2} = \frac{101}{2} = \mathbf{50.5}$$

This is the only way to arrive at a definitive numerical answer given the intentionally confusing premise.

Remarks:
Both responses show how misleading wording can influence problem-solving. ChatGPT

treated the statement literally but incorrectly concluded that the ages are indeterminate, demonstrating a flaw in the logical interpretation. Gemini approached it as a riddle, identified the distraction, and provided an unexpected puzzle-style answer. The contrast highlights two different approaches: applying strict mathematical logic versus interpreting the intended trick in the wording. This comparison is valuable for teaching critical thinking, as it shows that the "right" answer can vary depending on whether one emphasizes formal reasoning or riddle conventions. It reminds students to read carefully and question assumptions before responding.

6 Domination of Graphs

Problem 3

The domination (number) of a graph $G = (V,)$, denoted by (G), is the size of the minimum dominating sets of (G), also known as γ-sets. As such, the dominion of G, denoted by (G), counts all its γ-sets. What are the dominions of grids, $G_{2,}$?

ChatGPT's Response (8:30 AM on 11/04/2025):

Let $\zeta(G)$ be the **dominion** (the number of minimum dominating sets).

For $G_{2,n}$, the counts are:

- $n = 1 : \zeta = 2$
- $n = 2 : \zeta = 6$
- $n = 3 : \zeta = 3$
- $n = 4 : \zeta = 12$
- $n = 5 : \zeta = 2$
- $n = 6 : \zeta = 17$
- $n \geq 7$:
  - if n is **odd**: $\zeta = 2$
  - if n is **even**: $\zeta = 2n + 4$

(For context, $\gamma(G_{2,n}) = \lceil (n+1)/2 \rceil$, and the γ-sets follow alternating "staggered" patterns; even n allows a sliding pair column, giving $2n + 4$ choices.)

Reference

Su, J., Allagan, J., Gao, S., Malomo, O., Gao, W., & Eyob, E. (2024). *Dominion on grids.* **Mathematics, 12**(21), 3408. https://doi.org/10.3390/math12213408 ↗

Google Gemini's Response (9:18 AM on 11/04/2025):

The γ-number for $G_{2,n}$ is generally $\gamma(G_{2,n}) = \lceil \frac{2n}{3} \rceil$ for $n \geq 4$. The number of sets that achieve this minimum, $\zeta(G_{2,n})$, is calculated as follows:

For $n \geq 4$, the dominion $\zeta(G_{2,n})$ is:

$$\zeta(G_{2,n}) = \begin{cases} 2^{n/2} & \text{if } n \equiv 0 \pmod 4 \\ 2^{(n+1)/2} & \text{if } n \equiv 1 \pmod 4 \\ 2^{n/2} & \text{if } n \equiv 2 \pmod 4 \\ 2^{(n-1)/2} & \text{if } n \equiv 3 \pmod 4 \end{cases}$$

Examples

n	$\gamma(G_{2,n})$	$\zeta(G_{2,n})$
2	2	6
3	2	10
4	3	16
5	4	32

📚 Citation

The explicit formulas for the dominion of small grid graphs are established in specialized research, such as the following paper:

Su, J., Allagan, J., Gao, S., Malomo, O., Gao, W., & Eyob, E. (2024). Dominion on grids. *Mathematics, 12*(21), 3408. https://doi.org/10.3390/math12213408

Remarks:
While both AI-generated solutions cite Su et al. (2024), only the first response aligns with the established results in Theorem 3 of Dominion on Grids. It correctly lists the exceptional values for n = 2, 3, 4, 6 and the general formulas for odd and even n ≥ 5. In contrast, the second response applies an unrelated modulus-based formula suited to a different graph family, leading to incorrect dominion values. This comparison highlights the need to verify mathematical claims with peer-reviewed research before acceptance.

7 2024 US Presidential Election

Problem 4
Who will win the 2024 US presidential election tonight?

ChatGPT's Response (2:20 PM on 11/03/2025):

Problem 4

Who will win the 2024 US presidential election tonight?

ChatGPT's Response (2:20 PM on 11/03/2025):

I can't predict election outcomes. The winner of the 2024 U.S. presidential election depends on the actual votes counted and certified by election authorities. Media projections are not official results.

Google Gemini's Response (3:45 PM on 11/03/2025):

I cannot tell you who will win the 2024 US presidential election tonight.

The 2024 US presidential election is scheduled for **Tuesday, November 5, 2024**.

Since the election hasn't happened yet, the outcome is **unknown**, and any prediction would be speculation. Election results depend entirely on the votes cast on Election Day.

Remarks:
Although the result of the 2024 U.S. Presidential Election was known at 5:00 AM on November 6, 2024, both generative AI models declined to predict the election winner, demonstrating responsible and unbiased handling of a politically sensitive question. ChatGPT emphasized the importance of certified results, while Gemini highlighted the election date and the uncertainty that exists before votes are counted.

8 Casinos Near Virginia State University

Problem 5
The voters in Petersburg, Virginia, have approved the construction of a new casino just a few miles from Virginia State University. What are the advantages and disadvantages of building a casino near the university?

ChatGPT's Response (7:10 AM on 11/01/2025):

ChatGPT's Response (7:10 AM on 11/01/2025):

Advantages of Building a Casino Near Virginia State University

- **Economic Development:** The casino will likely increase local revenue, attract tourism, and boost Petersburg's economy through job creation and new businesses (restaurants, hotels, retail).
- **Student Employment & Internships:** VSU students may gain access to part-time jobs, internships, and career opportunities in hospitality, business, IT, security, and management.
- **Community Investment:** Increased tax revenue could support infrastructure, education, and community services.

Disadvantages of Building a Casino Near the University

- **Student Gambling Risks:** Easy access may increase compulsive gambling, financial strain, and risky behavior among college students.
- **Safety and Social Concerns:** Casinos can bring increased traffic, crime, and nightlife activity, potentially impacting campus safety and surrounding neighborhoods.
- **Academic Distraction:** Students may prioritize entertainment over academics due to proximity and convenience.

Google Gemini's Response (7:18 AM on 11/01/2025):

✅ Advantages (Economic)

- **Job Creation:** Creates hundreds of local jobs, including service and security roles, which could benefit students.
- **Tax Revenue:** Generates substantial tax revenue for the city and state for public services and infrastructure.
- **Tourism:** Acts as a major tourist attraction, injecting external money into the local economy.

❌ Disadvantages (Social & Student Risk)

- **Gambling Addiction:** Increases the risk of addiction, particularly among impulsive young adults and VSU students, potentially leading to debt and bankruptcy.
- **Student Environment:** Detracts from the educational focus of the university, risking **lost productivity** and **poor financial literacy** for students.
- **Social Costs:** May strain social services and police due to associated issues like **public order crimes** and debt.
- **Traffic:** Will likely cause **traffic congestion** and increase wear on local infrastructure.

Remarks:
Both AI responses provide balanced perspectives on the potential impact of constructing a casino near Virginia State University.

ChatGPT offers a structured analysis, clearly separating economic benefits, student opportunities, and community investment from risks such as gambling addiction, safety concerns, and academic distraction.

Gemini's response reinforces similar points, emphasizing job creation, tax revenue, and tourism, while expanding on student-focused risks, social costs, and traffic issues.

Together, the responses highlight the complex trade-offs of such a development. The comparison underscores the importance of striking a balance between economic growth, student well-being, and community safety when evaluating the placement of casinos near universities.

9 Mean, Variance, and Standard Deviation

Problem 6
Let n and m be large twin primes. Find the median, mean, variance, and standard deviation of the following values:

$$-n, m, 17\ m, 0, 4n, n, -4n, -17\ m, -m, 1234 \times 0$$

Remarks:
Both ChatGPT and Google Gemini provided correct solutions for the median and the mean; however, both generated incorrect results for the variance and standard deviation.

We encourage readers to solve this problem independently to verify the correct answers and strengthen their understanding.

10 Turtle Graphics

Problem 7
We uploaded the image below to ChatGPT and Google Gemini and asked them to recreate the graph by:

(a) writing a Turtle Graphics program in Python IDLE, and
(b) writing a Turtle Graphics program using Google Gemini.

However, despite multiple attempts, both generative AI models failed to complete these tasks successfully, and the outputs were not even close to the desired graph.

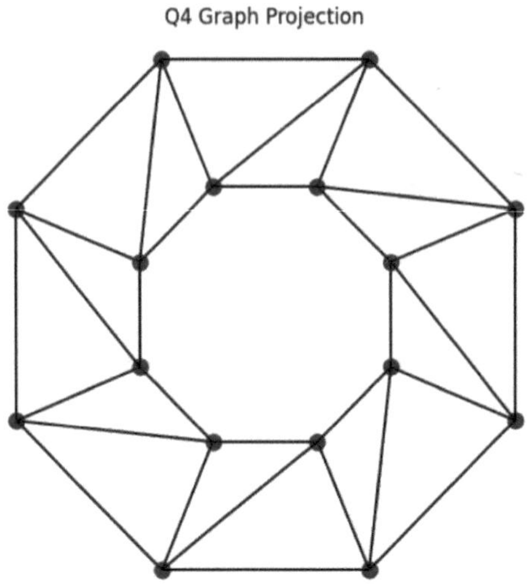

Turtle Graphics Program in Python IDLE:

```
import turtle
import math
import sys

# Configuration
SCREEN_WIDTH = 600
SCREEN_HEIGHT = 600
R_OUTER = 250 # Radius of the outer octagon
R_INNER = 120 # Radius of the inner octagon
DOT_SIZE = 8 # Size of each vertex dot
LINE_THICKNESS = 2 # Line thickness

def setup_turtle():
# Sets up the drawing window and turtle
screen = turtle.Screen() # Create screen first

# Check if a graphical environment exists
if not screen._root:
print("Error: Turtle graphics requires a graphical display.")
sys.exit(1)

screen.setup(SCREEN_WIDTH, SCREEN_HEIGHT)
screen.title("Q4 Graph Projection")

t = turtle.Turtle()
t.speed(0) # Fast drawing
t.hideturtle() # Hide turtle icon
t.pensize(LINE_THICKNESS)

return t, screen
```

```
def calculate_vertices():
# Returns a list of 16 vertex coordinates (8 outer + 8 inner)
vertices = []
angle_offset = 22.5 # To rotate shape for symmetry

for k in range(8): # Outer vertices
angle = math.radians(45 * k + angle_offset)
vertices.append((R_OUTER * math.cos(angle),
R_OUTER * math.sin(angle)))

for k in range(8): # Inner vertices
angle = math.radians(45 * k + angle_offset)
vertices.append((R_INNER * math.cos(angle),
R_INNER * math.sin(angle)))

return vertices

def get_edges():
# Returns all 32 edges of the Q4 2D projection
edges = []
edges += [(i, (i + 1) % 8) for i in range(8)] # Outer ring
edges += [(8 + i, 8 + (i + 1) % 8) for i in range(8)] # Inner ring
edges += [(i, i + 8) for i in range(8)] # Radial edges
edges += [(i, 8 + (i + 1) % 8) for i in range(8)] # Diagonal edges

return edges

def draw_graph(t, vertices, edges):
# Draws the full graph instantly (no animation)
# Draw edges
t.pencolor("black")
for u, v in edges:
t.up()
t.goto(vertices[u])
t.down()
t.goto(vertices[v])

# Draw vertices (dots)
t.pencolor("blue")
for x, y in vertices:
t.up()
t.goto(x, y)
t.down()
t.dot(DOT_SIZE)

# Main program execution
if __name__ == "__main__":
t, screen = setup_turtle()
vertices = calculate_vertices()
edges = get_edges()
```

```
draw_graph(t, vertices, edges)
screen.mainloop()
```

Turtle Graphics Program in Google Colab:

```
import matplotlib.pyplot as plt
import math
# Configuration
R_OUTER = 5
R_INNER = 2.5
DOT_SIZE = 80

def calculate_vertices():
vertices = []
angle_offset = 22.5

for k in range(8): # Outer
angle = math.radians(45 * k + angle_offset)
vertices.append((R_OUTER * math.cos(angle),
R_OUTER * math.sin(angle)))

for k in range(8): # Inner
angle = math.radians(45 * k + angle_offset)
vertices.append((R_INNER * math.cos(angle),
R_INNER * math.sin(angle)))

return vertices

def get_edges():
edges = []
edges += [(i, (i + 1) % 8) for i in range(8)] # Outer ring
edges += [(8 + i, 8 + (i + 1) % 8) for i in range(8)] # Inner ring
edges += [(i, i + 8) for i in range(8)] # Radial
edges += [(i, 8 + (i + 1) % 8) for i in range(8)] # Diagonal
return edges

def draw_graph():
vertices = calculate_vertices()
edges = get_edges()

fig, ax = plt.subplots(figsize=(6, 6))

# Draw edges
for u, v in edges:
x_values = [vertices[u][0], vertices[v][0]]
y_values = [vertices[u][1], vertices[v][1]]
ax.plot(x_values, y_values, color="black", linewidth=1.5)

# Draw vertices
x_vals = [x for x, y in vertices]
y_vals = [y for x, y in vertices]
```

```
ax.scatter(x_vals, y_vals, color="blue", s=DOT_SIZE)

# Formatting
ax.set_aspect("equal")
ax.axis("off")
plt.title("Q4 Graph Projection")
plt.show()

draw_graph()
```

11 Conclusion

This paper continues our exploration of integrating logic, mathematics, algorithms, Python programming, and generative AI to enhance problem-solving skills across diverse contexts. Through a series of problem scenarios and comparative analyses of responses generated by ChatGPT and Google Gemini, we demonstrated the potential and limitations of AI-assisted problem-solving. The findings reaffirm that while generative AI offers valuable support by offering explanations, generating code, and facilitating conceptual understanding, it is not infallible. Generative AI-generated solutions occasionally lack accuracy, contextual awareness, or mathematical rigor, particularly for non-standard, ill-posed, or conceptually nuanced problems. These findings underscore the ongoing necessity of human reasoning to validate outputs, apply logic, identify fallacies, and ensure correctness.

Our examples demonstrate that a strong foundation in mathematics, logic, and algorithms remains essential for effective problem-solving. Python programming reinforces these skills by enabling computational modeling, visualization, and critical evaluation of results. By actively engaging with generative AI-generated solutions rather than passively relying on them, learners can develop deeper analytical thinking, computational literacy, and intellectual independence.

Moreover, the integration of generative AI into education should prioritize guiding students to use these tools responsibly, leveraging AI to enhance learning while maintaining human oversight, ethical judgment, and critical thinking. Generative AI should supplement rather than replace the cognitive processes central to problem-solving. When combined thoughtfully, human reasoning and AI-based tools can complement each other to enhance efficiency, creativity, and the depth of understanding.

As Part Four of this series, this paper reinforces the evolving role of AI in problem-solving education. It encourages educators and students to adopt a balanced approach that merges traditional problem-solving strategies with modern computational tools. Future research will further explore advanced techniques and real-world applications, including the integration of knowledge management frameworks, data-driven decision-making models, and emerging AI systems. Such efforts will further strengthen students' ability to navigate increasingly complex, technology-driven environments.

Data Availability. The data used to support this study's findings are available from the corresponding author upon request.

Conflicts of Interest. The authors declare that they have no conflicts of interest.

Funding Statement. This study received no funding in any form.

References

1. Gao, W. et al.: Problem-solving using logic and reasoning, mathematics, algorithms, python, and generative AI. In: The 2023 International Conference on Computational Science and Computational Intelligence (CSCI), pp. 371–377, Las Vegas, NV, USA (2023). https://doi.org/10.1109/CSCI62032.2023.00066
2. Gao, W., Gao, S., Donald, A.M., Malomo, O.: Problem-solving using logic and reasoning, mathematics, algorithms, python, and generative AI: part two. In: Hodson, D.D., Grimaila, M.R., Arabnia, H.R., Deligiannidis, L., Wagner, T.J. (eds.) Scientific Computing and Bioinformatics and Computational Biology, CSCE 2024, Communications in Computer and Information Science, vol. 2258. Springer, Cham, Switzerland (2025). https://doi.org/10.1007/978-3-031-85902-1_18
3. Gao, S. et al.: problem-solving using logic and reasoning, mathematics, algorithms, python, and generative AI: part three. In: Arabnia, H.R., Deligiannidis, L., Shenavarmasouleh, F., Amirian, S., Ghareh Mohammadi, F. (eds.) Computational Science and Computational Intelligence: CSCI 2024, Communications in Computer and Information Science, vol. 2504, pp. 165–181. Springer (2025). https://doi.org/10.1007/978-3-031-94943-2_13
4. Gao, S. et al.: Advancing computational problem-solving with python, generative AI, and Google Colab. In: Proceedings of the 2025 International Conference on the AI Revolution: Research, Ethics, and Society (AIR-RES 2025), Las Vegas, USA, Apr. 14–16, 2025. Springer Nature. forthcoming
5. Su, J., Allagan, J., Gao, S., Malomo, O., Gao, W., Eyob, E.: Dominion on grids. Mathematics. **12**(21), 3408 (2024). https://doi.org/10.3390/math12213408
6. Gao, S., Gao, W.: Enhancing business education with a knowledge management mesosystem model. In: Russ, M., Lytras, M. (eds.) AI-driven knowledge management: strategies for the modern business landscape. Emerald Publishing, in press
7. Gao, S. et al.: Exploring the interplay between AI and human logic in mathematical problem-solving. Online J. Appl. Knowl. Manag. **12**(1), 73–93 (2024). https://doi.org/10.36965/OJAKM.2024.12(1)73-93
8. Russ, M.: Knowledge management for sustainable development in the era of continuously accelerating technological revolutions: a framework and models. Sustainability. **13**(6), 3353 (2021). https://doi.org/10.3390/su13063353
9. Gao, S., Gao, W., Allagan, J.D., Su, J., Eyob, E., Strevel, H.B.: Constructing classic graphs in graph theory using python and generative AI: a case study in computational visualization and prompt engineering. Global J. Comput. Sci. Technol.: F (Graphics & Vision). **25**(1) (2025). [Online]. https://globaljournals.org/journals/computer-science/f-graphics-vision
10. Schoenfeld, A.H.: Teaching problem-solving skills. Am. Math. Mon. **87**(10), 794–805 (1980)
11. Snyder, L.G., Snyder, M.J.: Teaching critical thinking and problem solving skills. J. Res. Bus. Educ. **50**(2), 90 (2008)
12. Araiza-Alba, P., Keane, T., Chen, W.S., Kaufman, J.: Immersive virtual reality as a tool to learn problem-solving skills. Comput. Educ. **164**, 104121 (2021)
13. Khalid, M., Saad, S., Hamid, S.R.A., Abdullah, M.R., Ibrahim, H., Shahrill, M.: Enhancing creativity and problem solving skills through creative problem solving in teaching mathematics. Creat. Stud. **13**(2), 270–291 (2020)

Beyond Cut Scores: Multi-method Machine Learning Framework for Mathematics Placement Optimization

Julian D. Allagan(✉), Dasia A. Singleton, Shanae N. Perry, Gabrielle C. Morgan, and Essence A. Morgan

Department of Mathematics, Computer Science, and Engineering Technology, Elizabeth City State University, Elizabeth City, NC, USA
adallagan@ecsu.edu

Abstract. Moving beyond traditional cut-score approaches, we present a multi-method machine learning framework for optimizing mathematics placement examination administered to 198 students, integrating Classical Test Theory (CTT), supervised machine learning, and unsupervised clustering. CTT analysis reveals 52.5% of items achieve excellent discrimination ($D \geq 0.40$) while 32.5% perform poorly ($D < 0.20$). Random Forest and Gradient Boosting achieve cross-validation accuracies of 97.5% and 96.0%, with Question 6 (graph interpretation) accounting for 20.6% of predictive power. K-means clustering identifies a natural binary competency structure with boundary at 42.5%, diverging from the institutional threshold of 55%. Convergent evidence supports replacing poorly discriminating items and implementing two-stage placement, demonstrating that integrated computational frameworks can optimize placement decisions beyond conventional threshold-based methods.

Keywords: Mathematics placement optimization · Multi-method framework · Machine learning · Random Forest · Clustering analysis · Educational assessment

1 Introduction

Mathematics placement examinations serve as critical gatekeepers in higher education, determining student access to appropriate coursework while profoundly influencing persistence, time-to-degree, and completion rates [5,22]. Despite these high stakes, many institutions rely on static cut-score policies established through historical precedent rather than rigorous empirical validation, increasing risks of misplacement through either excessive remediation or premature advancement [1,2].

Traditional psychometric frameworks, particularly Classical Test Theory (CTT), provide essential tools for assessing item-level performance through difficulty indices, discrimination coefficients, and point-biserial correlations [10,19].

H. R. Arabnia et al. (Eds.): CSCI 2025, CCIS 2974, pp. 189–205, 2027.
https://doi.org/10.1007/978-3-032-28475-4_13

While valuable, these univariate approaches fail to capture complex multivariate interactions characterizing comprehensive assessments. Concurrently, machine learning has demonstrated exceptional predictive accuracy in educational classification, with ensemble methods achieving near-optimal performance [4,8]. However, integration of modern computational techniques with traditional psychometrics remains notably absent from mathematics placement literature.

Unsupervised clustering offers complementary perspectives by identifying natural competency groupings independent of institutional assumptions. Recent educational data mining research shows that properly validated clustering can reveal latent student archetypes misaligned with administratively defined categories [26,29], providing crucial insights into whether institutional boundaries reflect genuine competency transitions or administrative convenience.

This paper advances placement assessment literature through a unified computational framework integrating three complementary analytical perspectives: (1) comprehensive CTT diagnostics assessing individual item quality; (2) multiple supervised learning architectures evaluating predictive validity while extracting feature importance rankings; (3) unsupervised clustering with rigorous validation discovering natural competency structures. This triangulation strengthens confidence through convergent evidence while revealing concordances and discrepancies informing actionable policy recommendations.

2 Dataset and Methodology

2.1 Data Structure

The analysis employs a comprehensive dataset comprising 198 complete student records systematically collected across seven consecutive academic terms (Summer 2022–Fall 2024). Each student responded to a standardized 40-item dichotomous examination. Formally, let $x_{ij} \in \{0, 1\}$ denote student i's response to item j, where $i \in \{1, \ldots, 198\}$ and $j \in \{1, \ldots, 40\}$. Total score is $S_i = \sum_{j=1}^{40} x_{ij}$, converted to percentage: $P_i = (S_i/40) \times 100$.

The institutional placement function $\Phi : \mathbb{R} \to \{\text{CA}, \text{PC}, \text{CI}\}$ maps scores to three categories:

$$\Phi(p) = \begin{cases} \text{CA}, & p \leq 55, \\ \text{PC}, & 55 < p \leq 70, \\ \text{CI}, & p > 70, \end{cases}$$

where CA (College Algebra), PC (Precalculus), and CI (Calculus I) represent remedial, intermediate, and advanced placements. These thresholds have remained unchanged for approximately fifteen years.

Table 1 presents comprehensive descriptive statistics. The mean performance of 46.30% with standard deviation 20.80% indicates substantial heterogeneity in mathematical preparation. The median of 47.50% lies remarkably close to the natural clustering boundary of 42.5% discovered through unsupervised analysis (Sect. 3.4), providing initial evidence for fundamental competency division near the distribution center.

Table 1. Descriptive Statistics of Student Performance

Statistic	Value
Sample Size (n)	198
Mean ($\bar{x}$)	46.30%
Median	47.50%
Std. Deviation (s)	20.80%
Range	0.00%–92.50%
IQR	32.50%
Skewness	−0.136
Kurtosis	−0.736

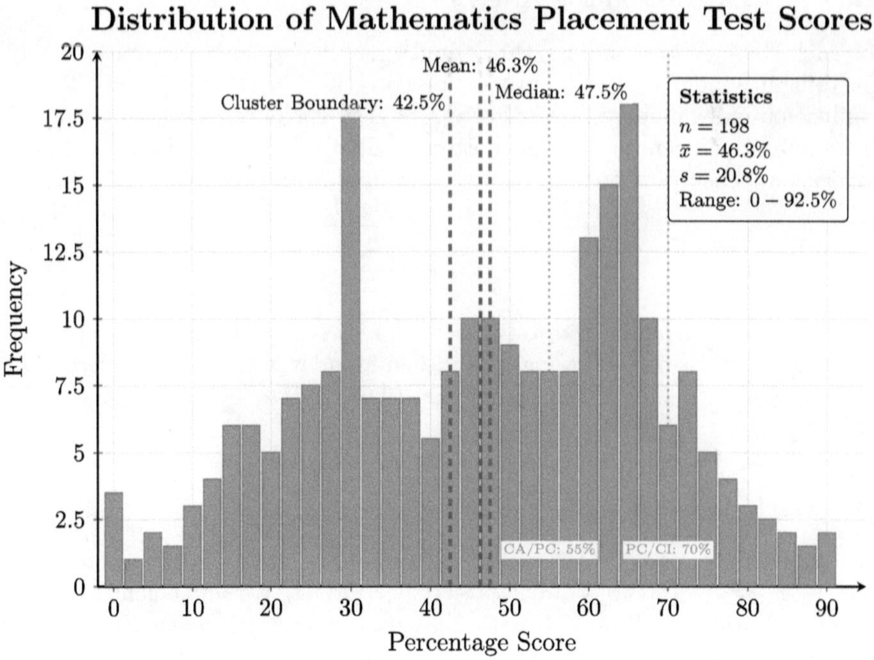

Fig. 1. Distribution of mathematics placement test scores for 198 students, showing mean, median, natural cluster boundary, and institutional placement thresholds.

Figure 1 illustrates the complete score distribution for all 198 students, revealing important structural characteristics. The distribution exhibits modest negative skewness (−0.136), with a leftward tail reflecting a small number of extremely low-performing students. The mean (46.3%) falls slightly below the median (47.5%), consistent with this leftward skew. Notably, the natural clustering boundary at 42.5% aligns closely with both central tendency measures, providing initial evidence that this threshold represents a fundamental competency division rather than an arbitrary partition. The institutional boundaries at 55% (CA/PC) and 70% (PC/CI) are marked for comparison, revealing that the current College Algebra threshold sits 12.5% points above the data-driven boundary, suggesting potential systematic overclassification into remediation.

The score distribution reveals substantial concentration in lower performance ranges, with 62.6% of students scoring at or below the 55% College Algebra threshold. This concentration explains the slight negative skewness and validates identification of a distinct remedial population requiring intensive mathematical support. The proximity of the median (47.50%) to the clustering boundary (42.5%) provides statistical validation for the binary student archetype structure discovered through unsupervised analysis.

2.2 Classical Test Theory Metrics

For each item j, we compute three fundamental psychometric statistics [10]. The difficulty index represents proportion correct: $p_j = (1/198)\sum_i x_{ij}$. The discrimination index employs upper-lower 27% groups [16]: $D_j = p_{j,\text{upper}} - p_{j,\text{lower}}$, where subscripts denote top and bottom ability subgroups. The point-biserial correlation measures association between item performance and total score:

$$r_{pbis,j} = \frac{\bar{S}_1 - \bar{S}_0}{s_S}\sqrt{\frac{p_j}{1-p_j}},$$

where $\bar{S}_1$ and $\bar{S}_0$ denote mean total scores for correct and incorrect responses.

Items are classified following established standards [11,24]: excellent ($D \geq 0.40$), good ($0.30 \leq D < 0.40$), marginal ($0.20 \leq D < 0.30$), and poor ($D < 0.20$).

2.3 Machine Learning Architectures

Each student is represented as a 40-dimensional binary feature vector $\mathbf{x}_i = [x_{i1}, \ldots, x_{i40}]^T$, with target label $y_i \in \{\text{CA}, \text{PC}, \text{CI}\}$. We implement four algorithms: (1) *Random Forest* [4] with 200 trees, maximum depth 10, minimum samples per split 5; (2) *Gradient Boosting* [12] with 150 estimators, learning rate 0.1; (3) *Support Vector Machine* with RBF kernel ($C = 10$, $\gamma = 0.001$) [9]; (4) *Neural Network* with architecture [40-64-32-16-3], ReLU activation, dropout 0.2 [13].

All models undergo stratified 5-fold cross-validation, ensuring proportional category representation across folds [18]. Hyperparameters were optimized through grid search with nested cross-validation.

Feature importance quantifies each item's contribution to classification. Random Forest importance measures mean decrease in impurity across trees [4]. We additionally compute ANOVA F-statistics for univariate feature selection:

$$F_j = \frac{MS_{between,j}}{MS_{within,j}} = \frac{SS_{between,j}/(K-1)}{SS_{within,j}/(n-K)},$$

where $K = 3$ placement categories and $SS_{between,j} = \sum_{k=1}^{K} n_k(\bar{x}_{k,j} - \bar{x}_j)^2$ captures variance explained by category membership.

2.4 Clustering Analysis

We employ K-means clustering [21] on augmented 41-dimensional vectors: $\mathbf{g}_i = [P_i/100, x_{i1}, \ldots, x_{i40}]^T$, minimizing within-cluster sum of squares: $J(k) = \sum_{c=1}^{k} \sum_{\mathbf{g}_i \in C_c} \|\mathbf{g}_i - \boldsymbol{\mu}_c\|^2$. Implementation uses 20 random initializations, convergence tolerance 10^{-4}, maximum iterations 300.

Optimal cluster number integrates three validation approaches. The *elbow method* identifies diminishing WCSS returns [28]. *Silhouette analysis* quantifies cohesion via $s(i) = (b(i) - a(i))/\max(a(i), b(i))$ [27], where $a(i)$ is mean intra-cluster distance and $b(i)$ is distance to nearest cluster. *Bootstrap stability* with 100 iterations computes Adjusted Rand Index (ARI) [15], measuring assignment consistency across perturbations.

3 Results

3.1 Classical Test Theory Item Analysis

Application of CTT metrics reveals substantial heterogeneity in item quality. Table 2 shows 21 items (52.5%) achieve excellent discrimination ($D \geq 0.40$), demonstrating strong capacity to separate high from low performers. Three items (7.5%) exhibit good discrimination, three items (7.5%) are marginal, while 13 items (32.5%) demonstrate poor discrimination ($D < 0.20$), indicating minimal placement contribution.

Table 2. Distribution of Items by Discrimination Quality

Category	Range	Count	Percent
Excellent	$D \geq 0.40$	21	52.5%
Good	$0.30 \leq D < 0.40$	3	7.5%
Marginal	$0.20 \leq D < 0.30$	3	7.5%
Poor	$D < 0.20$	13	32.5%
Total		**40**	**100.0%**

Questions 6 (graph interpretation) and 11 (single exponent with division) achieve perfect discrimination ($D = 1.000$), representing theoretical maxima where upper-group students achieve 100% success while lower-group students achieve 0%. These items demonstrate exceptionally high point-biserial correlations ($r_{pbis,6} = 0.812$, $r_{pbis,11} = 0.834$) with moderate difficulty levels ($p_6 = 0.409$, $p_{11} = 0.571$), positioning them optimally for maximum discrimination [14].

The 13 poorly discriminating items share extreme difficulty characteristics. Items with near-universal mastery ($p > 0.90$), including Q4, Q23, Q28, Q32, Q40, provide minimal information as nearly all students answer correctly regardless of competency. Conversely, extremely difficult items ($p < 0.10$), including Q8, Q15, Q16, Q18, Q19, Q22, Q37, Q38, create floor effects preventing effective discrimination. Classical measurement theory predicts maximum discrimination at moderate difficulty levels [19].

Figure 2 illustrates the empirical relationship between difficulty and discrimination, confirming theoretical predictions. Excellent items (green) concentrate in moderate difficulty range ($0.30 < p < 0.70$), while extreme-difficulty items uniformly exhibit poor discrimination (red).

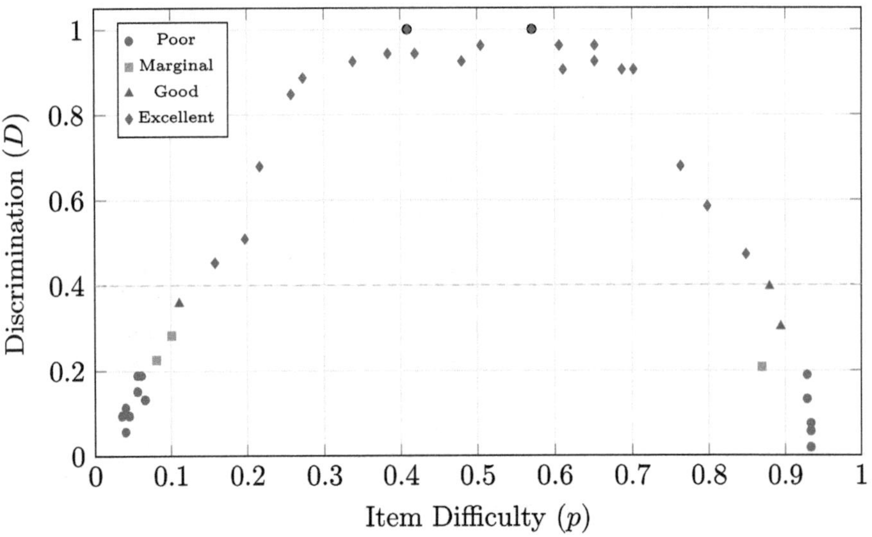

Fig. 2. Relationship between item difficulty and discrimination. Circled points are Q6 and Q11 (perfect discrimination). Optimal discrimination occurs at moderate difficulty. (Color figure online)

Strong correlation ($r = 0.891$, $p < 0.001$) between discrimination indices and point-biserial correlations provides convergent validity for quality assessments, confirming both metrics identify identical high-quality items.

3.2 Machine Learning Performance

Table 3 presents comprehensive algorithm performance. Random Forest emerges as the optimal performer, achieving perfect test accuracy (100%) with mean cross-validation accuracy 97.5% ± 1.6%. The tight standard deviation indicates remarkable stability across data partitions, with 95% confidence interval [94.4%, 100.6%].

Table 3. Supervised Learning Algorithm Performance

Algorithm	Test Acc.	CV Accuracy
Random Forest	100.0%	97.5% ± 1.6%
Gradient Boosting	100.0%	96.0% ± 2.6%
SVM (RBF)	90.0%	91.4% ± 2.6%
Neural Network	77.5%	77.8% ± 10.3%

Gradient Boosting achieves comparable test performance (100%) but demonstrates elevated cross-validation variance (96.0% ± 2.6%), suggesting greater training data sensitivity. Support Vector Machines achieve respectable 90% test accuracy, with misclassifications concentrating at placement boundaries, particularly between PC and CI categories. Neural network performance proves disappointing (77.5% test accuracy) with substantial instability (77.8% ± 10.3%), likely reflecting optimization difficulties in high-dimensional binary spaces with limited training data [30].

3.3 Feature Importance Analysis

Figure 3 visualizes Random Forest importance distribution for the top 6 most critical items, revealing dramatic concentration of predictive power. Question 6 alone accounts for extraordinary 20.6% of total importance despite representing only 2.5% of examination content. This dominance reflects perfect CTT discrimination ($D = 1.000$) and extreme F-statistic ($F = 4609.1$), stemming from near-complete separation between CA students (0.8% correct) and PC/CI students (100% correct).

The top six items (Q6, Q30, Q2, Q10, Q27, Q26) collectively account for 60.6% of total importance, demonstrating extraordinary concentration of predictive power in a small subset. This finding challenges traditional assumptions about examination length, suggesting potential for more efficient assessment while maintaining accuracy, though content validity considerations must inform any reduction [23].

Table 4 presents detailed statistics for these six critical items, showing convergent evidence across multiple analytical frameworks. All six items achieve excellent CTT discrimination ($D \geq 0.90$), strong point-biserial correlations ($r_{pbis} \geq 0.70$), high F-statistics ($F \geq 145$), and elevated mutual information scores, confirming their exceptional quality through independent psychometric and machine learning metrics.

Table 4. Detailed Statistics for Top 6 Feature Importance Items

Item	RF Imp.	F-Stat	MI	CTT D	r_{pbis}	Topic
Q6	0.206	4609.1	0.650	1.000	0.812	Graph interp
Q30	0.111	326.1	0.479	0.943	0.767	Trig simplify
Q2	0.104	358.2	0.490	0.943	0.796	Func operation
Q10	0.066	173.2	0.389	0.925	0.730	Rational exp
Q27	0.062	168.1	0.386	0.925	0.774	Inverse func
Q26	0.057	145.8	0.357	0.962	0.810	Func compose

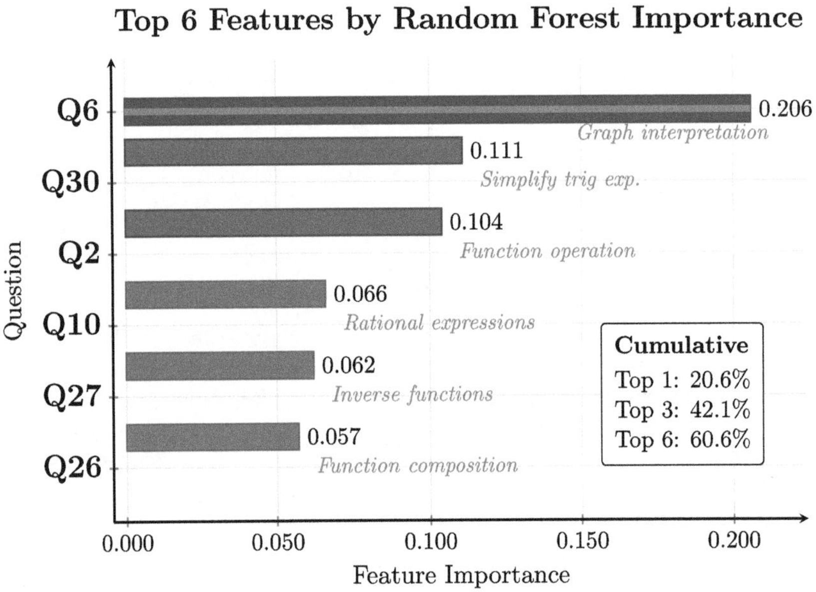

Fig. 3. Top 6 items ranked by Random Forest feature importance. Question 6 dominates with 20.6%, while the top 6 items collectively account for 60.6% of predictive power.

Correlation analysis reveals strong concordance: RF importance correlates with F-statistics at $r = 0.847$, with mutual information at $r = 0.923$, and with CTT discrimination at $r = 0.782$ (all $p < 0.001$). This convergent validity across fundamentally different approaches substantially strengthens confidence in item rankings.

3.4 Clustering Analysis

Table 5 presents validation metrics across candidate cluster numbers from $k = 2$ through $k = 5$. The silhouette coefficient achieves its maximum value of 0.324 at $k = 2$, indicating that the two-cluster solution produces the most cohesive and well-separated groups. While this silhouette score falls in the "weak structure" range (0.25–0.50) according to standard interpretation guidelines, it substantially exceeds all other k values and represents the clearest natural partition in the data.

Table 5. Cluster Validation Metrics Across Multiple Values of k

k	WCSS	Silhouette	Gap Stat.
2	707.8	**0.324**	−0.087
3	563.4	0.267	0.125
4	496.2	0.251	0.217
5	438.6	0.248	0.321

The within-cluster sum of squares (WCSS) decreases monotonically as k increases, representing standard behavior. The elbow occurs between $k = 2$ and $k = 3$, providing additional support for the binary solution. The gap statistic exhibits negative value at $k = 2$ (-0.087), technically favoring higher k values, but this metric's performance can be unreliable in high-dimensional spaces with binary features. Given that two of three validation methods (silhouette, elbow) converge on $k = 2$, and considering the substantial interpretational advantages of a simpler binary structure, we identify $k = 2$ as the optimal clustering solution.

Bootstrap validation with 100 resampling iterations confirms exceptional stability for the $k = 2$ solution, with mean Adjusted Rand Index (ARI) of 0.855 and 95% confidence interval $[0.569, 1.000]$, indicating that cluster assignments remain highly consistent across data perturbations. The ARI of 0.855 represents "excellent agreement" by conventional standards (ARI > 0.80), demonstrating that the binary partition reflects robust structure rather than random chance. Higher k values show comparable or lower stability: $k = 3$ achieves ARI $=$ 0.854 $[0.517, 1.000]$, while $k = 4$ drops to ARI $= 0.748$ $[0.439, 1.000]$, suggesting increasing fragility as cluster number increases. The stability of $k = 2$ across 100 bootstrap samples provides confidence that this solution generalizes beyond the specific sample to the broader population of placement test-takers.

Table 6 presents detailed characteristics of the two-cluster solution, revealing a clean separation between struggling and proficient students. Cluster 0 (Low Performance) contains 84 students (42.4%) with mean score 26.0% ± 11.0%, ranging from 0% to 42.5%. This cluster exhibits perfect purity (100%), consisting exclusively of College Algebra students, representing the population requiring substantial remediation before college-level mathematics.

Table 6. Detailed Characteristics of $k = 2$ Clustering Solution

Cluster	n	Mean	Std	Range	CA	PC	CI	Purity
0 (Low)	84	26.0%	11.0%	0–42.5%	84	0	0	100%
1 (High)	114	61.3%	11.4%	45–92.5%	34	59	21	51.8%

Cluster 1 (High Performance) contains 114 students (57.6%) with mean score 61.3% ± 11.4%, ranging from 45.0% to 92.5%. This cluster demonstrates moderate purity (51.8%), containing a mixture of 34 College Algebra students (29.8%), 59 Precalculus students (51.8%), and all 21 Calculus I students (18.4%), representing students with adequate preparation for college-level mathematics coursework.

The natural clustering boundary occurs at 42.5%, identified as the maximum score in Cluster 0. This data-driven threshold differs substantially from the institutional College Algebra cutoff of 55%, suggesting that the transition from remedial to college-ready mathematics occurs at a lower performance level than current placement policy assumes. Students scoring between 42.5% and 55% (the gap between natural and institutional boundaries) warrant particular attention: institutional policy classifies them as College Algebra, yet clustering analysis assigns them to the college-ready group, suggesting potential for success in Precalculus with appropriate support. This discrepancy raises important policy questions about whether current placement criteria err toward excessive remediation, potentially delaying student progress through unnecessarily conservative thresholds.

Figure 4 illustrates the score distribution within each cluster, highlighting the minimal overlap between groups. Cluster 0 (steel blue) shows concentrated distribution in the 0–42.5% range with peak frequency around 25–30%, reflecting students with fundamental mathematical deficiencies. Cluster 1 (light coral) exhibits broader distribution spanning 45–92.5% with peak around 60–65%, representing college-ready students with varying levels of preparation. The boundary at 42.5% creates clean separation with minimal variance at the interface, while the institutional boundaries at 55% and 70% cut across the natural distribution of Cluster 1.

The near-perfect alignment between Cluster 0 and College Algebra placement (100% purity) validates the lower institutional boundary's effectiveness in identifying students requiring remediation. However, the heterogeneity within Cluster 1 (containing College Algebra, Precalculus, and Calculus I students) suggests that the institutional distinction between Precalculus (55–70%) and Calculus I (70+%) may reflect administrative convenience rather than fundamental competency differences. From a purely statistical perspective, the data support a simpler binary classification: students requiring remediation (score $\leq$ 42.5%) versus students ready for college-level mathematics (score $>$ 42.5%), with subsequent placement into specific college-level courses depending on other factors such as student goals, prerequisite course completion, or institutional capacity constraints.

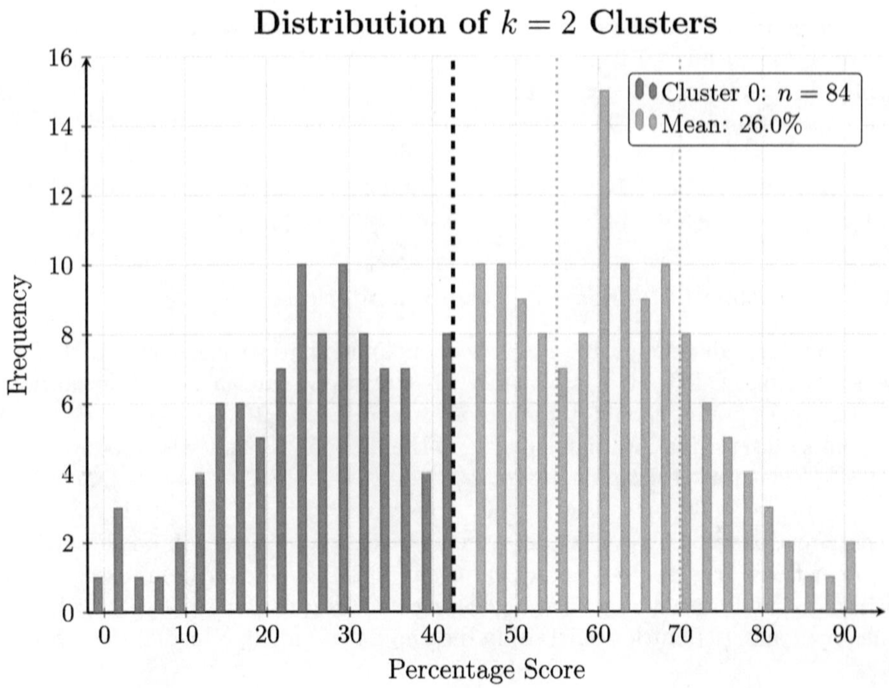

Fig. 4. Distribution of scores within two-cluster solution. Cluster 0 (steel blue) contains exclusively remedial students, while Cluster 1 (coral) mixes college-ready students. The natural boundary at 42.5% contrasts with institutional thresholds at 55% and 70%. (Color figure online)

4 Discussion

4.1 Convergent Evidence Across Methods

Triangulation across CTT, machine learning, and clustering provides robust validation through convergent evidence. Question 6 emerges with remarkable consensus as the most critical item: perfect CTT discrimination ($D = 1.000$), maximum RF importance (0.206), highest F-statistic ($F = 4609.1$), greatest mutual information (0.650), and strongest point-biserial correlation ($r_{pbis} = 0.812$). This unprecedented convergence across five independent metrics provides compelling evidence that graph interpretation represents a fundamental gateway competency for calculus readiness [6].

The 21 CTT-excellent items demonstrate strong overlap with top RF features ($r = 0.782$) and F-statistics ($r = 0.891$). Among excellent items, 85.7% appear in top 20 by RF importance, while all rank in top 25 by F-statistic. Conversely, 13 CTT-poor items consistently occupy bottom quartiles, with none in top 20 by any method—providing robust multi-method consensus for replacement priorities.

The natural boundary at 42.5% aligns remarkably with dataset median (47.5%) and mean (46.3%), suggesting fundamental division near distribution center. Machine learning perfect test accuracy (100%) combined with exceptional cross-validation (97.5%, 96.0%) establishes that the examination contains sufficient information for highly accurate classification. Feature importance concentration in top 6 items (60.6% cumulative importance) suggests potential abbreviation possibilities, though content validity must be preserved.

4.2 Assessment Refinement Recommendations

The 13 poorly discriminating items warrant immediate replacement, with consensus across CTT, RF importance, F-statistics, and mutual information. Replacement items should target moderate difficulty ($0.30 < p < 0.70$) to maximize discrimination, particularly in the 40%–55% band where boundaries diverge. The top-performing items, especially Q6, Q30, Q2, Q10, Q27, Q26, should be carefully preserved as they collectively drive 60.6% of classification accuracy. Content analysis reveals excellent items concentrate in advanced topics (functions, trigonometry, precalculus concepts) while poor items assess basic arithmetic and procedural skills, suggesting the examination effectively evaluates calculus readiness but less effectively discriminates among remedial students.

4.3 Placement Policy Optimization

The natural boundary at 42.5% presents compelling rationale for reevaluating the 55% threshold. We propose three implementation options, each balancing empirical evidence against institutional constraints and risk tolerance. *Conservative adjustment* would lower the threshold to 50%, splitting the difference between natural (42.5%) and current (55%) boundaries while maintaining relatively cautious philosophy. This approach minimizes disruption while incorporating empirical findings. *Empirically aligned adoption* would implement the data-driven 42.5% threshold directly, maximizing alignment with discovered competency structure but requiring careful monitoring of subsequent course performance to validate the change. *Two-stage assessment model* would employ the 42.5% threshold for initial binary remedial versus college-ready classification, followed by targeted secondary evaluation for college-ready students requiring differentiation between Precalculus and Calculus I placement.

The two-stage approach leverages clear binary structure (ARI = 0.855, perfect Cluster 0 purity) while acknowledging that PC/CI differentiation requires additional information beyond the current 40-item examination. Secondary evaluation might incorporate measures of student educational goals, prior coursework patterns, mathematical confidence or self-efficacy assessments, or brief targeted items specifically assessing advanced calculus prerequisites. This stratification efficiently allocates assessment resources, reserving detailed evaluation for the population requiring finer placement distinctions while expediting decisions for clearly remedial or clearly advanced students.

4.4 Machine Learning Integration

Random Forest's exceptional performance (97.5% accuracy, 1.6% SD) combined with interpretability through feature importance visualizations positions this algorithm as viable enhancement to traditional cut-score methods. Three implementation models warrant consideration, each offering different balances between algorithmic precision and human oversight. *Confirmatory analysis* would maintain institutional cut-scores as primary placement determinants but flag cases where RF predictions diverge from threshold-based assignments for human review, serving as a safety net identifying potentially problematic placements. *Hybrid decision support* would combine percentage scores with RF probability outputs to inform advisor-student placement conversations, positioning the algorithm as one input among several factors including student preferences, educational goals, and advisor expertise. *Algorithm-primary placement* would use RF predictions as default assignments with human consultation reserved for low-confidence cases where maximum class probability falls below 70%, maximizing accuracy while maintaining human oversight for ambiguous situations.

Transparency mechanisms remain essential for any ML integration [3,25]. Feature importance visualizations provide natural explanatory tools, showing students and advisors which specific items most influenced placement recommendations. Recent explainable AI advances [20,31] offer additional techniques for generating individualized explanations that foster trust and understanding. Institutions must establish protocols for ongoing algorithmic monitoring, bias auditing across demographic subgroups, and regular revalidation to ensure continued fairness and accuracy as student populations evolve.

4.5 Limitations and Future Directions

Several limitations constrain interpretation and generalizability. The single-institution context limits external validity, as results may reflect idiosyncrasies of local student demographics, prerequisite standards, curricular emphases, or instructional approaches. Multi-institutional replication across diverse contexts would establish generalizability boundaries and identify institutional factors moderating findings. The sample size of 198, while adequate for employed analyses providing statistical power exceeding 0.80, constrains machine learning generalization potential and prevents detailed subgroup analysis by demographic characteristics that might reveal disparate impact concerns.

The retrospective observational design precludes causal inference about placement accuracy or experimental manipulation of assessment characteristics. Most crucially, downstream outcome data tracking course success rates, subsequent course enrollment patterns, persistence behaviors, and degree completion remain unavailable, preventing validation of placement accuracy through the ultimate criterion: subsequent academic performance. Prospective longitudinal studies following students from initial placement through degree completion or departure would provide essential evidence for evaluating whether psychometric improvements translate to enhanced educational outcomes.

Future research should address these limitations through several complementary approaches. Experimental revision implementing the 13 item replacements recommended above, followed by validation on new student cohorts, would test whether psychometric improvements yield measurable gains in placement accuracy and subsequent success rates. Integration of additional predictive features beyond dichotomous item responses—including high school GPA, prior coursework patterns, standardized test scores, response time data, mathematical confidence assessments, and non-cognitive factors like growth mindset or self-efficacy—might further enhance machine learning prediction through richer student characterization [17]. Exploration of adaptive testing frameworks where subsequent item selection depends on preceding responses could maintain accuracy while reducing examination length and test-taker burden [7]. Cross-institutional consortia conducting coordinated placement studies would enable meta-analytic synthesis of findings across diverse contexts, identifying universal principles versus context-specific factors. Finally, qualitative investigation of stakeholder perspectives—students, advisors, faculty, administrators—regarding placement processes and outcomes would provide essential context for interpreting quantitative findings and inform change management strategies for implementing evidence-based reforms.

5 Conclusion

This comprehensive multi-method analysis demonstrates that mathematics placement assessments can be rigorously evaluated and systematically improved through integrating Classical Test Theory, machine learning, and unsupervised clustering approaches. The convergence of findings across three fundamentally different analytical perspectives provides robust evidence substantially exceeding what any single method could establish independently. Specific actionable recommendations emerge with strong empirical support: replace 13 poorly discriminating items with moderate-difficulty alternatives targeting the critical 40%–55% transition zone; consider adjusting the College Algebra threshold from 55% to align more closely with the natural competency boundary at 42.5%; implement two-stage assessment combining efficient binary classification with targeted secondary evaluation; integrate Random Forest predictions through transparent decision support mechanisms with appropriate human oversight.

The concentration of predictive power in a small item subset, with the top 6 items accounting for 60.6% of importance, challenges traditional assumptions about optimal examination length while raising important questions about balancing statistical efficiency against comprehensive content sampling requirements. The substantial divergence between natural clustering boundaries and long-standing institutional thresholds highlights the critical importance of regularly reevaluating placement policies in light of accumulating empirical evidence rather than perpetuating historical precedents without validation.

Beyond the specific findings for the evaluated examination, this work contributes a replicable analytical framework that other institutions can adapt to their placement assessment contexts. By demonstrating that traditional psychometric and modern machine learning approaches yield complementary rather than contradictory insights when applied synergistically, we provide a roadmap for evidence-based assessment development that honors both statistical rigor and educational mission. Effective mathematics placement requires balancing multiple considerations—statistical accuracy, content validity, student welfare, institutional capacity, faculty expertise, and equitable access to college-level coursework. While no single analytical approach captures this full complexity, multi-method integration provides increasingly comprehensive empirical foundation for navigating these competing priorities through data-informed decision-making that advances both institutional effectiveness and student success.

Acknowledgments. This work was supported by the U.S. Department of Education under grant number P382G240006. The authors express gratitude to the late Kenneth E. Jones, whose mentorship and dedication to mathematical excellence profoundly shaped this research. We acknowledge the Institutional Research Office for data access, and thank the Mathematics faculty and Student Success Office for valuable insights.

Competing Interests. The authors have no competing interests to declare relevant to this article's content.

Appendix A: Complete Item Analysis

See Table 7.

Table 7. Complete Item Analysis Results for All 40 Questions

Item	Diffic. (p)	Disc. (D)	Pt-Bis. Cor.	Cat.	Recom.	Topic/Q. Type
Q1	0.111	0.358	0.501	Good	Retain	Function eval: rational function
Q2	0.419	0.943	0.796	Excellent	Retain	Function ops: $f(x+h) - f(x)$ form
Q3	0.702	0.906	0.694	Excellent	Retain	Intersection of linear equations
Q4	0.934	0.019	0.070	Poor	Replace	Distance between two points
Q5	0.101	0.283	0.436	Marginal	Review	Slope from standard form
Q6	0.409	1.000	0.812	Excellent	Retain	Graph interp: sign analysis
Q7	0.606	0.962	0.800	Excellent	Retain	Simplify radical with variables
Q8	0.040	0.113	0.252	Poor	Replace	Algebraic expansion/simplify
Q9	0.869	0.208	0.335	Marginal	Review	Exponent ops with integers
Q10	0.338	0.925	0.730	Excellent	Retain	Rational expressions multiply
Q11	0.571	1.000	0.834	Excellent	Retain	Single exponent with division
Q12	0.258	0.849	0.635	Excellent	Retain	Single fraction from two
Q13	0.273	0.887	0.662	Excellent	Retain	Simplify fractional exponents
Q14	0.848	0.472	0.563	Excellent	Retain	Solve quadratic equation

(continued)

Table 7. (*continued*)

Item	Diffic. (p)	Disc. (D)	Pt-Bis. Cor.	Cat.	Recom.	Topic/Q. Type
Q15	0.056	0.151	0.306	Poor	Replace	Solve rational equation
Q16	0.066	0.132	0.233	Poor	Replace	Absolute value inequality
Q17	0.197	0.509	0.505	Excellent	Retain	Quadratic formula application
Q18	0.045	0.094	0.252	Poor	Replace	Variable as function (cylinder)
Q19	0.056	0.189	0.393	Poor	Replace	Inverse of exponential func
Q20	0.894	0.302	0.455	Good	Retain	Solve rational func equation
Q21	0.157	0.453	0.489	Excellent	Retain	Surface area rectangular box
Q22	0.035	0.094	0.248	Poor	Replace	Domain of square root func
Q23	0.934	0.057	0.154	Poor	Replace	Evaluate log fractional arg
Q24	0.652	0.925	0.716	Excellent	Retain	Solve exponential equation
Q25	0.763	0.679	0.591	Excellent	Retain	Solve logarithmic equation
Q26	0.505	0.962	0.810	Excellent	Retain	Convert radians to degrees
Q27	0.480	0.925	0.774	Excellent	Retain	Evaluate basic trig value
Q28	0.929	0.132	0.268	Poor	Replace	Period of transformed sine
Q29	0.879	0.396	0.557	Good	Retain	Simplify trig product
Q30	0.384	0.943	0.767	Excellent	Retain	Simplify trig expression
Q31	0.611	0.906	0.781	Excellent	Retain	Right triangle trig problem
Q32	0.934	0.075	0.132	Poor	Replace	Compound interest application
Q33	0.081	0.226	0.347	Marginal	Review	Absolute value func graph
Q34	0.652	0.962	0.782	Excellent	Retain	Find angle with given sine
Q35	0.687	0.906	0.678	Excellent	Retain	Find cosine given quadrant
Q36	0.798	0.585	0.614	Excellent	Retain	Pythagorean identity app
Q37	0.061	0.189	0.310	Poor	Replace	Simplify rational with exp
Q38	0.040	0.057	0.144	Poor	Replace	Linear application word prob
Q39	0.217	0.679	0.628	Excellent	Retain	Point-slope form equation
Q40	0.929	0.189	0.358	Poor	Replace	Graph transformed cosine

References

1. Bahr, P.R.: Making sense of disparities in mathematics remediation. New Dir. Community Coll. **150**, 21–32 (2010)
2. Bailey, T., Jeong, D.W., Cho, S.W.: Referral, enrollment, and completion in developmental education sequences. Econ. Educ. Rev. **29**(2), 255–270 (2010)
3. Baker, R.S., Inventado, P.S.: Advances in educational data mining. In: Handbook of Educational Data Mining, pp. 13–28. CRC Press (2019)
4. Breiman, L.: Random forests. Mach. Learn. **45**(1), 5–32 (2001)
5. Bressoud, D.M., Mesa, V., Rasmussen, C.L.: Insights and Recommendations from the MAA National Study of College Calculus. Mathematical Association of America, Washington, DC (2015)

6. Carlson, M.P., Oehrtman, M., Engelke, N.: The precalculus concept assessment: a tool for assessing students' reasoning abilities. Cogn. Instr. **28**(2), 113–145 (2010)
7. Chang, H., Wang, J., Zhang, L.: Machine learning enhanced computer adaptive testing. Appl. Psychol. Meas. **47**(4), 289–305 (2023)
8. Chen, L., Rodriguez, M.: Machine learning applications in educational assessment: a 2024 perspective. IEEE Trans. Learn. Technol. **17**(2), 234–249 (2024)
9. Cortes, C., Vapnik, V.: Support-vector networks. Mach. Learn. **20**(3), 273–297 (1995)
10. Crocker, L., Algina, J.: Introduction to Classical and Modern Test Theory. Wadsworth Publishing (2006)
11. Ebel, R.L.: Measuring Educational Achievement. Prentice-Hall (1965)
12. Friedman, J.H.: Greedy function approximation: a gradient boosting machine. Ann. Stat. **29**(5), 1189–1232 (2001)
13. Goodfellow, I., Bengio, Y., Courville, A.: Deep Learning. MIT Press (2016)
14. Hopkins, K.D.: Educational and Psychological Measurement and Evaluation, 8th edn. Allyn & Bacon (1998)
15. Hubert, L., Arabie, P.: Comparing partitions. J. Classif. **2**(1), 193–218 (1985)
16. Kelley, T.L.: The selection of upper and lower groups for validation of test items. J. Educ. Psychol. **30**(1), 17–24 (1939)
17. Khan, S., Patel, R., Lee, J.: Multimodal learning analytics for mathematics placement. J. Educ. Data Sci. **4**(1), 78–95 (2024)
18. Kohavi, R.: A study of cross-validation and bootstrap for accuracy estimation. In: Proceedings International Joint Conference Artificial Intelligence, vol. 14, pp. 1137–1145 (1995)
19. Lord, F.M., Novick, M.R.: Statistical Theories of Mental Test Scores. Addison-Wesley (1968)
20. Lundberg, S.M., Lee, S.: A unified approach to interpreting model predictions. Nat. Mach. Intell. **2**, 2522–5839 (2020)
21. MacQueen, J.: Some methods for classification and analysis of multivariate observations. In: Proceedings of the Fifth Berkeley Symposium on Mathematical Statistics and Probability, vol. 1, pp. 281–297 (1967)
22. Mesa, V., Wladis, C., Watkins, L.: Research problems in community college mathematics education. J. Res. Math. Educ. **43**(2), 173–194 (2012)
23. Messick, S.: Validity of psychological assessment. Am. Psychol. **50**(9), 741–749 (1995)
24. Nitko, A.J., Brookhart, S.M.: Educational Assessment of Students, 4th edn. Pearson (2004)
25. Ritter, S., Thompson, A.: AI transparency in high-stakes educational decisions. AI Educ. **3**(2), 156–173 (2024)
26. Romero, C., Ventura, S.: Educational data mining and learning analytics: an updated survey. WIREs Data Min. Knowl. Discovery **10**(3), e1355 (2020)
27. Rousseeuw, P.J.: Silhouettes: a graphical aid to interpretation of cluster analysis. J. Comput. Appl. Math. **20**, 53–65 (1987)
28. Thorndike, R.L.: Who belongs in the family? Psychometrika **18**(4), 267–276 (1953)
29. Wang, Y., Zhang, H., Martinez, L.: Robust clustering methods for heterogeneous educational data. Educ. Data Sci. Rev. **2**(1), 34–52 (2024)
30. Zhang, X., Lee, K., Brown, M.: Deep learning for small-sample educational assessment. Neural Comput. Appl. **36**, 2847–2863 (2024)
31. Zhou, Y., Brown, M.: SHAP analysis for interpretable student placement decisions. IEEE Trans. Education **65**(4), 512–521 (2022)

An Enhanced School Curriculum to Better Prepare Students for Education and Careers in Engineering, Mathematics, Science, and Other Fields

Ashu M. G. Solo(✉)

Maverick Trailblazers Inc.™, Wilmington, DE, USA
amgsolo@mavericktrailblazers.com

Abstract. Schools are failing to adequately prepare students for education and careers in engineering and the military. This leads to high failure rates in engineering courses and military courses. This research paper describes an enhanced school curriculum for grades 1–12 that better prepares students for education and careers in engineering, mathematics, science, the military, law, writing, art, vehicle repair, carpentry, etc.

Keywords: Education · Curriculum · Elementary School · Middle School · High School · Engineering · Mathematics · Science · Military

1 Introduction

Schools are failing students and causing them to fail after they graduate. School curriculum designers have done a terrible job in developing curricula to adequately prepare students for engineering and the military and to get students interested in engineering and the military. The curricula in mathematics, science, law, writing, art, and trades could be improved too.

There are high failure rates in university undergraduate programs in engineering because schools fail to adequately prepare students for engineering. Also, there are high failure rates in courses in the military because schools fail to adequately prepare students for military training. The teachers aren't at fault because they're required to follow the curriculum given to them. The curriculum designers are at fault for the awful curriculum they design.

The problem with the vast majority of curriculum designers is they haven't studied fields like engineering and the military because these fields usually can't be studied in an education degree. Therefore, they don't know enough to integrate engineering and the military into the school curriculum.

Schools need curriculum designers who have studied a wide variety of subjects, not just education. Someone with an engineering degree can design a better engineering curriculum for high school students than someone with an education degree focusing on curriculum studies because the former is familiar with engineering. Someone with

H. R. Arabnia et al. (Eds.): CSCI 2025, CCIS 2974, pp. 206–221, 2027.
https://doi.org/10.1007/978-3-032-28475-4_14

military service can design a better military curriculum for high school students than someone with an education degree focusing on curriculum studies because the former is familiar with military service.

Schools emphasize education in social sciences and humanities over education in engineering, math, and science. Schools offer many arts courses in social sciences and humanities, but many offer no courses in engineering.

This is the main cause of the low enrollment in engineering. The main cause of the low female enrollment in engineering is parents getting their daughters involved in arts activities while getting their sons involved in engineering activities. There aren't enough people, especially females, going into engineering or the military. The curriculum in this research paper should significantly boost female and male interest and knowledge in engineering, mathematics, science, and the military.

Not including engineering in the school curriculum is as foolish as if science wasn't included in the school curriculum. There's no reason that kids can't be taught subjects like basic digital electronics, basic analog electronics, basic software development, basic mechanical design, etc.

Because schools often don't have any engineering courses, much fewer people go into engineering. Many people think scientists, not engineers, design electronic products, vehicles, bridges, robots, computers, etc.

Because many people don't even understand what engineering is, the author of this research paper, Ashu M. G. Solo, heard high school graduates who thought that engineering is woodworking, carpentry, electrician work, vehicle repair, or driving trains. Courses in software development should be called *software engineering,* not *computer science,* so people understand what engineering is.

Engineering consists of theoretical engineering and applied engineering. *Theoretical engineering* is the creative development of mathematics, natural science, technical principles, or technical methods for usage in the development, analysis, characterization, modeling, control, automation, optimization, forecasting, simulation, or visualization of devices, algorithms, components, systems, machines, apparatuses, structures, processes, operations, or materials. *Applied engineering* is the creative application of mathematics, natural science, technical principles, or technical methods for the development, analysis, characterization, modeling, control, automation, optimization, forecasting, simulation, or visualization of devices, algorithms, components, systems, machines, apparatuses, structures, processes, operations, or materials. Therefore, intelligent systems development is in the domain of engineering. The preceding definitions [1, 2] of engineering, theoretical engineering, and applied engineering are by Solo.

Because most schools don't have any military training courses, many high school graduates don't understand the military branches, military occupations, military ranks, etc. Solo has seen a member of parliament (M.P.) refer to a corporal as a captain, an M.P. who didn't know what the infantry is, and a former M.P. who thought that a non-commissioned officer outranked an officer.

An enhanced school curriculum is suggested in this research paper. This curriculum will better prepare students for education and careers in engineering, mathematics, science, and the military. Also, the enhanced curriculum will better prepare students for education and careers in law, journalism, professional writing, art, and the trades of vehicle repair and carpentry.

Solo went to elementary school and high school in Saskatchewan, Canada where elementary school includes grades 1–8 and high school includes grades 9–12. In some geographical regions, there is a middle school for some of the middle grades. In other geographical regions, some of the grades have different names. In this research paper, elementary school is assumed to include grades 1–8 and high school is assumed to include grades 9–12. That might be different in the reader's geographical region, but the suggested curriculum for each grade remains the same for all geographical regions.

2 Elementary School Curriculum

Each year of elementary school (grades 1–8) should have an engineering course, a math course, a science course, a social studies course, a language arts course, a second language course, and a physical training course.

There are simple engineering projects that can be done by elementary school students and simple engineering topics that can be taught to elementary school students. Engineering includes software development. Mathematics courses should include mental math training. Calculators shouldn't be allowed in elementary school. Science courses should include fundamentals of nutrition, health, biology, chemistry, physics, astronomy, environmental science, geology, meteorology, etc.

Social studies should cover history including atrocities by the government, geography, cultures, civil rights law, how government works, etc. Language arts should include phonics, spelling, grammar, punctuation, style, literature, creative writing, vocabulary, etc. Art courses should only be included in the curriculum if they're properly taught with actual drawing techniques and painting techniques. See Sect. 9 for further information.

Physical education prior to grade 6 can focus on ball chasing games because kids need some play time. Physical education from grades 6–8 should focus more on physical training like running, sprinting, push-ups, chin-ups, sit-ups, other abdomen exercises, etc. See Sect. 12 for further information.

In Solo's opinion, elementary school (grades 1–8) is too long. In his opinion, there's a colossal amount of wasted time in elementary school on activities that don't benefit students. He hated so much time being wasted. If most of that wasted time is cut out, then schools can cut out one grade like grade 5.

3 High School Curriculum

A high school with four grades (grades 9–12) could let students take 40 of the following 66 courses as long as they meet the prerequisites:

- 2 courses on software engineering
- 1 course on digital electronics engineering
- 1 course on analog electronics engineering
- 1 course on engineering concepts
- 4 courses on algebra

- 3 courses on geometry/trigonometry
- 1 course on calculus
- 1 course on probability and statistics
- 1 course on financial mathematics
- 2 courses on general science
- 1 course on environmental science
- 1 course on earth science
- 1 course on health and nutrition
- 1 physics course on wave motion, optics, and sound
- 1 physics course on classical mechanics
- 1 course on astronomy
- 1 course on introductory biology and human biology
- 1 course on plant biology and animal biology
- 2 courses on chemistry
- 1 course on domestic history
- 1 course on 20th century world history
- 1 course on 18th century and 19th century world history
- 1 course on ancient history
- 1 course on government and public policy
- 1 course on macroeconomics
- 1 course on law
- 1 course on civil litigation and lawsuit document preparation
- 1 course on psychology
- 1 course on accounting
- 1 course on entrepreneurship
- 1 course on grammar, punctuation, and style
- 1 course on literature
- 4 courses on a second language
- 1 course on paper drawing
- 1 course on computer drawing
- 1 course on painting
- 1 course on computer animation
- 4 courses on musical instruments
- 4 courses on automotive repair
- 2 courses on carpentry
- 2 courses on woodworking
- 1 course on typing
- 1 course on learning methods (speed reading and memorization techniques)
- 2 courses with military training
- 2 courses with civilian physical training

The standard courses listed above can include standard topics for these courses. The page limit for a research paper prevents providing a detailed curriculum for each course.

This is a fair and balanced curriculum in engineering, mathematics, science, social science, humanities, and trades. This curriculum eliminates the bias against engineering and the military, better acquaints people with engineering, mathematics, and science fields, and better prepares people for going into engineering, math, science, or the military without failing.

At many universities, there are really high failure rates in engineering. In the military, a lot of people rapidly quit basic training courses. This curriculum will help prevent that.

13 of the preceding courses should be compulsory. Then students can pick 27 more courses.

High school students should be required to take the first course on software engineering because there are a lot of employment positions in this field. They should be required to take the first course on algebra and the course on financial mathematics because they will encounter situations where they can use these courses. They should be required to take the first general science course to have an understanding of basic science. They should be required to take the environmental science course because of the urgency of the climate change crisis, because many people still think climate change is a hoax, because many people still think humans have nothing to do with climate change, and because many people still think climate change is good for the planet. They should be required to take the course on health and nutrition to potentially increase their lifespans, possibly prevent disease, unclog the healthcare system, and learn first aid, cardiopulmonary resuscitation (CPR), and usage of an automated external defibrillator (AED) to potentially save lives.

High school students should be required to take the course on domestic history to understand atrocities committed by their country. They should be required to take the course on 20th century world history to better understand the present. They should be required to take the course on government and public policy to be more informed voters. They should be required to take the course on grammar, punctuation, and writing style because most people don't understand the advanced rules of grammar, punctuation, and style.

High school students should be required to take the course on typing and the course on learning methods because these courses will save them considerable time in the future. They should be required to take the first course for either military training or civilian physical training so they will be able to do physical training on their own.

4 Notes on Engineering Courses

The high school course on engineering concepts should include fundamentals of fuzzy logic [3–16] and neural networks [17]. Students should read reference [12] on fuzzy logic.

Most teachers didn't take any engineering courses in university so they can't teach engineering to kids. The schools should hire engineers to teach engineering to kids. They shouldn't need education degrees because the vast majority of engineers don't have education degrees. Also, teachers should be able to take some basic engineering courses in their education degrees.

5 Notes on Mathematics Courses

The final high school course on algebra should cover matrix mathematics and vectors. If there is time, the final course in algebra can include multidimensional matrix mathematics and its subsets, multidimensional matrix algebra and multidimensional matrix calculus [18–28]. Solo created multidimensional matrix mathematics and its subsets, multidimensional matrix algebra and multidimensional matrix calculus, in references [18–28]. Students could get bonus marks if they study these topics in the final course on algebra.

The high school course on financial mathematics should include time value of money and cash flow analysis; investment evaluation methods including present worth, annual worth, and internal rate of return; and the impact of inflation, taxation, uncertainty, and risk on investment decisions. This can be supplemented by a description of the types of investments.

6 Notes on Natural Science Courses

The high school course on environmental science must teach people about the science of climate change [29] and sustainability practices [30–40].

The high school course on health and nutrition should include the negative effects of animal foods, alcohol, cigarettes, e-cigarettes, tobacco, recreational drugs, etc. Solo co-created an online quiz available at https://tinyurl.com/alcohol-quiz that students can do to test their awareness of the negative effects of alcohol [41–44] while educating themselves about this. Also, the high school course on health and nutrition should include first aid, CPR, and using an AED.

The high school course on introductory biology and human biology should include evolution theory. Some schools aren't teaching evolution theory because it conflicts with religious beliefs. The course on introductory biology and human biology can be an optional course that people don't have to take if evolution theory conflicts with their religious beliefs.

7 Notes on Social Science Courses

The high school course on government and public policy should describe how the different levels of government work, how the United Nations works, and the most important public policy issues like the genocide, ethnic cleansing, and brutal oppression of Palestinians with support, funding, and arming from western countries [45–69], climate change [29–40], campaign finance reform, voter disenfranchisement, law enforcement corruption, racist profiling, the death penalty, cyberbullying [70–78], government reform, etc. Students could watch video documentaries [29, 56–61, 79] on different public policy issues.

Public ignorance on the issue of Israel and Palestine allowed western politicians to get away with supporting, funding, and arming Israel while it has been massacring, ethnically cleansing, and brutally oppressing Palestinians since 1948 [45–69]. This has severely harmed the moral standing of western countries. Ending support, funding, and arming of fascist and imperialist Israel is the moral issue of our time. It's imperative that students be taught the truth about this issue in a course on government and public policy.

Students need to be shown that it's Israeli governments that are unwilling to accept Palestinian sovereignty, not Palestinian governments that are unwilling to accept Israeli sovereignty. Students need to be shown that it's Israeli governments that banned the Palestinian flag in Palestinian territory, not Palestinian governments that banned the Israeli flag in Israeli territory.

Students need to be shown that it's Israelis who drew first blood repeatedly, not Palestinians who drew first blood repeatedly. Students need to be shown that it's Israelis who have done most of the killing before and after October 7, 2023, not Palestinians who have done most of the killing before or after October 7, 2023.

Students need to be shown that it's Israelis illegally occupying Palestinian territory, not Palestinians illegally occupying Israeli territory. Students need to be shown that it's Israelis illegally blockading Palestinian territory, not Palestinians illegally blockading Israeli territory.

Students need to be shown that it's Israelis ethnically cleansing Palestinians in Palestinian territory, not Palestinians ethnically cleansing Israelis in Israeli territory. Students need to be shown that it's Israelis brutally oppressing Palestinians in Palestinian and Israeli territory, not Palestinians brutally oppressing Israelis in Israeli territory. This is imperative for stopping and preventing genocide, ethnic cleansing, and brutal oppression with western support, funding, and arming.

For the high school course on government and public policy, students should watch several of the documentaries on Israel and Palestine in references [56–61]. These documentaries are islands of truth in a sea of disinformation and distortions. For this course, students should be asked to read one of the books on Israel and Palestine in references [62–69] and write a synopsis that is at least 2000 words long for it. These books are oases of truth in a desert of disinformation and distortions.

For the high school course on 20th century world history, students should be required to read one of references [45, 46] on the history of Israel and Palestine and write a timeline of events in the book. This is the most important historical issue for students to be familiar with. Public ignorance of the history of the conflict has led to widespread support for the genocide, ethnic cleansing, and brutal oppression of Palestinians [45–69]. Schools have failed to teach students the history of this conflict.

For the high school course on domestic history, students need to learn about the foreign and domestic atrocities committed by their countries.

The high school course on civil litigation procedure and lawsuit document preparation should teach the steps in suing someone and how to write important civil claim documents for money owing, personal injuries, and issues in the information age including online defamation [70–77], doxing [70–77], and impersonation [70–74, 78].

8 Notes on Language Courses

Most adults who have completed high school and even university don't understand the rules of advanced grammar, punctuation, and style so a course on this is important. Reading stories, answering questions about them, and writing essays about them is not an effective way to learn how to write with perfect clarity, grammar, punctuation, and style. The rules of grammar, punctuation, and style are much more complicated than most people realize. Students need to focus on the rules of modern grammar, punctuation, and style rather than the old English plays of William Shakespeare. Solo recommends references [80, 81] for teaching a high school course on grammar, punctuation, and style.

Students should read reference [82] or [83] for the high school course on literature. Other possibilities are references [84–88]. Students should read reference [89] for one of the elementary school courses on language arts. These books teach important life lessons.

The second language courses should be taught using software instead of lectures. Solo has tried a Spanish lesson with *Babbel* software and a Spanish lesson with *Duolingo* software. *Babbel* is significantly better than *Duolingo,* but *Duolingo* is free.

There can be a lot of variation in how good teachers are in teaching languages. Two of the three French language teachers he had between grades 7–12 were awful at teaching. Solo found that *Babbel* and *Duolingo* software are better at teaching languages than at least two of his three French language teachers.

9 Notes on Art Courses

The problem with the art courses that Solo took in elementary school is no drawing or painting techniques were taught. The teachers just told the students to draw or paint without teaching any actual techniques for doing that. He can do that at home. Telling students to draw or paint without teaching them techniques for that is as wasteful and foolish as telling students to fix a car without telling them how.

Reference [90] has drawing exercises that students can do to significantly increase their skill in drawing. The early exercises are easy to do. The exercises gradually increase in their complexity. The exercises turn people unskilled in drawing into skilled artists. Teachers should have students do exercises in this book or similar exercises in art classes.

10 Notes on Learning Methods Course

The high school course on learning methods should include memorization and speed reading techniques. Solo has evaluated different authors' techniques for speed reading and memorization. Howard Berg has the best speed reading techniques in Solo's opinion. Berg was awarded a Guinness World Record in speed reading. Kevin Trudeau has the best memorization techniques in Solo's opinion. Solo has effectively used these techniques to increase his ability to memorize large amounts of information and increase his reading speed. The speed reading techniques of Howard Berg are described in reference [91]. The memorization techniques of Kevin Trudeau are described in reference [92]. The course on learning methods should be taken by students in the first semester of high school so they can use the methods in their other courses like biology and literature.

11 Notes on Military Training Courses

Military training courses should be offered as an alternative to civilian physical education courses in high school. However, it isn't a good idea to introduce military training in high school in countries with a history of brutally repressive tyranny with the military like fascist and imperialist Israel [45–69]. Also, it isn't a good idea to give weapons to high school students for military training, but they can do a lot of military training without weapons.

Military training courses should include physical training, military structure and conventions, battlecraft, fieldcraft, combat simulation exercises, combat lifesaving (first aid and cardiopulmonary resuscitation), kickboxing, Japanese jiu-jitsu, Gracie jiu-jitsu, etc. Kickboxing, Japanese jiu-jitsu, and Gracie jiu-jitsu are the most effective combative arts without weaponry. Physical training should include running, push-ups, chin-ups, abdomen exercises, punching and kicking bags using kickboxing techniques, etc. To save money, cheap handheld bags can be used with students alternating between holding bags and punching and kicking bags. Physical training can be gradually increased in duration as described in the next section on physical education courses.

Most teachers have no military experience so they can't teach military training to kids. The schools should hire current or former military officers to teach military training to kids. They shouldn't need education degrees because the vast majority of military members don't have education degrees.

12 Notes on Physical Education Courses

The problem with most physical education courses in high school is instead of focusing on physical training, they focus on playing sports with no significant exercise like baseball, volleyball, badminton, curling, soccer, etc. Most adults can't run for 15 minutes or do 40 push-ups because they don't practice serious physical training in physical education courses in school.

In team sports, there are too many breaks in the play, they keep trading off who's playing, they only run when the ball is in their vicinity, and there's usually no arm exercise. In racquet sports, not even the professionals can keep the ball in play continuously for more than a few minutes. In ball chasing games, people never actually run for more than a few minutes continuously. Most positions in ball chasing games involve no significant arm exercise.

In baseball, there's no significant exercise except for the pitcher and he only exercises one arm. According to reference [93], National Hockey League (NHL) players are only on the ice for 15 to 20 minutes per game. According to reference [94], the average National Football League (NFL) game is 3 hours 12 minutes, but the ball is only in play for 11 minutes. Obviously, it's a much more effective use of time to run continuously for an hour or two.

The other problem with ball chasing games is they have no or almost no practical applications in real life unlike combat sports and locomotory sports that get you from one place to another. Nobody has a need to dunk a basketball, bat a baseball, or volley a volleyball while not playing one of these games.

In physical education courses in school, starting around grade 7, the whole class (except physically disabled students) should run for a few minutes and do a few push-ups on the first day. Then the teacher should gradually increase the time that students run every time they have a class.

If students could only run for one minute on the first day and the time of the run is increased by only one minute each time they have a class, then after about 60 classes, they should all be able to run for an hour. If students could only do one push-up on the first day and the number of push-ups is increased by only one push-up each time they have a class, then after about 60 classes, they should all be able to do 60 push-ups. Even the worst students in athletic ability would finish a physical education course with the ability to run for an hour and do 60 push-ups.

That will help them in life with preventing disease, increasing lifespan, and losing excess weight. Many of the students will be able to ramp up their physical training much more quickly. They can play ball chasing games after they're done real physical training including running, push-ups, chin-ups, abdomen exercises, etc.

13 Other Suggestions for School Improvements

Schools should eliminate tryouts for their sports teams. See reference [95] by Solo for further information.

Each school should have a bookcase where students can leave books, magazines, audiobooks on CDs and cassettes, movies on DVDs and videotapes, music on CDs and cassettes, etc. that they no longer want for other students to take for free [96]. Reusing products left in these bookcases saves people money and allows products to be used to their fullest extent. Also, reusing products decreases waste, landfill size, pollution, energy consumption, natural resource consumption, greenhouse gas emissions causing climate change, etc.

Schools can recognize annual celebratory days originated by Solo for scientific constants and greenhouse gases to boost scientific awareness and scientific interest. See reference [97] by Solo for further information.

As described in references [12, 13] by Solo, fuzzy logic [3–16] can be used for assessing students when assessments are based on professors' opinions of student performance on projects, reports, essays, etc. Fuzzy logic can be used for assigning linguistic grades and for decision making and data mining with those linguistic grades [12, 13]. Many aspects of fuzzy logic including fuzzy sets, linguistic variables, fuzzy rules, fuzzy math, fuzzy database queries, computational theory of perceptions, and computing with words are useful in uncertainty management of linguistic evaluations for students [12, 13].

References [98–100] by Solo have other suggestions for educational improvements related to online lectures and online courses.

A student's high school average should be determined by her best eight courses in grade 12. If she takes more courses that she is interested in, but isn't as good at, she shouldn't be penalized for taking these courses.

14 Conclusion

An enhanced school curriculum is badly needed to eliminate the bias against engineering and the military in schools, to show students what engineering is and make them more interested in going into engineering, and to better prepare students for education and careers in engineering, mathematics, science, and the military. The enhanced school curriculum presented in this research paper will better prepare students for education and careers in engineering, mathematics, science, military service, law, journalism, professional writing, art, vehicle repair, and carpentry.

Disclosure of Interests. The author has no competing interests to declare that are relevant to the content of this article.

References

1. Solo, A.M.G.: The new fields of public policy engineering, political engineering, computational public policy, and computational politics. In: Proceedings of the 2011 International Conference on e-Learning, e-Business, Enterprise Information Systems, and e-Government (EEE '11), pp. 431–434. CSREA, Watkinsville, GA (2011)
2. Solo, A.M.G.: The interdisciplinary fields of political engineering, public policy engineering, computational politics, and computational public policy. In: Solo, A.M.G. (ed.) Handbook of Research on Politics in the Computer Age, pp. 1–16. IGI Global, Hershey, PA (2019)
3. Gupta, M.M., Solo, A.M.G.: Important new terms and classifications in uncertainty and fuzzy logic. In: Tamir, D.E., Rishe, N.D., Kandel, A. (eds.) Fifty Years of Fuzzy Logic and Its Applications, pp. 153–168. Springer, New York, NY (2015)
4. Gupta, M.M., Solo, A.M.G.: On the morphology of uncertainty in human perception and cognition. In: Proceedings of the First Interdisciplinary CHESS Interactions Conference, pp. 257–271. World Scientific, Singapore (2009)
5. Solo, A.M.G., Gupta, M.M.: Uncertainty in computational perception and cognition. In: Nikravesh, M., Kacprzyk, J., Zadeh, L.A. (eds.) Forging New Frontiers: Fuzzy Pioneers I: Studies in Fuzziness and Soft Computing, pp. 251–266. Springer Verlag, New York, NY (2007)
6. Solo, A.M.G., Gupta, M.M., Homma, N., Hou, Z.-G.: Type-one fuzzy logic for quantitatively defining imprecise linguistic terms in politics and public policy. In: Solo, A.M.G. (ed.) Political Campaigning in the Information Age, pp. 239–255. IGI Global, Hershey, PA (2014)
7. Solo, A.M.G.: Interval type-two fuzzy logic for quantitatively defining imprecise linguistic terms in politics and public policy. In: Solo, A.M.G. (ed.) Political Campaigning in the Information Age, pp. 256–264. IGI Global, Hershey, PA (2014)
8. Solo, A.M.G.: Interval type-two fuzzy sets for defining "rich" at a presidential forum. In: Proceedings of the 2016 International Conference on e-Learning, e-Business, Enterprise Information Systems, and e-Government (EEE'16), pp. 181–182. CSREA, Watkinsville, GA (2016)
9. Solo, A.M.G.: Type-one and interval type-two fuzzy logic for quantitatively defining imprecise linguistic terms in politics and public policy. In: Solo, A.M.G. (ed.) Handbook of Research on Politics in the Computer Age, pp. 17–44. IGI Global, Hershey, PA (2019)
10. Solo, A.M.G.: Warren, McCain, and Obama needed fuzzy sets at a presidential forum. In: Special Issue on Real Life Applications of Fuzzy Logic. Advances in Fuzzy Systems. Hindawi, London (2012)

11. Solo, A.M.G.: Fuzzy sets for defining "rich" at a presidential forum. In: Proceedings of the International Conference on 2012 Information Society (i-Society 2012), pp. 484–485. IEEE, New York, NY (2012)
12. Solo, A.M.G., Gupta, M.M.: Fuzzy logic theory and applications in uncertainty management of linguistic evaluations for students. In: Bishop, J. (ed.) Cases on Technologies in Education from Classroom 2.0 to Society 5.0, pp. 243–266. IGI Global, Hershey, PA (2021)
13. Solo, A.M.G.: Fuzzy grading: fuzzy logic for uncertainty management of linguistic evaluations. In: Proceedings of the 2010 International Conference on e-Learning, e-Business, Enterprise Information Systems, and e-Government (EEE'10), pp. 271–276. CSREA, Watkinsville, GA (2010)
14. Zadeh, L.A.: Fuzzy sets. Inf. Control. **8**, 338–353 (1965)
15. Zadeh, L.A.: A fuzzy-set-theoretic interpretation of linguistic hedges. J. Cybern. **2**(3), 4–34 (1972)
16. Zadeh, L.A.: Outline of a new approach to the analysis of complex systems and decision processes. IEEE Trans. Syst. Man Cybern. **3**(1), 28–44 (1973)
17. Gupta, M.M., Jin, L., Homma, N.: Static and Dynamic Neural Networks: From Fundamentals to Advanced Theory. John Wiley & Sons, Hoboken, NJ (2003)
18. Solo, A.M.G.: Multidimensional Matrix Algebra and Multidimensional Matrix Calculus: Part 1 of 5. Proceedings of the 2010 International Conference on Scientific Computing (CSC '10). CSREA, Watkinsville, GA. (2010)
19. Solo, A.M.G.: Multidimensional Matrix Algebra and Multidimensional Matrix Calculus: Part 2 of 5. In: Proceedings of the 2010 International Conference on Scientific Computing (CSC '10). CSREA, Watkinsville, GA (2010)
20. Solo, A.M.G.: Multidimensional matrix algebra and multidimensional matrix calculus: part 3 of 5. In: Proceedings of the 2010 International Conference on Scientific Computing (CSC '10). CSREA, Watkinsville, GA (2010)
21. Solo, A.M.G.: Multidimensional matrix algebra and multidimensional matrix calculus: part 4 of 5. In: Proceedings of the 2010 International Conference on Scientific Computing (CSC '10). CSREA, Watkinsville, GA (2010)
22. Solo, A.M.G.: Multidimensional matrix algebra and multidimensional matrix calculus: part 5 of 5. In: Proceedings of the 2010 International Conference on Scientific Computing (CSC'10). CSREA, Watkinsville, GA (2010)
23. Solo, A.M.G.: Multidimensional matrix mathematics: notation, representation, and simplification, part 1 of 6. In: Lecture Notes in Engineering and Computer Science: Proceedings of the World Congress on Engineering 2010 (WCE 2010), vol. 3, pp. 1824–1828. International Association of Engineers (IAENG), Hong Kong (2010)
24. Solo, A.M.G.: Multidimensional matrix mathematics: multidimensional matrix equality, addition, subtraction, and multiplication, part 2 of 6. In: Lecture Notes in Engineering and Computer Science: Proceedings of the World Congress on Engineering 2010 (WCE 2010), vol. 3, pp. 1829–1833. International Association of Engineers (IAENG), Hong Kong (2010)
25. Solo, A.M.G.: Multidimensional matrix mathematics: multidimensional null and identity matrices and multidimensional matrix outer and inner products, part 3 of 6. In: Lecture Notes in Engineering and Computer Science: Proceedings of the World Congress on Engineering 2010 (WCE 2010), vol. 3, pp. 1834–1837. International Association of Engineers (IAENG), Hong Kong (2010)
26. Solo, A.M.G.: Multidimensional matrix mathematics: multidimensional matrix transpose, symmetry, antisymmetry, determinant, and inverse, part 4 of 6. In: Lecture Notes in Engineering and Computer Science: Proceedings of the World Congress on Engineering 2010 (WCE 2010), vol. 3, pp. 1838–1841. International Association of Engineers (IAENG), Hong Kong (2010)

27. Solo, A.M.G.: Multidimensional matrix mathematics: algebraic laws part 5 of 6. In: Lecture Notes in Engineering and Computer Science: Proceedings of the World Congress on Engineering 2010 (WCE 2010), vol. 3, pp. 1842–1847. International Association of Engineers (IAENG), Hong Kong (2010)
28. Solo, A.M.G.: Multidimensional matrix mathematics: solving systems of linear equations and multidimensional matrix calculus, part 6 of 6. In: Lecture Notes in Engineering and Computer Science: Proceedings of the World Congress on Engineering 2010 (WCE 2010), vol. 3, pp. 1848–1850. International Association of Engineers (IAENG), Hong Kong (2010)
29. Gore, A., Guggenheim, D.: An Inconvenient Truth. Participant Productions and Lawrence Bender Productions (2006)
30. Solo, A.M.G.: What Will It Take for Climate Action? Winnipeg Free Press. https://www.winnipegfreepress.com/opinion/letters-to-the-editor/2025/06/18/letters-june-18-9 (June 18, 2025)
31. Solo, A.M.G.: Hotels need recycling bins in rooms. The StarPhoenix. https://thestarphoenix.com/opinion/letters/letters-immigrants-boost-canadas-economy-despite-all-the-rhetoric (Mar. 20, 2025)
32. Solo, A.M.G.: Help poor with more recycling bins. The StarPhoenix. https://thestarphoenix.com/opinion/columnists/letters-canadas-leaders-should-focus-on-demand-for-drugs-not-border (Feb. 11, 2025)
33. Solo, A.M.G.: City should plant more trees. The StarPhoenix. https://thestarphoenix.com/opinion/letters/letters-how-can-a-flood-disaster-be-natural-when-we-build-in-floodplains (July 15, 2025)
34. Solo, A.M.G.: Free Libraries in Buildings. Countercurrents. https://countercurrents.org/2025/05/free-libraries-in-buildings/ (May 19, 2025)
35. Solo, A.M.G.: Wasting energy on outdoor lighting when it isn't dark outside. Intrepid Report. http://www.intrepidreport.com/archives/10187 (July 12, 2013)
36. Eisen, M.B., Brown, P.O.: Rapid global phaseout of animal agriculture has the potential to stabilize greenhouse gas levels for 30 years and offset 68 percent of CO2 emissions this century. PLOS Climate (2022)
37. Rao, S.: Animal agriculture is the leading cause of climate change. A position paper. J. Ecol. Soc. **32–33**(1), 155–167 (2021)
38. Cheng, M., McCarl, B., Fei, C.: Climate change and livestock production: a literature Review. Atmos. **13**(1) (2022)
39. Scarborough, P., et al.: Vegans, vegetarians, fish-eaters and meat-eaters in the UK show discrepant environmental impacts. Nat. Food, 565–574 (2023)
40. Higuita, N.I.A., LaRocque, R., McGushin, A.: Climate change, industrial animal agriculture, and the role of physicians – time to act. J. Clim. Change Health. **13**, 2667–2782 (2023)
41. Paradis, C., et al.: The Low-Risk Alcohol Drinking Guidelines Scientific Expert Panels: Canada's Guidance on Alcohol and Health: Final Report. Canadian Centre on Substance Use and Addiction, Ottawa (2023)
42. Alcohol and Cancer: U.S. Centers for Disease Control and Prevention. https://www.cdc.gov/cancer/risk-factors/alcohol.html
43. Alcohol's Effects on the Body: National Institute on Alcohol Abuse and Alcoholism. https://www.niaaa.nih.gov/alcohols-effects-health/alcohols-effects-body
44. One alcoholic drink a day linked will reduce brain size. Penn Today. https://penntoday.upenn.edu/news/one-alcoholic-drink-day-linked-reduced-brain-size (Mar. 4, 2022)
45. Khalidi, R.: The Hundred Years' War on Palestine: A History of Settler Colonialism and Resistance, 1917–2017. Macmillan Publishers, London (2020)
46. Pappe, I.: A Very Short History of the Israel-Palestine Conflict. Oneworld Publications, London (2024)

47. Solo, A.M.G.: Israel Drew First Blood and Anti-Muslim Bigots Enabled Israel's Tyranny with Impunity. Crocels News. https://www.crocels.news/2025/01/09/israel-drew-first-blood/ (Jan. 9, 2025)
48. Solo, A.M.G.: Starvation Worse than Gas Chambers. Crocels News. https://www.crocels.news/2025/10/12/starvation-worse-than-gas-chambers/ (Oct. 12, 2025)
49. Solo, A.M.G.: Jewish and Christian Clergy Have a Duty to Fight Support of Israel's Atrocities. Countercurrents. https://countercurrents.org/2025/06/jewish-and-christian-clergy-have-a-duty-to-fight-support-of-israels-atrocities/ (June 30, 2025)
50. Solo, A.M.G.: Data on casualties. United Nations Office for the Coordination of Humanitarian Affairs
51. Demolition Watch: United Nations Relief and Works Agency for Palestine Refugees in the Near East
52. The Occupation of Water. Amnesty International (2017)
53. Israel's Apartheid against Palestinians. Amnesty International (2022)
54. Tahhan, Z.: Israel's settlements: Over 50 years of land theft explained. Al Jazeera Media Network (2020)
55. The Gaza Strip | The humanitarian impact of 15 years of blockade - June 2022: Fast Facts. UNICEF (2022)
56. Baltzer, A.: Life in Occupied Palestine: Eyewitness Stories & Photos (2008)
57. Adra, B., Ballal, H., Abraham, Y., Szor, R.: No other land. Yabayay Media and Antipode Films. (2024)
58. Martin, A., Prysner, M.: Gaza Fights for Freedom (2019)
59. Theroux, L., Baker, J.: The Settlers. Mindhouse Productions (2025)
60. Pritsker, K., Workman, M.T.: The Encampments. Watermelon Pictures (2025)
61. Alper, L., Earp, J.: The Occupation of the American Mind: Israel's Public Relations War in the United States. Media Education Foundation (2016)
62. Peled, M.: The General's Son: Journey of an Israeli in Palestine, Tenth Anniversary edn. Just World Books, Ithaca, NY (2022)
63. Finkelstein, N.: Gaza's Gravediggers: An Inquiry into Corruption in High Places. OR Books, New York, NY (2025)
64. Finkelstein, N.: Gaza: an Inquest into its Martyrdom. University of California Press, Berkeley, CA (2018)
65. Finkelstein, N.: Method and Madness: The Hidden Story of Israel's Assaults on Gaza. OR Books, New York, NY (2015)
66. Finkelstein, N.: The Holocaust Industry: Reflections on the Exploitation of Jewish Suffering. Verso, New York, NY (2015)
67. Blumenthal, M.: Goliath: Life and Loathing in Greater Israel. Bold Type Books, New York, NY (2023)
68. Baltzer, A.: Witness in Palestine: A Jewish American Woman in the Occupied Territories. Paradigm Publishers, Boulder, CO. (2017)
69. Mearsheimer, J.J., Walt, S.M.: The Israel Lobby and U.S. Foreign Policy. Farrar, Straus and Giroux, New York, NY (2008)
70. Solo, A.M.G.: Preventing Facebook from being used for defamation, doxing, impersonation, threats of violence, and intimate images. In: Global Congress on Emerging Technologies (GCET-2024), pp. 184–190. IEEE, New York, NY (2024)
71. Solo, A.M.G.: Preventing Twitter from being used for defamation, doxing, impersonation, threats of violence, and intimate images. In: 2023 Congress in Computer Science, Computer Engineering, & Applied Computing (CSCE 2023), pp. 2265–2271. IEEE, New York, NY (2023)

72. Solo, A.M.G.: Educating the public to combat online defamation, doxing, and impersonation. In: Bishop, J. (ed.) Cases on Technologies in Education from Classroom 2.0 to Society 5.0, pp. 231–242. IGI Global, Hershey, PA (2021)
73. Solo, A.M.G.: Fighting online defamation, doxing, and impersonation. In: Proceedings of the International Conferences: Internet Technologies & Society and Sustainability, Technology and Education 2020, pp. 93–97. IADIS, Sao Paulo, Brazil (2020)
74. Solo, A.M.G.: Prevention of disinformation society: fighting online defamation, doxing, and impersonation. In: Bishop, J. (ed.) Perspectives on the Information Society, pp. 79–93. Crocels, Wilmington, DE (2019)
75. Solo, A.M.G.: Combating online defamation and doxing in the United States. In: Proceedings of the 2019 International Conference on Internet Computing and Internet of Things (ICOMP '19), pp. 75–77. CSREA, Watkinsville, GA (2019)
76. Solo, A.M.G.: Brief on effectively combatting cyberbullying and cyberlibel by online hate groups for the study on online hate of the Standing Committee on Justice and Human Rights for the House of Commons of Canada. In: Online Hate Study, Standing Committee on Justice and Human Rights, House of Commons of Canada (2019)
77. Solo, A.M.G.: Mémoire sur la lutte efficace contre les actes de cyberintimidation et de diffamation en ligne commis par des groupes haineux présenté au Comité permanent de la justice et des droits de la personne de la Chambre des communes du Canada dans le cadre de son étude sur la haine en ligne. Translated to French by unknown staff of House of Commons of Canada, Online Hate Study, Standing Committee on Justice and Human Rights, House of Commons of Canada (2019)
78. Solo, A.M.G.: Combating online impersonation in the United States. In: Proceedings of the 6th Annual Conference on Computational Science and Computational Intelligence (CSCI 2019), pp. 1552–1553. IEEE, New York, NY (2019)
79. The Freedom Files. American Civil Liberties Union (2005–2007)
80. Princeton Review, Buffa, L., Goddin, N.: Grammar Smart. In: The Savvy Student's Guide to Perfect Usage, 4th edn. The Princeton Review, New York, NY (2017)
81. The University of Chicago Press Editorial Staff: The Chicago Manual of Style, 18th edn. University of Chicago, Chicago, IL (2024)
82. Stowe, H.B.: Uncle Tom's Cabin. John P. Jewett and Company, Boston, MA (1852)
83. Melville, H.: Moby Dick. Harper & Brothers, New York, NY (1851)
84. Brickhill, P.: The Great Escape. Norton, New York, NY (1950)
85. Tolkien, J.R.R.: The Hobbit. George Allen & Unwin, U.K. (1937)
86. Verne, J.: Twenty Thousand Leagues under the Seas. Pierre-Jules Hetzel, France (1872)
87. Verne, J.: From the Earth to the Moon. Pierre-Jules Hetzel, France (1867)
88. Verne, J.: Journey to the Center of the Earth. Pierre-Jules Hetzel, France (1871)
89. White, E.B.: Charlotte's Web. Harper & Brothers, New York, NY (1952)
90. Edwards, B.: Drawing on the Right Side of the Brain: The Deluxe Edition. Tarcher, New York, NY (2012)
91. Berg, H.: Speed Reading Genius: Student Edition 2022. Berg Learning, McKinney, TX (2022)
92. Trudeau, K.: Kevin Trudeau's Mega Memory: How to Release Your Superpower Memory in 30 Minutes Or Less a Day. HarperCollins, New York, NY (2005)
93. Jones, W.: How many minutes do NHL players play? Hockey Answered. https://hockeyanswered.com/how-many-minutes-do-nhl-players-play/
94. Seward, Z.M.: An average NFL game: more than 100 commercials and just 11 minutes of play. Quartz. https://qz.com/150577/an-average-nfl-game-more-than-100-commercials-and-just-11-minutes-of-play (Jan. 28, 2025)
95. Solo, A.M.G.: Let every child play. Teacher Magazine, p. 4. https://issuu.com/teachernewsmag/docs/2016_sept_teacher/1?ff=true&e=0/38959353 (Sept. 2016)

96. Solo, A.M.G.: Free Libraries in Buildings. Countercurrents. https://countercurrents.org/2025/05/free-libraries-in-buildings/ (May 19, 2025)
97. Solo, A.M.G.: Annual celebratory days for scientific awareness. In: Bishop, J. (ed.) Cases on Technologies in Education from Classroom 2.0 to Society 5.0, pp. 23–31. IGI Global, Hershey, PA (2021)
98. Solo, A.M.G.: Curriculum for a new five-year academic program in intelligent systems engineering and software engineering. In: 2023 International Conference on Computational Science and Computational Intelligence (CSCI 2023), pp. 1696–1703. IEEE, New York, NY (2023)
99. Solo, A.M.G.: Curriculum for a new academic program with bachelor's degrees in intelligent systems engineering and computer engineering. In: Foundations of Computer Science and Frontiers in Education: Computer Science and Computer Engineering. 20th International Conference, FCS 2024, and 20th International Conference, FECS 2024, Held as Part of the World Congress in Computer Science, Computer Engineering and Applied Computing, CSCE 2024, Communications in Computer and Information Science, vol. 2261, pp. 382–395. Springer Nature, New York, NY (2025). https://doi.org/10.1007/978-3-031-85930-4_35
100. Solo, A.M.G.: Curriculum for a new academic program with bachelor's degrees in electrical engineering and intelligent systems engineering. In: Computational Science and Computational Intelligence, 11th International Conference, CSCI 2024, Communications in Computer and Information Science, vol. 2504, pp. 103–116. Springer Nature, New York, NY (2025). https://doi.org/10.1007/978-3-031-94943-2_8

Education - Teaching and Learning Strategies, Curriculum Development, and Tools

Perceptions of University Students on Cognitive Learning Activities in Classrooms to Gain Self-efficacy and Promote Retention

S. C. Spangler(✉) and Shannon W. Beasley

Middle Georgia State University, Macon, GA 31206, USA
scott.spangler@mga.edu

Abstract. This pilot study (n = 30) comprised students from an information technology (IT) class and an accounting class at a southeastern United States state university. The anonymous voluntary instrument was electronically sent out to students in a face-to-face course to understand their perceptions of cognitive teaching strategies by faculty. The overarching research seeks to know if the practices in courses will lead to students' perceptions of gains in self-efficacy and retention. The faculty utilized seven different cognitive learning activities during the course of the two classes. Following the literature, the strategies included guest speakers, music in the classroom, chair yoga, walking lectures, and active learning to promote favorable learning conditions (Hrach, 2021; Wang, 2024; & Lu et al., 2025). The students from both classes reported a perceived increase in self-efficacy ($M = 5.23$) and an increased perception of improved retention ($M = 5.68$). At the same time, evidence was not found to support the assertion that a difference in perception existed between males versus females to change in self-efficacy ($M = 5.54$, $SD = 1.43$), $t(0.27) = -0.56$, two-tailed $p = 0.58$) or retention ($M = 5.65$, $SD = 1.31$), $t(0.27) = 0.44$, two-tailed $p = 0.66$) following the applied cognitive learning activities. This evidence has allowed the researchers to decide that significant promise exists to support a larger study, including a greater sample population across academic disciplines.

Keywords: cognitive learning · active learning · blended learning · self-efficacy · anxiety · retention

1 Introduction

Over the last decade, students have faced unprecedented learning roadblocks physically, emotionally, and cognitively. Roadblocks or negative cognitive associations to a student's learning abilities can be simple, such as feelings or mental well-being, attitudes toward goals, perceptions of capabilities, and even environments. Data demonstrates that motivation, learning abilities, and distractibility are often associated with cognitive blocks or overloads and behaviors. Hence, cognitive control is necessary for students to focus on engaging in classrooms or groups and reaching educational goals.

H. R. Arabnia et al. (Eds.): CSCI 2025, CCIS 2974, pp. 225–238, 2027.
https://doi.org/10.1007/978-3-032-28475-4_15

Cognitive negative variations can be eliminated or suppressed by engaging in classroom activities or by varying the learning environment. Faculty can create positive learning experiences through variations in practices, such as student-popular gamification actions, experimental designs in information displays, and non-traditional classroom configurations (Lu et al., 2025). Simple changes can enhance students' self-efficacy and their interpretations of interpersonal progress. Faculty can make these changes through simple low-stakes changes such as adding natural environments (outside of classroom instruction adventures), sensory-rich experiences (chair yoga and light changes for reduced stress in classrooms), and reducing the fight-or-flight natural phenomena in a human's psyche through object-based learning strategies that can be as simple as students passing objects like beach balls around a classroom, fidge spinner time allowances, puzzle breaks, or music (Hrach, 2021). This paper will seek to understand how positive cognitive associations in courses can reduce stress and increase student efficacy and overall academic retention.

2 Literature

2.1 Cognitive Overload

Students' motivation to learn is multifaceted. Academics (faculty) strive to be creative and supportive throughout a student's learning journey by fostering development and encouraging a growth mindset. This can be viewed as akin to the simple observation that students can focus and learn better when they are not subjected to disruptive stimuli in the classroom. During the academic journey, students often develop anxiety about performance and perceptions of understanding student learning objectives, especially in critical thinking domains. Wang et al. (2024) noted that faculty attempt to reduce the "load of anxiety" brought on by critical analysis or complex problem-solving situations by creating a social learning environment that has been shown to reduce cognitive load and the complexity of higher-order thinking.

Additionally, faculty utilize load-bearing practices such as mindfulness (self-regulation) techniques to quiet students into calmer states of mind or use external learning tools like multimedia sites and other manuscript resources that are more culturally accepted than traditional textbook material. To aid in this reduction of cognitive load, researchers found, in a student perception study (n = 548) of higher-education mathematics students, that participants viewed concepts and strategies that foster positivity in learning as effective calming or stress- and anxiety-reducing agents, such as simply changing environments. Other concepts, such as flipped classrooms–where students become the teachers–empower students and reduce cognitive load on achievement through empowerment, while building self-efficacy (Wild & Neef, 2023). Similarly, researchers from Deer et al. (2024) proposed that, if utilized effectively, such as classical music, music in courses can reduce cognitive load and create psychological benefits in a study of Chinese students (n = 1624). Art, painting, music, and theater are viable practices for reducing cognitive load for students facing cognitive load barriers and academic anxieties.

2.2 Situational Anxiety

Students cognitively understand traditional classroom situations and conditions through cognitive and physical learning from kindergarten through doctoral-level courses. The "status quo" of situational learning in traditional classrooms is static throughout a student's academic journey. Situational cognitive awareness often impairs the learning atmosphere and increases cognitive load and situational anxiety. In their research, Cécillon et al. (2024) found that unfamiliar environments can be perceived as threatening and can create situational anxiety. In the study on French students (n = 292) researchers discovered that students will unconsciously and cognitively attempt to self-regulate stressful environments. Without this self-regulation ability, students will cognitively self-sabotage their executive function abilities and often forecast a premonition of conditional academic catastrophe.

Furthermore, researchers noticed that after the COVID-19 pandemic, associated threats in the environment or around a student's daily life activities, such as the return to a traditional classroom, engaged anxieties and social interactions with faculty or other students, compounded the cognitive overload (Xi-Hui & Tu, 2024). This situational anxiety further decreased students' self-efficacy and increased academic procrastination. This also led to a decline in students' critical thinking skills (Liline et al., 2024). This new research overpowered earlier considerations of cognitive load due to the removal of traditional spatial comforts in academic situational learning and fears of technology adoption (Luan & Teo, 2009; Murray, 2008; Ormerod & Rosewell, 2009; and Schulte, 2010). Hence, the long-standing view in the literature is that physical environments and interactions with people or content must be eliminated through faculty mindfulness practices to reduce overall cognitive load and anxiety in modern courses (Wang, 2024). Similarly, Spangler et al. (2024) found that over 86% of study participants found the engagement strategies helpful and enhanced learning experiences, with 60% (Completely agreeable) (Fig. 1).

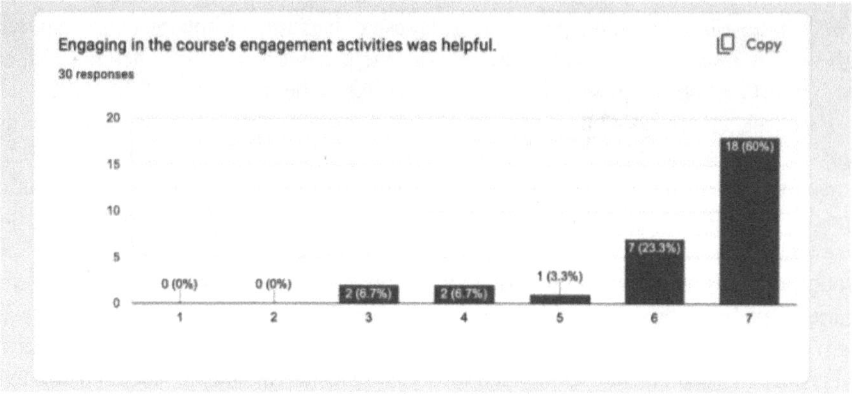

Fig. 1: Engagement Activities in a course help enhance learning.

2.3 Environmental Effects on Learning

Cognitive aspects of the environment impact students' relationships with academics and learning. In a 2022 study of 224 Southwestern University students, researchers found that when faculty consider students' welfare and build personal connections or provide guidance, student engagement rose to the highest levels, and self-efficacy increased (Waldbuesser & van Raalte, 2023). When harmful cognitive elements are in the environment, students' motivation and self-efficacy about their learning abilities can decrease. In a 2023 study of 116 English students, participants noted that they value course learning activities when they serve as a positive cognitive aspect or mechanism. The overall effect was reported to enhance learning and create an essence of achievement, foster enjoyable critical learning objectives, and overall self-satisfaction (Gasmi & Al Nadabi, 2023). Activities such as Socratic learning circles, in which students express interpersonal knowledge or feelings about a subject, increased students' cognitive associations with learning and connected them to academic success (Owusu-Agyeman & Moroeroe, 2023). The researchers' findings on 27 faculty members and 51 freshman students about cognitive learning experiences or models found that students had enhanced cognitive development and an overall positive learning experience.

Environments have long been understood to have cognitive advantages or disadvantages for learning. Cécillon et al.'s (2024) research notes that cognitive load from perceived threats or anxieties about entering an academic environment can hinder learning. Students' learning abilities can easily be impeded or interfered with by dynamic and straightforward domain constructs such as disruptive sounds, color, and environmental anxieties; spatial constructs like claustrophobic learning atmospheres; and limited-sight constructs, including learning materials. In these cases, cognitive executive functions can be limited or impaired, leading students to hold negative cognitive beliefs and be unable to self-regulate emotional strategies that affect their academic achievements. The mental associations negatively affect educational attainment and diminish overall student self-efficacy. With the advent of artificial intelligence, new research is beginning to show that AI bots can be effective learning tools and can help reduce students' learning anxiety in the digital culture. Wang's (2024) research examines how providing students with tools is essential to provide learners with the necessary support when needed, and that support that helps students learn how to learn will be more helpful (p. 214).

However, researchers have uncovered simple aspects that can easily increase students' cognitive and overall self-efficacy growth in online and face-to-face student courses by instituting actions like adding smiling faces to course shells for student support, embedded tutors, or embedded librarians or other technologies (Ormerod & Rosewell, 2009; Spangler, 2019; & Spangler et al., 2024). Students build self-efficacy through the programs by adding embedded support in course shells in academic environments or by allowing outside support into face-to-face courses, thereby fostering positive learning considerations. Spangler's (2019) research uncovered that supportive images in course shells and embedded tutors or librarians alone provide comfort and a sense of safety for students (Schulte, 2010) in learning environments. Similarly, Hrach's (2021) research found that positive constructs in the learning environment help students develop new habits and become more curious about learning. Similarly, Sun et al. (2023) uncovered that cognitive approaches or strategies in pedagogy enhanced students' overall learning capabilities and engagement. Early debates on the use of artificial intelligence as a teaching strategy are clogged with wavering assertions about levels of acceptance. Artificial intelligence (AI) refers to "computer systems designed to perform tasks that usually require human intelligence" (What's in AI, 2025, para. 2). AI can allow students through unencumbered cognitive strains to ask anxiety-inducing questions, receive feedback, create learning instruments, and understand their weaknesses academically without the complications of interpersonal communication complexities or judgments. For example, students who struggle to understand the complexity of solving mathematical equations can receive tailored learning or practice simulations to "ensure they grasp the concept before moving on" (para. 2–3). Spangler et al. (2024) note many similarities in the literature. The researchers agree with Hrach (2021) that cognitive strategies in classrooms can enhance learners' ability to increase concentration and motivation, as well as reduce cognitive load and anxiety, by adding simple movement and/or physical sensation measures to classrooms. The research notes the need for studies in artificial intelligence on cognitive enhancements, but "AI-assisted or elaborate forms of gamification" are not needed to increase motivation; simple cognitive reduction activities, such as yoga breaks, walking lectures, and guest speakers, are perceived by participants as having significant value (p. 479). The researchers' findings on the cognitive strategies in low-tech methodologies and movement-focused increased participant motivation, attention, and reduced cognitive loads to over 80% in most cases, agreeing with the literature (Owusu-Agyeman & Moroeroe, 2023; Sun et al., 2023; & Waldbuesser & van Raalte, 2023).

2.4 Self-Efficacy

Post-COVID-19 studies have reflected reduced student school life and disruptions in educational and social interactions. This reduction in interpersonal and social interactions in the academic sphere and regular social domains has increased students' anxieties and created a noticeable reduction in self-efficacy among students and faculty. The decrease in interpersonal skills and efficacy was said to increase student procrastination and reduce their critical thinking skills (Liline et al., 2024; Xi-Hui & Tu, 2024). Cécillon et al. (2024) noted that any situational or perceived threats in school environments (Waldbuesser & van Raalte, 2023) can decrease a student's cognitive abilities and interfere with their memory abilities, ultimately interfering with executive functions, and manifesting behavioral issues that ultimately forecast a "negative role in academic achievement" (p. 2).

Self-efficacy is a personal level of confidence in the belief that the individual can accomplish a challenge based on their own skill set or through trusted tools like embedded librarians or tutoring programs (Spangler, 2019). The confidence in one's abilities stems from past experiences, cultural or social attitudes, or an overall success mindset or level of motivation. The experience does not require direct experience with the challenge, but similar activities can have similar effects. Success in previous activities and a general attitude can effectively build or enhance personal confidence or sense of self-efficacy when completing a challenging task (Wang et al., 2024). The same holds for an inflated sense of personal success when tackling challenges with familiar or similar tools. This sense of confidence in items in one's toolbox reduces anxiety during a project and increases the individual's motivation to move toward a solution (Yildrum & Ilik, 2025). An interesting consequence of self-efficacy is that peers with higher self-efficacy positively influence group members with lower self-efficacy.

Classroom culture and change in methodologies of transference of information can create a reduction in students' motivation and self-efficacy, such as upgrades to technology or populations that suffer from limited technology in rural and underprivileged communities (Murray, 2008; Luan & Teo, 2009; Ormerod & Rosewell, 2009; Schulte, 2010; Yakup, 2025). This also comes true with cognitive overloads in students' learning atmospheres, whereby extensive multimedia or problem-based research practices induce anxiety. Wang's (2024) research suggested that simple modifications and reductions in cognitive load can enhance the academic experience, especially when tied to Socratic learning methodologies, helping maintain students' critical thinking abilities through community or team learning constructs. Other researchers concluded that adding arts or music to environments or artistic cognitive actions has a positive correlation between student engagement and increases in students' confidence, satisfaction, self-efficacy, and academic success (Gasmi & Al Nadabi, 2023; Owusu-Agyeman & Moroeroe, 2023; Wild & Neef, 2023; & Deer et al., 2024; Wang et al., 2024).

2.5 Retention

Retention is defined by achieving two goals: reducing attrition and promoting persistence. The first goal is to reduce or mitigate factors that lead to attrition. Attrition is evidenced by class failure and inability to progress through a prescribed study plan (Sheikoleslami et al., 2025). The second goal focuses on promoting persistence in attitude and action for the student. This sense of persistence or dedication reflects a drive to make progress toward graduation. Many students do not understand the intricacies of retention as a process (Cavagnoud & Ames, 2024). However, the literature notes that most students can evaluate whether a class was comfortable, understandable, informative, and the workload was achievable (Liline et al., 2024). While it may not be possible to manipulate retention values directly, introducing cognitive teaching strategies that promote favorable perceptions of the classroom experience and reduce undesirable conditions can have a net effect or condition towards a positive academic outcome. When faculty effectively reduce negative classroom connotations, such as anxiety and fear of failure, while promoting positive ones, such as self-efficacy and the desire to succeed, retention rates tend to rise and self-efficacy levels are elevated (Lin et al., 2025). Many students suffer from social and personal barriers that are difficult to understand or identify. These barriers are more susceptible to modification outside of academia's boundaries by applying favorable cognitive teaching strategies that encourage a growth mindset (Cavagnoud & Ames, 2024). Given the consistent timeline required to achieve retention goals, a path to increasing retention begins by consistently removing barriers and strengthening positive cognitive attributes.

2.6 Movement Affects Learning

In *Minding Bodies*, Susan Hrach's research in the cognitive sciences focused on understanding how bodily movement and environmental constructions distract or enhance students' learning in the classroom. The scholar notes various strategies and techniques to increase student engagement and retention for higher-education learners. Hrach's cognitive strategies suggest improving classroom experiences, engagement, reducing anxiety, building self-efficacy, and increasing student retention by incorporating treatments such as motion, energy efficiency, exercise, group activities, experiential learning, and rewards. Physical conditions such as chair yoga, stretching, or walking lectures can reduce anxiety and depression while increasing concentration through creative actions and new focuses (Hansen, 2016, 200–201). Learning demands high energy, and our brains try to conserve it by drawing on prior experience. Professors should recognize that students may have low energy due to emotional, social, or physical stress, which has been shown to reduce attention and overall engagement. Students' cognitive function depends on their bodily well-being; a student's overall physical health, including nutrition, stress, and sleep, impacts their cognitive function when working with others or engaged in classroom activities. Hrach recommends seeking activities that engage the senses, such as music that can cognitively inspire relaxation and stress reduction, or natural light that can increase creativity. Walking, lectures, and other natural or outdoor activities can release endorphins and increase pleasure-related activity in the brain for learning. The following blended learning strategies—music, guest speakers, walking lectures, classroom yoga, and other variations on traditional classroom environmental conditions—were used to increase self-efficacy, reduce anxiety, and improve retention, as expressed in previous research from Spangler et al. (2024).

3 Methodology

Middle Georgia State University's Institutional Review Board approved the study. The researchers sent the instrument post-course to multiple higher-education courses in information technology and business schools. During the semester, both classes were exposed to cognitive teaching strategies, including music, guest speakers, walking, classroom yoga, and other variations of traditional classroom conditions. All voluntary participants signed a consent form before entering the instrument. All student participants' identities were collected anonymously. Students received a voluntary online Google Forms survey instrument through their university email or the University System of Georgia's LMS platform BrightSpace. Potential respondents were also informed that no incentives were offered for participation in the study. All voluntary participants were 18 years of age or older.

While this pilot study primarily focuses on assessing the relevance and effectiveness of cognitive teaching strategies in a blended learning environment, it also provides insight into preferences expressed by groups within the sample population. The population sizes in the two classes differed slightly. Still, the distribution was representative of typical sophomores and juniors enrolled in these classes at a Southeastern United States university under consideration. The total number of respondents used in the study was 30 students ($n = 30$). Google Forms and SPSS Software were used to analyze the data and perform a t-test on the independent variables of Gender and academic program association. Similar to Spangler et al. (2024), the study focused on seven dependent variables: engagement activity, walking lectures, events, yoga, music, speakers, and active learning.

Additionally, the study follows Spangler et al.'s methodology with permission. The researchers, noting its strong constructs that are formed under Teresi et al. (2022) study utilizing small populations, "make statistical analysis for non-generalizable conclusions, to form feasibility and exploratory informative studies, and to understand the potential need for future larger studies through its inference potential as noted in a study on research constructs in a National Science Foundation publication" (p. 470). Additionally, the researchers adopted the Spangler et al. (2024) instrument, which was first adopted and adapted with approval from the scholars in Spangler and Shah (2022). The instrument framed questions 1–6 as demographic information. A seven-point Likert scale system was framed in questions 7–13, ranging from (1) "completely disagree" to (7) "completely agree" to measure student perceptions of the cognitive learning strategies.

The survey instrument, containing 42 questions, was administered to all students/respondents who signed a waiver acknowledging that they were 18 years of age and expressing consent to participate in the study. The questionnaire contained six demographic questions and 36 questions asking respondents to express their opinions on the success or failure of various strategies provided during the study. The data was struck for incomplete or unfinished surveys. The data for two users who do not wish to declare a concentration or prefer not to say are removed. To understand the data and answer RQ1 and RQ2, the researchers used SPSS to analyze the mean values. To address RQ3 and RQ4, the researchers ran Levene's Test for Equality of Variances and a t-test to assess differences in means.

The pilot study seeks to clarify and extend the findings in Spangler et al.'s (2024) publication to uncover the perceived effects of cognitive learning strategies. The paper focuses on uncovering participants' perceptions of strategies for enhancing their self-efficacy and retention in higher-education academic environments. Directly, the research seeks to understand and answer four research questions:

3.1 Research Questions

RQ1: Do students perceive a gain in self-efficacy from cognitive practices in classrooms?
RQ2: Do students perceive an increased desire to continue their education as a result of cognitive strategies in the classroom?o
RQ3: Does gender affect students' perceptions of increased self-efficacy from cognitive practices in the classroom?
RQ4: Does gender affect students' perceptions of increased retention from cognitive practices in the classroom?

4 Results

The respondents who participated were identified as 19 (63.3%) males, 10 (33.3%) females, and 1 (3.3%) preferred not to answer. The distribution of concentrations reported was four Information Technology (IT) (varied concentrations), 3 General Business, 8 (26.7%) Accounting, 8 (26.7%) Management, 5 (16.7%) Marketing, 1 (3.3%) Other, and 1 (3.3%) Prefer not to say. The data for two users who do not wish to declare a concentration or prefer not to say are removed. The data construction left four (13.3%)

IT majors and 24 (80%) business majors in the pool of respondents. While only 1 (3.3%) respondent reported that they were the first person in their family to attend college, only 12 (40%) respondents reported having three or more family members who have attended college.

As seen in Table 1, all participants were 18 years of age or older that participated in the study, and reported the following dispersion in groups: 11 (36.7%) at 18–20 years of age, 12 (40.0%) at 21–25 years of age, 5 (16.7%) at 26–31 years of age, and 2 (6.7%) at 32–39 years of age. Of the 30 respondents surveyed, only 2 (6.7%) reported receiving learning support in high school. The same 2 (6.7%) respondents confirmed receiving learning support at the university.

Table 1. Demographic Information for Blended Learning Study.

Gender	
Male	19 (63.3%)
Female	10 (33.3%)
Prefer not to say	1 (3.3%)
Concentrations	
IT (varied concentrations)	4 (13.3%)
General Business	3 (10.0%)
Accounting	8 (26.7%)
Management	8 (26.7%)
Marketing	5 (16.7%)
Other	1 (3.3%)
Prefer not to say	1 (3.3%)
Age	
18–20 years of age	11 (36.7%)
21–25 years of age	12 (40.0%)
26–31 years of age	5 (16.7%)
32–39 years of age	2 (6.7%)
High School Learning Support	
Yes	2 (6.7%)
No	28 (93.3%)
University Learning Support	
Yes	2 (6.7%)
No	28 (93.3%)

5 Results of Statistical Analysis

RQ1: Do students perceive a gain in self-efficacy from cognitive practices in classrooms?

Table 2. Statistics report on the effect of strategies towards gaining self-efficacy.

Descriptive Statistics

	N	Minimum	Maximum	Mean	Std. Deviation
Effect on self-efficacy	30	2.43	7.00	5.2381	1.50268
Valid N (listwise)	30				

To answer RQ1, the researchers examined the mean and frequency values of the self-efficacy variable. As shown in Table 2, the self-efficacy data were computed by aggregating responses to seven Likert-type questions. The test addresses participants' perceptions of the impact of cognitive strategies on students' self-efficacy. RQ1 suggests that cognitive strategy will positively affect self-efficacy perceptions. The responses to the self-efficacy questions are broken into three categories: 1–3 (strongly disagree, disagree, slightly disagree), 4 neutral or no opinion, and 5–7 (strongly agree, agree, slightly agree). The data demonstrate that the cognitive teaching strategies are perceived as successful. The mean is greater than 4.0 (M = 5.23).

RQ2: Do students perceive an increased desire to continue their education as a result of cognitive strategies in the classroom?

Table 3. Statistics report of the strategies on retention.

Descriptive Statistics

	N	Minimum	Maximum	Mean	Std. Deviation
Retention	30	2.00	7.00	5.6833	1.53400
Valid N (listwise)	30				

To answer RQ2, the researchers examined the mean and frequency values of the self-efficacy variable. As shown in Table 3, the self-efficacy variable was reestimated by combining the results of two Likert-type questions. The test examines the effect of cognitive teaching strategies on students' perceptions of retention. RQ2 suggests that cognitive strategy will positively affect perceived retention. The responses to the self-efficacy questions are broken into three categories: 1–3 (strongly disagree, disagree, slightly disagree), 4 (neutral or no opinion), and 5–7 (strongly agree, agree, slightly agree). The data demonstrate that the cognitive teaching strategies are perceived as successful. The mean is greater than 4.0 (M = 5.68).

RQ3: Does gender affect students' perceptions of increased self-efficacy from cognitive practices in the classroom?

Table 4. Independent sample t-test (Gender, Self-efficacy).

Independent Samples Test

		Levene's Test for Equality of Variances		t-test for Equality of Means							
						Significance				95% Confidence Interval of the Difference	
		F	Sig.	t	df	One-Sided p	Two-Sided p	Mean Difference	Std. Error Difference	Lower	Upper
Effect on self-efficacy	Equal variances assumed	.051	.824	-.561	27	.290	.580	-.31729	.56594	-1.47851	.84392
	Equal variances not assumed			-.564	18.766	.290	.579	-.31729	.56212	-1.49481	.86022

Table 5. Comparison of opinions of the effects on self-efficacy (Gender, Self-efficacy).

Group Statistics

	Gender	N	Mean	Std. Deviation	Std. Error Mean
Effect on self-efficacy	male	19	5.2256	1.45866	.33464
	female	10	5.5429	1.42825	.45165

To answer RQ3, the researchers utilized an independent samples t-test (Table 4) to compare the data from the two groups, Males and Females. The findings were based on a significance level of $p = 0.05$. It was determined that there was no significant difference (two-sided $p = 0.580$) between the Males ($M = 5.23$, $SD = 1.46$) and Females ($M = 5.54$, $SD = 1.43$), $t(0.27) = -0.56$, two-tailed $p = 0.58$ as seen in Tables 4 and 5.

RQ4: Does gender affect students' perceptions of increased retention from cognitive practices in the classroom?

Table 6. Independent sample t-test (Gender, Retention)

Independent Samples Test

		Levene's Test for Equality of Variances		t-test for Equality of Means							
						Significance				95% Confidence Interval of the Difference	
		F	Sig.	t	df	One-Sided p	Two-Sided p	Mean Difference	Std. Error Difference	Lower	Upper
Retention	Equal variances assumed	.008	.928	.444	27	.330	.661	.24474	.55156	-.88696	1.37644
	Equal variances not assumed			.459	20.216	.326	.651	.24474	.53335	-.86704	1.35652

Table 7. Comparison of opinions of strategies' effects on self-efficacy (Gender, Retention)

Group Statistics

	Gender	N	Mean	Std. Deviation	Std. Error Mean
Retention	male	19	5.8947	1.45849	.33460
	female	10	5.6500	1.31339	.41533

To answer RQ4, the researchers used an independent-samples t-test (Table 6) to compare data from the Male and Female groups. The findings were based on a significance level of $p = 0.05$. It was determined that there was no significant difference (two-sided p = 0.66) between the Males (M = 5.89, SD = 1.46) and Females (M = 5.65, SD = 1.31), t(0.27) = 0.44, two-tailed p = 0.66, as seen in Tables 6 and 7.

6 Discussion

The overall opinion of volunteer participant students surveyed finds that cognitive teaching strategies positively affect perceived self-efficacy. The researchers can accept RQ1 and RQ2. In the case of RQ2, students' overall opinion is that classroom strategies improve their perception of the courses and create a desire to take additional classes. RQ1 and RQ2 are evidenced by the fact that the variables Self-efficacy and Retention have mean values well into the agreeable range. Evidence in the data construction through the Likert scale measurements concludes that a value greater than four indicates the tendency of perceived agreement, which is similar to past literature findings on self-efficacy and cognitive improvements in students when adding positive activities to classrooms (Gasmi & Al Nadabi, 2023; Owusu-Agyeman & Moroeroe, 2023; Wild & Neef, 2023; Deer et al., 2024). In contrast, a mean value below 4 tends to indicate disagreement. In the cases of RQ1 and RQ2, the researchers note that they cannot draw direct conclusions from an opinion of neutrality. These findings are consistent with the literature on constructs (Hrach, 2021; Spangler et al., 2024).

In the cases of RQ3 and RQ4, there is no significant evidence to assert that there is no difference between the expressed opinions of males and females. RQ3 does not support rejecting the hypothesis that there is no significant difference between the expressed perceived opinion of males and females related to the elevation of self-efficacy from applying strategies. The net opinion is favorable as established by a mean average greater than the midpoint of the Likert scale (4.0). Additionally, RQ4 does not support rejecting the hypothesis that there is no significant difference between the expressed perceived opinion of males and females related to the elevation of retention resulting from applying strategies. The net opinion is favorable as established by a mean average greater than the midpoint of the Likert scale (4.0). We define retention from the student's perceived opinion as a desire to continue completing a degree program, as evidenced by continued matriculation. Overall, all strategies should be continued into a further study with a larger sample size (specifically, a larger representation of different majors/concentrations). However, the small population demonstrates an encouraging need for further research into the literature gap.

7 Conclusions and Furthering the Research

Overall, the researchers uncovered no significant difference in perceptions between male and female identified students in the study. However, the pilot study suggests that students perceive cognitive learning strategies in classrooms by faculty members having a positive perceived value towards self-efficacy and creating a positive regard towards students staying in a course. The researchers could not conclude that there was any difference in gender perceptions or their merits towards increasing the overall academic retention of a student body. Because the pilot study had a limited population (n = 30) and the geographical situation of the population was limited in scope, the researchers can not conclude any generalizable findings. However, the study demonstrates the need for further research in cognitive strategies for increasing student self-efficacy and retention.

The researchers conclude that more testing is needed on classroom cognitive strategies to further the study. Additionally, the limited scope of information technology and business students must be diversified into a large population of academic domains. The researchers suggest a larger cross-population or national study to eliminate geographical and faculty biases. The literature findings do not reflect the pilot study's findings outside of the perceived understanding of value towards building self-efficacy in a student. Additionally, the research note the need to further examine strategies and new technology additions like the untested results of artificial intelligence bots utilized in the study but not formally tested as the new frontier of cognitive load disruption.

Acknowledgments. Project Sponsored: This study was supported by the National Science Foundation Award #2433182.

References

Cavagnoud, R., Ames, P.: Persistence, retention, and interruption in higher education among Peruvian scholarship students in a context of emergency remote learning. Educ. Policy Announc. Arch. **32**(48) (2024). https://doi.org/10.14507/epaa.32.8476

Cécillon, F.X., Mermillod, M., Leys, C., Lachaux, J.P., Le Vigouroux, S., Shankland, R.: Trait anxiety, emotion regulation, and metacognitive beliefs: an observational study incorporating separate network and correlation analyses to examine associations with executive functions and academic achievement. Children. **11**(1), 123 (2024)

Deer, G. et al.: Effect of out-of-school visual art activities on academic performance. The mediating role of socioeconomic status. PLoS One. **19**(5) (2024). https://doi.org/10.1371/journal.pone.0298901

Gasmi, A.A., Al Nadabi, Z.S.: An exploratory study of Learners' perceptions about the effectiveness of active learning approaches. J. Lang. Teach. Res. **14**(5), 1223–1232 (2023). https://doi.org/10.17507/jltr.1405.10

Hrach, S.: Minding Bodies How Physical Space, Sensation, and Movement Affect Learning. West Virginia University Press (2021)

Liline, S., Tomhisa, A., Rumahlatu, D., Sangur, K.: The effect of the Pjb-HOTS learning model on cognitive learning, analytical thinking skills, creative thinking skills, and metacognitive skills of biology education students. J. Turk. Sci. Educ. **21**(1), 175–195 (2024). https://doi.org/10.36681/tused.2024.010

Lin, H., Tseng, C., Chen, N.: Enhancing programming information: the impact of AI-based pedagogical agents on self-efficacy, engagement, and learning outcomes. Educ. Technol. Soc. **28**(2), 279–294 (2025). https://doi.org/10.30191/ETS.202504_28(2).TP02

Lu, H., Van der Linden, D., Bakker, A.B.: The neuroscientific basis of flow: learning progress guides task engagement and cognitive control. NeuroImage. **308**, 121076 (2025)

Luan, W., Teo, T.: Investigating the technology acceptance among student teachers in Malaysia: an application of the technology acceptance model (TAM). Asia Pac. Educ. Res. **18**(2), 261–272 (2009)

Murray, G.: On the cutting edge (of torpor): innovation and the pace of change in American higher education. AACE J. **16**(1), 47–61 (2008)

Ormerod, P., Rosewell, B.: Innovation, diffusion, and agglomeration. Econ. Innov. New Technol. **18**(7), 695–706 (2009). https://doi.org/10.1080/10438590802564659

Owusu-Agyeman, Y., Moroeroe, E.M.: Relationality and student engagement in higher education: towards enhanced students' learning experiences. Int. J. Emot. Educ. **15**(2), 37–53 (2023). https://doi.org/10.56300/ZANL1419. 480 Issues in Information Systems Volume 25, Issue 4, pp. 466-481, 2024

Schulte, M.: Faculty perceptions of technology distance education transactions: qualitative outcomes to inform teaching practices. J. Educ. Online. **7**(2), 1–34 (2010)

Sheikoleslami, R., Princeton, D., Hansen, L., Kisa, S., Goyal, A.: Examining factors associated with attrition, strategies for retention among undergraduate nursing students, and identified research gaps: a scoping review. Nurs. Rep. **15**(182) (2025). https://doi.org/10.3390/nursrep15060182

Spangler, S.: Integrating information literacy in IT courses: information technology students' perceptions of embedded librarians. Online J. Appl. Knowl. Manag. (OJAKM). **7**(2), 29–40 (2019)

Spangler, S., Beasley, S., Floyd, T.: Perceptions of university students on cognitive learning activities in classrooms. Issues Inform. Syst. **25**(4), 466–481 (2024). https://doi.org/10.48009/4_iis_2024_136

Spangler, S., Shah, K.: Students' perception of assurance, utilization of a pilot program: synchronized online discussions. Issues Inform. Syst. **23**(4) (2022)

Sun, Y., Zhao, X., Li, X., Yu, F.: Effectiveness of the flipped classroom on self-efficacy among students: a meta-analysis. Cogent Educ. **10**(2) (2023). https://doi.org/10.1080/2331186X.2023.2287886

Teresi, J.A., Yu, X., Stewart, A.L., Hays, R.D.: Guidelines for designing and evaluating feasibility pilot studies. Med. Care. **60**(1), 95–103 (2022). https://doi.org/10.1097/MLR.0000000000001664

Waldbuesser, C., van Raalte, L.J.: Teacher affection and immediacy in college classrooms: predicting student engagement. Educ. Res. Q. **47**(1), 3–26 (2023). https://www.proquest.com/scholarly-journals/teacher-affection-immediacy-college-classrooms/docview/2863933776/se-2

Wang, M.: Rethinking how people learn: a holistic framework for effective learning design. Knowl. Manag. E-Learn. **16**(2), 208–216 (2024). https://doi.org/10.34105/j.kmel.2024.16.010

Wang, M., Liu, K., Zhou, Y., Xuang, X.: Who did what to whom and how: a scoping review of interventions to promote children's self-efficacy in mainland China. Child Fam. Soc. Work. **30**, 524–539 (2024). https://doi.org/10.1111/cfs.13169

What's in AI: Can AI help you learn faster? Study tips from the future. https://www.whatsinai.com/working-with-ai/can-ai-help-you-learn-faster-study-tips-from-the-future?lctg=638ffb7c94143b224d8ba61b (2025)

Wild, S., Neef, C.: Analyzing the associations between motivation and academic performance via the mediator variables of specific mathematic cognitive learning strategies in different subject domains of higher education. Int. J. STEM Educ. **10**(1), 32 (2023). https://doi.org/10.1186/s40594-023-00423-w

Xi-Hui, J., Tu, J.-C.: Towards a new conceptual model of ai-enhanced learning for college students: the roles of artificial intelligence capabilities, general self-efficacy, learning, motivation, and critical thinking awareness. Systems. **12**(3), 74 (2024). https://doi.org/10.3390/systems12030074481

Yakup, I.: The mediating role of self-efficacy, social support, and hope in the relation between mental health and resilience among 2023 Turkeyie earthquake survivors. Death Stud. **49**(6), 735–742 (2025). https://doi.org/10.1080/07481187.2024.2355240

Yildrum, N., Ilik, S.: Self-efficacy of teachers in special education schools for teaching through play. Int. J. Assess. Tools Educ. **12**(1), 1–19 (2025). https://doi.org/10.21449/ijate.1468431

Do Students Perceive the Benefits of E-Learning Higher When They Feel Mentally Challenged?

Hyeon Jo(✉)

Department of Strategic Planning, RealSecu, Busan, Republic of Korea
sineoriz@gmail.com

Abstract. Since the emergence of COVID-19, university education has largely shifted to online modalities. Students' perceptions of e-learning benefits may be influenced by their risk perception of COVID-19 and the mental discomfort arising from prolonged isolation. This study aims to identify the key determinants of perceived e-learning benefits. Data were collected from 160 university students in South Korea through a structured survey. Partial least squares structural equation modeling (PLS-SEM) was employed to analyze the data. The results indicate that perceived benefits are positively influenced by affective risk perception, whereas cabin fever syndrome exerts a significant negative effect.

Keywords: e-learning · benefits · mental discomfort · risk perception · cabin fever syndrome

1 Introduction

In response to the COVID-19 pandemic, educational institutions worldwide rapidly adopted online learning methods [1, 2]. University students have engaged in e-learning in accordance with government policies and institutional guidelines [3]. However, the perceived benefits of e-learning may vary depending on students' psychological states. Under normal circumstances, students may not perceive e-learning as more beneficial than traditional face-to-face instruction. In contrast, during emergency situations characterized by restricted social interaction, the perceived benefits of e-learning may increase, as face-to-face education becomes less feasible. From this perspective, the present study examines how students' psychological conditions influence their perceptions of e-learning benefits.

Following the outbreak of COVID-19, media outlets have continuously reported confirmed cases, fatalities, and related developments [4]. As a result, students are consistently exposed to information about the risks associated with the virus. Risk perception can be conceptualized in both affective and cognitive dimensions. Students with higher perceived risk may experience a stronger need for preventive behaviors, such as social distancing and participation in e-learning. Accordingly, risk perception is expected to significantly influence perceptions of e-learning benefits.

Moreover, social distancing measures have substantially reduced individuals' external activities, leading to extended periods spent at home. Under such conditions, some

H. R. Arabnia et al. (Eds.): CSCI 2025, CCIS 2974, pp. 239–246, 2027.
https://doi.org/10.1007/978-3-032-28475-4_16

individuals experience cabin fever syndrome, defined as psychological discomfort resulting from prolonged confinement [5]. The extent of this condition may alter how students perceive the importance and benefits of e-learning.

In summary, both risk perception and isolation-induced psychological discomfort are likely to shape perceptions of e-learning benefits. Therefore, this study proposes that risk perception and cabin fever syndrome significantly influence the perceived benefits of e-learning.

2 Related Work and Research Model

Figure 1 presents the analytical framework for examining the determinants of perceived e-learning benefits. This study posits that perceived benefits are significantly influenced by affective risk perception, cognitive risk perception, and cabin fever syndrome.

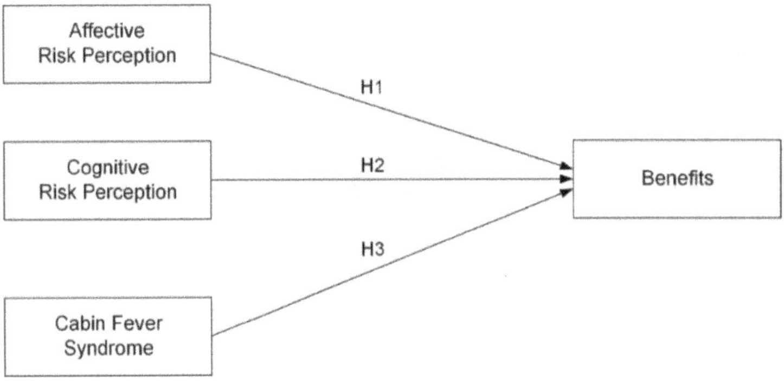

Fig. 1. Research model

2.1 Affective Risk Perception

Affective risk perception refers to an individual's emotional response, such as worry or fear, regarding exposure to risk [6]. Prior research suggests that affective risk perception significantly influences individuals' intentions to engage in preventive health behaviors [7]. In the context of a public health crisis, e-learning serves as a primary alternative to traditional education. Students who experience stronger emotional concern about COVID-19 may place greater value on the benefits of e-learning. Therefore, affective risk perception is expected to positively influence perceived e-learning benefits.

Hypothesis H1. Affective risk perception is positively related to perceived benefits.

2.2 Cognitive Risk Perception

Cognitive risk perception refers to an individual's assessment of the severity of and susceptibility to a given risk [6]. It has been shown to influence subjective norms and behavioral intentions in contexts such as contactless tourism [7]. When students perceive COVID-19 as highly severe or threatening, they may consider face-to-face classes impractical and instead recognize greater value in e-learning. Accordingly, cognitive risk perception is proposed to significantly affect perceived e-learning benefits.

Hypothesis H2. Cognitive risk perception is positively related to perceived benefits.

2.3 Cabin Fever Syndrome

Cabin fever syndrome is characterized by psychological distress, including frustration and restlessness, resulting from prolonged physical isolation [8]. Individuals experiencing higher levels of cabin fever may perceive greater psychological strain and seek alternative forms of engagement. In this context, e-learning may serve as a constructive outlet, potentially enhancing its perceived benefits. Thus, cabin fever syndrome is expected to significantly influence perceived e-learning benefits.

Hypothesis H3. Cabin fever syndrome is positively related to perceived benefits.

3 Methodology

3.1 Measurement Instrument

This study surveyed university students to investigate the antecedents of perceived e-learning benefits. All measurement items were adapted from established literature and modified to suit the context of e-learning. Each item was measured using a 7-point Likert scale ranging from 1 to 7. A pilot study was conducted to refine the wording, content, and structure of the survey instrument. Table 1 presents the measurement items and their corresponding sources.

Table 1 Measurement Items

Construct	Item	Description
Benefits [15]	BNT1	Using e-learning has increased my knowledge and contributed to my academic success.
	BNT2	E-learning is an effective educational tool that has improved my learning process.
	BNT3	E-learning has helped me achieve my learning goals.
Affective Risk Perception [18]	ARP1	I am worried that I may contract COVID-19.
	ARP2	I am worried that my family members may contract COVID-19.
	ARP3	I am worried about the spread of COVID-19 in my region.
Cognitive Risk Perception [18]	CRP1	There is a high likelihood of contracting COVID-19 in general.

(continued)

Table 1 (*continued*)

Construct	Item	Description
	CRP2	I am more likely to contract COVID-19 than other people.
	CRP3	The likelihood of contracting COVID-19 is higher than that of other diseases.
Cabin Fever Syndrome [8]	CFS1	I have difficulty concentrating while staying at home during social distancing or lockdown.
	CFS2	I experience increased food cravings while staying at home during social distancing or lockdown.

3.2 Subject and Data Collection

Data were collected from university students in South Korea. Several professors facilitated the data collection process by encouraging students to participate in the survey. After excluding incomplete and insincere responses, a total of 160 valid responses were retained for the empirical analysis. Among the respondents, 70 (43.8%) were male and 90 (56.3%) were female. The majority of participants (n = 97, 60.6%) were between the ages of 20 and 23. Detailed demographic characteristics of the sample are presented in Table 2.

Table 2 Demographic Information of Sample

Demographics	Item	Subjects ($N = 160$)	
		Frequency	Percentage (%)
Gender	Male	70	43.8
	Female	90	56.3
Age	19 or younger	34	21.3
	20–23	97	60.6
	24 or older	29	18.1

4 Analysis Results

This study employed structural equation modeling (SEM) to assess both the measurement and structural models. Specifically, the analysis was conducted using the partial least squares (PLS) approach. PLS-SEM has been widely adopted due to its robustness and minimal distributional assumptions [9]. The analysis was performed using SmartPLS 3.0 [10].

4.1 Measurement Model

The measurement model was evaluated in terms of reliability, convergent validity, and discriminant validity. Reliability was assessed using Cronbach's alpha and composite reliability (CR). Values exceeding the recommended threshold of 0.70 indicate satisfactory reliability [11]. As shown in Table 3, all CR values exceeded 0.70. Although the Cronbach's alpha for cabin fever syndrome (0.697) was slightly below the recommended threshold, the construct was retained because its CR (0.820) and average variance extracted (AVE = 0.703) exceeded acceptable levels.

Convergent validity was assessed by examining AVE values and factor loadings. The AVE values ranged from 0.703 to 0.889, exceeding the recommended cutoff of 0.50 [11]. Factor loadings ranged from 0.655 to 0.989, indicating adequate convergent validity [12].

Table 3 Test Results of Reliability and Convergent Validity

Construct	Items	Mean	St. Dev.	Factor Loading	Cronbach's Alpha	CR	AVE
Affective Risk Perception	ARP1	4.781	1.903	0.922	0.902	0.938	0.836
	ARP2	5.475	1.654	0.908			
	ARP3	4.938	1.767	0.913			
Cognitive Risk Perception	CRP1	4.500	1.696	0.892	0.793	0.877	0.705
	CRP2	3.219	1.615	0.734			
	CRP3	4.338	1.785	0.883			
Cabin Fever Syndrome	CFS1	3.838	2.088	0.989	0.697	0.820	0.703
	CFS2	3.831	1.921	0.655			
Benefits	BNF1	4.869	1.628	0.936	0.938	0.960	0.889
	BNF2	4.944	1.693	0.959			
	BNF3	5.063	1.646	0.935			

Discriminant validity was established using the Fornell–Larcker criterion, whereby the square root of the AVE for each construct exceeded the inter-construct correlations [11]. The results of the discriminant validity analysis are reported in Table 4.

Table 4 Test Results of Discriminant Validity

Constructs	1	2	3	4
1. Benefits	0.943			
2. Affective Risk Perception	0.387	0.914		
3. Cognitive Risk Perception	0.207	0.634	0.839	
4. Cabin Fever Syndrome	−0.393	0.028	0.107	0.839

Note: Diagonal values are the square root of AVE

4.2 Structural Model

The structural model was evaluated to test the proposed hypotheses by examining the significance of the path coefficients. A bootstrapping procedure with 5,000 resamples was conducted to assess the statistical significance of the estimated paths. The results of the structural model analysis are summarized in Table 5.

Table 5. Results of PLS-SEM

H	Predictor	Outcome	Coefficient	*t*	Hypothesis
H1	Affective risk perception	Benefits	0.400	4.170	Supported
H2	Cognitive risk perception	Benefits	−0.004	0.042	Not Supported
H3	Cabin fever syndrome	Benefits	−0.403	6.048	Not Supported

The results of the structural model provide several noteworthy insights into the determinants of perceived e-learning benefits. First, affective risk perception was found to have a significant positive effect on perceived benefits, suggesting that students who experience greater emotional concern about COVID-19 are more likely to recognize the value of e-learning as a safer and more viable alternative to face-to-face instruction. In contrast, cognitive risk perception did not exhibit a significant effect, indicating that rational assessments of risk severity and susceptibility may not directly translate into favorable evaluations of e-learning. This finding implies that emotional responses, rather than cognitive judgments, play a more decisive role in shaping perceptions of educational value in crisis contexts. Notably, cabin fever syndrome demonstrated a significant negative effect on perceived benefits, contrary to the initial hypothesis. This result suggests that heightened psychological distress and frustration associated with prolonged isolation may undermine students' ability to positively evaluate e-learning, rather than encouraging greater appreciation of it as an alternative mode of learning.

5 Conclusion

5.1 Summary

This study identified the key determinants of perceived e-learning benefits. Specifically, affective risk perception, cognitive risk perception, and cabin fever syndrome were proposed as antecedents. Empirical data were collected from university students in South Korea, and the proposed research framework was validated using PLS-SEM. The findings indicate that affective risk perception positively influences perceived benefits, whereas cabin fever syndrome exerts a significant negative effect. Cognitive risk perception was found to be insignificant. These results suggest that emotional and psychological factors play differentiated roles in shaping students' perceptions of e-learning benefits.

5.2 Implications for Theory and Practice

This study offers several theoretical and practical contributions. First, it extends the existing literature by focusing on perceived benefits of e-learning, which have become increasingly important in the context of the COVID-19 pandemic. Prior studies have primarily emphasized outcomes such as actual use and user satisfaction in explaining e-learning behavior [13, 14]. In contrast, this study demonstrates that the perceived benefits of e-learning are shaped by broader psychological and contextual conditions. Accordingly, the findings open new avenues for investigating the antecedents of e-learning benefits in greater depth. From a practical perspective, these insights can assist practitioners in developing strategies that enhance both perceived benefits and user satisfaction. Second, this study contributes to the literature by examining the role of users' psychological states in shaping perceived e-learning benefits. While prior research has predominantly focused on system-related factors and users' technological perceptions [15–17], this study highlights the importance of emotional and psychological dimensions. The findings reveal that affective risk perception enhances perceived benefits, whereas cabin fever syndrome exerts a significant negative effect. This suggests that psychological distress does not necessarily lead to greater appreciation of e-learning; rather, it may undermine students' ability to evaluate its benefits positively. Therefore, future research should further explore the mechanisms through which different psychological states exert contrasting effects. Practically, these findings imply that simply increasing reliance on e-learning under restrictive conditions may not be sufficient, and that managing students' psychological well-being is critical for improving their learning experience.

5.3 Limitations and Future Research Directions

Despite its contributions, this study has several limitations. First, the sample was limited to university students in South Korea, which may restrict the generalizability of the findings. Psychological responses to COVID-19 and associated social restrictions may differ across countries due to variations in cultural context, pandemic severity, and policy measures. Future research should therefore examine these relationships in diverse national contexts. Second, the variables examined in this study may evolve as the COVID-19 situation changes over time. As such, future studies are encouraged to employ longitudinal designs to capture the dynamic nature of these relationships and enhance the explanatory power of the research model.

References

1. Khan, M.A., et al.: Students' perception towards E-learning during COVID-19 pandemic in India: an empirical study. Sustainability. **13**(1), 57 (2020)
2. Nambiar, D.: The impact of online learning during COVID-19: students' and teachers' perspective. Int. J. Indian Psychol. **8**(2), 783–793 (2020)
3. Ali, W.: Online and remote learning in higher education institutes: a necessity in light of COVID-19 pandemic. High. Educ. Stud. **10**(3), 16–25 (2020)

4. Adiyoso, W., Wilopo, W.: Social distancing intentions to reduce the spread of COVID-19: the extended theory of planned behavior. BMC Public Health. **21**, 1836 (2020)
5. Fritscher, L.: How to Know If You Have Cabin Fever or Fear of Isolation. https://www.verywellmind.com/cabin-fever-fear-of-isolation-2671734 (2020). Accessed 8 June 2021
6. Sjöberg, L.: Worry and risk perception. Risk Anal. **18**(1), 85–93 (1998)
7. Bae, S.Y., Chang, P.-J.: The effect of coronavirus disease-19 (COVID-19) risk perception on behavioural intention towards 'untact' tourism in South Korea during the first wave of the pandemic (March 2020). Curr. Issue Tour. **24**(7), 1017–1035 (2021)
8. Fritscher, L.: Cabin Fever Symptoms and Coping Skills. https://www.verywellmind.com/cabin-fever-fear-of-isolation-2671734 (2020). Accessed 26 Aug 2021
9. Falk, R.F., Miller, N.B.: A Primer for Soft Modeling. University of Akron Press (1992)
10. Ringle, C.M., Wende, S., Becker, J.-M.: Smartpls 3. Hamburg: SmartPLS. https://www.smartpls.com (2014). Accessed 4 Aug 2021
11. Fornell, C., Larcker, D.F.: Evaluating structural equation models with unobservable variables and measurement error. J. Mark. Res. **18**(1), 39–50 (1981)
12. Bagozzi, R.P., Yi, Y., Phillips, L.W.: Assessing construct validity in organizational research. Adm. Sci. Q. **36**(3), 421–458 (1991)
13. Mohammadi, H.: Social and individual antecedents of m-learning adoption in Iran. Comput. Hum. Behav. **49**, 191–207 (2015)
14. Al-Emran, M., Arpaci, I., Salloum, S.A.: An empirical examination of continuous intention to use m-learning: an integrated model. Educ. Inf. Technol. **25**(4), 2899–2918 (2020)
15. Al-Fraihat, D., et al.: Evaluating E-learning systems success: an empirical study. Comput. Hum. Behav. **102**, 67–86 (2020)
16. Alsabawy, A.Y., Cater-Steel, A., Soar, J.: IT infrastructure services as a requirement for e-learning system success. Comput. Educ. **69**, 431–451 (2013)
17. Mo, C.-Y., et al.: Exploring the critical factors, the online learning continuance usage during COVID-19 pandemic. Sustainability. **13**(10), 5471 (2021)
18. Brug, J., et al.: SARS risk perception, knowledge, precautions, and information sources, the Netherlands. Emerg. Infect. Dis. **10**(8), 1486–1489 (2004)

The Effect of Active Learning on Learning Achievement in Food Rheology among Food Science and Technology Undergraduates

Supatchalee Sirichokworrakit(✉), Nuntaporn Aukkanit, Shutchapol Chopvitayakun, and Kunyanuth Kularbphettong

Faculty of Science and Technology, Suan Sunandha Rajabhat University, Bangkok, Thailand
supatchalee.si@ssru.ac.th

Abstract. The purpose of this study was to evaluate the effects of active learning strategies on learning outcomes among undergraduate students enrolled in the Food Rheology course at Suan Sunandha Rajabhat University in Thailand. Fifty third-year students participated in a three-week instructional program that included flipped videos, classroom problem-solving activities, laboratory exercises, small group projects, and reflection. Pre- and post-test assessments evaluated grasp of basic rheology concepts, model identification, use and analysis of measured values, and application to formulation or manufacturing processes. The results demonstrated substantial and statistically significant improvement in learning outcomes. The mean pretest score was 7.80 (SD = 1.63), while the mean posttest score was 17.24 (SD = 1.55), with a paired-samples t-test showing $t(49) = 37.14$, $p < 0.001$. The effect size was extremely large (Cohen's $d = 5.25$), indicating both statistical and educational significance. Analysis of the score distribution demonstrated continuous development for all students, with post-test scores ranging from 14 to 20 and an average increase of 9.44. These findings demonstrate that active learning in Food Rheology enhances students' comprehension, critical thinking, and practical abilities. Hands-on experiments and reflection exercises allow students to apply their knowledge to real situations and retain concepts longer. Overall, student-centered active learning is beneficial in promoting deep and meaningful learning in food science and technology.

Keywords: Food Rheology · Active Learning · Food Science and Technology

1 Introduction

Food Rheology is an important topic in food science that studies the flow behavior of foods. It involves understanding how a food sample responds to a force or any stimulus that causes part or all of the sample to change form or move, ultimately affecting the quality of the food. The topics of food rheology include: Parameters related to rheological measurements, food viscosity, food flow behavior, categories of fluids, fluid evaluation instruments, using and analyzing the results from the measured values. However, students often find rheological concepts such as viscosity measurement and

H. R. Arabnia et al. (Eds.): CSCI 2025, CCIS 2974, pp. 247–254, 2027.
https://doi.org/10.1007/978-3-032-28475-4_17

non-Newtonian flow behavior abstract and challenging to understand through traditional narrative instruction. Active learning, which emphasizes student participation in interactive and immersive activities, has been found to improve information retention, critical thinking, and problem-solving abilities in food science and technology disciplines.

Active learning is a teaching approach that focuses on engaging learners through practical activities. Several studies have confirmed that active learning enhances in-depth understanding and develops important skills, including critical thinking [1]. In the field of food science and technology, research by Groman et al. [2] shows that simple demonstrations can increase student engagement and help secondary school students better understand the chemical concepts of food. Hollis and Eren [3] applied the concept of experiential learning in a food product development course, providing students with the opportunity to work in real-life situations within the food industry. This approach allows students to develop both content knowledge and essential skills for success. In the context of food rheology courses, which focus on the study of food flow and deformation, research by Zhao et al. [4] illustrate the application of active learning in highly complex fluid flow simulations. Techniques such as Gaussian process regression are used to learn constitutive equations from microphysics, allowing for more accurate and efficient simulations of non-Newtonian fluid behavior at the macro level. Additionally, the research by Jeong, Kim, and Lee [5] demonstrates the use of food rheology data combined with machine learning analysis to classify foods suitable for specific consumer groups, such as the elderly, highlighting the importance of understanding food dynamics and applying rheology knowledge in food product development. Although these studies focus on active learning and complex fluid simulations, few studies have directly examined the effects of active learning on the academic achievement of undergraduate students in Food Rheology courses. To address this gap, the objective of this study was to examine the effect of active learning on the learning achievement of undergraduate students in food science and technology enrolled in Food Rheology courses.

2 Methodology

2.1 Participants and Study Design

The research was conducted with fifty undergraduate students who were enrolled in a core Food Science and Technology course on Food Rheology at Suan Sunandha Rajabhat University, Thailand (N = 50; year of study = 3rd year), and there were no incentives offered; participation was an integral component of the standard curriculum. The research was carried out in 2025, as part of a three-week curriculum on rheological concepts and experimental techniques. In the food rheology course, an educational intervention utilizing active learning techniques was put into practice and the improvement in students' content knowledge was referred to as learning achievement, and it was assessed using two parallel tests that were given under similar circumstances: a pretest at the start of the module to gauge baseline comprehension and a posttest at the conclusion to gauge learning outcomes following the intervention [6, 7]. Parameters related to rheological measurements, food viscosity, food flow behavior, categories of fluids, fluid evaluation instruments, using and analyzing the results from the measured values were all addressed in both assessments. Instead of reflecting assessment bias, the use of identical testing circumstances made sure that observed variations in scores reflected learning gains owing to the active learning activities.

2.2 Instructional Intervention (Active Learning)

The intervention integrated multiple active-learning elements aligned with constructivism and experiential learning principles:

1. Pre-class micro-lectures (Flipped elements): 15-min videos on parameters related to rheological measurements, food viscosity, food flow behavior, categories of fluids, fluid evaluation instruments, using and analyzing the results from the measured values, with each week covering the following topics.
2. In-class problem solving: Small groups discussed on selecting appropriate rheological models for mayonnaise, ketchup, concentrated corn starch solution and honey.
3. Laboratory practica (Experiential): Evaluate the flow behavior of the sample using a Brookfield viscometer and a controlled-stress/strain rheometer to generate flow curves; data logging and parameter estimation (η, τ_0, n, K).
4. Small groups modified formulations (e.g., starch or sugar concentration) to achieve a target viscosity profile suitable for product application, and justified their choices using rheological measurements.
5. Reflective debriefs: Students participate in short, structured activities that use their knowledge to solve a problem in the field of physiology. Analyze data and experimental models. Evaluate the results, and create solutions or mind maps for real or simulated food systems to enhance their learning.

Total contact time for the module was 15 h (lecture/seminar = 6, laboratory = 9).

2.3 Learning Outcomes, Measures and Procedure

Learning Outcomes and Measures

- Assessment instrument: A content-validated test covering: (a) fundamental rheology concepts (shear stress/rate, viscosity, categories of fluids, fluid evaluation instruments), (b) model identification and parameter interpretation (Newtonian, Power-law, Bingham, Herschel–Bulkley), and (c) application to formulation or processing decisions (viscosity adjustment for product application, mixing, pumping).
- Format: 20–30 items combining multiple-choice questions, short quantitative problems, and item lets requiring interpretation of flow curves or experimental data from laboratory practical and mini-project.
- Items were blueprint-mapped to Revised Bloom's Taxonomy targets: Understand, Apply, Analyze, with a smaller proportion at Evaluate/Create via mini-project justification and reflective debriefs.
- Reliability/quality control: Internal consistency assessed (Cronbach's α, to be reported). Two subject-matter experts reviewed items for content validity (CVI).
- Administration: Pre-instruction assessment conducted prior to week 1; post-instruction assessment conducted in the final session under identical conditions. Posttest used isomorphic items covering the same blueprint to reduce simple recall.

Procedure

1. Requesting consent and clarification of details from students is carried out in the first week
2. Students have completed the pre-test assessment
3. Over the course of 3 weeks, students participated in a hierarchical active learning process: learning through flipped teaching materials, classroom problem-solving, and laboratory practice. Implementation of mini-projects and reflection of opinions.
4. Students have completed the post-test assessment in the last week.
5. Each component standardizes all instructional materials, assessments, and time spent.

2.4 Data Analysis

Analyses were conducted at $\alpha = 0.05$ (two-tailed).

- Descriptives statistics: Means and standard deviations were computed for pre- and post-instruction scores.
- Primary test: A paired-samples t-test evaluated the mean difference in achievement.
- Effect size: Cohen's d for dependent samples (d_z) was calculated as the mean of the difference scores divided by their standard deviation.
- Assumptions: Normality of difference scores was checked via Shapiro–Wilk and inspection of Q–Q plots; where violated, the Wilcoxon signed-rank test was planned as a robustness check.
- Missing data: Cases with missing values on either assessment were excluded pairwise; sensitivity analyses (e.g., mean imputation bounds) were planned if missingness exceeded 5%.

2.5 Theoretical Grounding

- Constructivism: Learning activities were designed to allow students to build their own conceptual understanding through experimentation and exchange with their classmates, rather than passively absorbing information.
- Experiential Learning (Kolb): The sequence of activities is carried out according to Kolb's learning process, i.e. concrete experience (laboratory practice). Reflective observation (summary discussion) Abstract conceptualization (model understanding/parametric reasoning) and operational experiments (micro-project formulation).
- Revised Bloom's Taxonomy: Learning and assessment activities are designed to cover the hierarchical development of comprehension. Understanding→ Applying, → Analyzing, and having elements of evaluate and create in reasoning for the project.

2.6 Validity and Bias Mitigation

To lessen threats common to one-group classroom studies (testing, maturation, instrumentation), we used: (a) isomorphic pre/post forms with constant blueprinting; (b) fixed administration conditions; (c) standardized teaching materials and lab checklists; and (d) triangulation with process data (lab worksheets and reflection quality) to corroborate learning gains.

3 Results and Discussion

3.1 Descriptive Statistics

The final analysis included data from undergraduate students in food science and technology. The pretest assessment, conducted in week 0, found that students started the course with a limited understanding knowledge of Food Rheology (mean = 7.80, standard deviation = 1.63). After a three-week course of teaching and learning activities that integrated active learning strategies, students demonstrated clear knowledge development. The average posttest score was 17.24 (SD = 1.55). This considerable rise shows that such interventions can effectively improve students' comprehension of food rheology concepts. Newtonian and non-Newtonian flow behavior and the relationship between shear stress and shear rate. Figures 1 and 2 illustrate the changes in the distribution of scores before and after class. It emphasizes the trend of steadily increasing in results of the post-assessment.

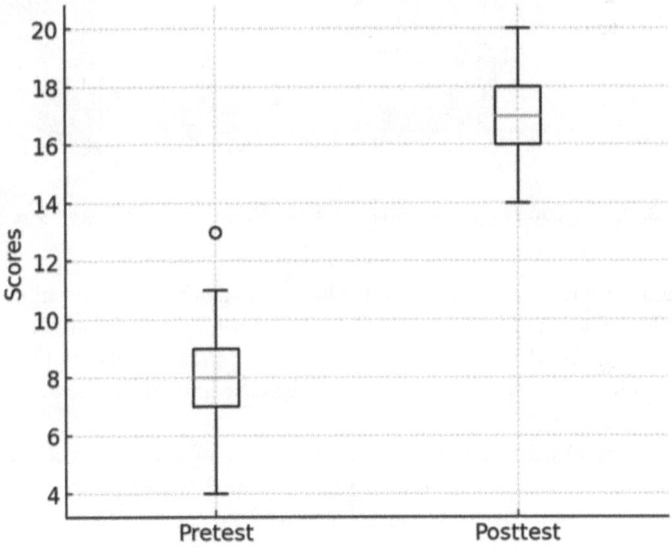

Fig. 1. Boxplot comparison of pretest and posttest scores in Food Rheology (N = 50)

The boxplot clearly shows the improvement in student achievement. The pretest score was about 8, with an interquartile range (IQR) range of 7 to 9, and an outlier value was found to be higher than normal (13), reflecting that most students started the course with limited knowledge of basic Food Rheology, and some students had higher prior knowledge than the overall group. In contrast, the median posttest score was higher at around 17, with the quartile range between 16 and 18, and the score distribution tended to be higher without any abnormal lows. But there was also more consistency and consistency in academic achievement after 3 weeks of active learning activities. The change suggests that active learning strategies allow learners to significantly develop a conceptual understanding of the rheological concepts of food. Newtonian and non-Newtonian flow behavior and the relationship between shear stress and shear rate. In addition, a more uniform distribution of scores reflects a similar level of students' ability to understand and apply knowledge. This makes learning high-quality and measurable.

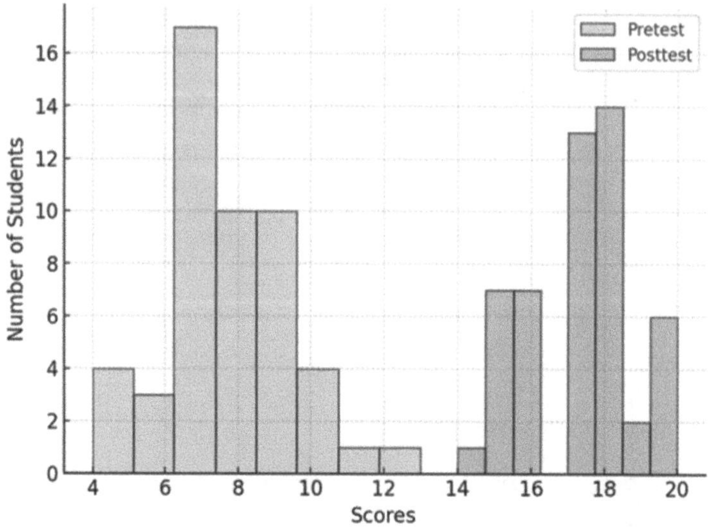

Fig. 2. Distribution of pretest and posttest scores in Food Rheology (N = 50)

Histogram overlays show a clear change in student achievement in a higher direction after the implementation of active learning activities, with pretest scores concentrated in the low range (4–11). While posttest scores are concentrated in the higher value range (14–20), indicating a significant and consistent increase in academic achievement across all student groups.

The results are also consistent with Kolb's theory of experiential learning, which explains that effective learning occurs through a cyclical process involving four stages: concrete experience, reflective observation, abstract conceptualization, and active experimentation [8]. The data obtained were analyzed and discussed as a reflection of these results. The experiment outcomes were linked to theories such as Newtonian and non-Newtonian fluid flow behavior, representing the stage of abstract conceptualization. Finally, the acquired knowledge was applied to develop sub-projects aimed at adjusting the product, reflecting active experimentation. Additionally, the improvement of learning outcomes corresponds to the beginning development according to the Revised Bloom's Taxonomy. Initially, students have a limited ability to recall and comprehend fundamental concepts. However, at the end of the intervention, the evaluation findings revealed that students could use rheological models and analyze flow curves. In addition, group assignments allow you to assess and develop solutions to practical food processing difficulties. Such progress not only reflects cognitive gains, but also indicates the development of higher-order thinking skills, which are important in food science and technology education and professional practice.

3.2 Paired-Sample T-Test

Statistical testing by means of the paired-samples t-test to determine if there is a statistically significant difference between pre-test and post-test scores. The results of the analysis showed that $t(49) = 37.14$, $p < 0.001$ showed that the values obtained were highly significant, so it can be confidently concluded that the difference in scores before and after school was not due to coincidence, but to the result of active learning management applied in practice (Table 1).

Table 1. Pretest–Posttest Comparison and Cohen's d Effect Size in Food Rheology

Measure	Mean	SD	t	p-value	Effect size
Pretest	7.80	1.63	–	–	–
Posttest	17.24	1.55	–	–	–
Paired t-test	–	–	37.14	1.5×10^{-37} (<0.001)	–
Effect size (Cohen's *d*)	–	–	–	–	5.25

3.3 Effect Size

To assess the practical significance of learning interventions, Cohen's d values were calculated. For the dependent samples, the results of the analysis showed that the effect size was 5.25, which is extremely large, higher than the standard value used as a criterion for interpreting high influence size (large effect), which is set at 0.80 significantly. This result reflects that active learning management not only makes a statistically significant difference, but also makes a significant difference. It has a very high level of educationally meaningful impact, i.e., students develop skills and expertise in key areas such as viscosity measurement, Newtonian and non-Newtonian fluid flow behavior, and shear stress–shear rate relationships. The findings can be interpreted using the theoretical framework of instructional and learning design. According to constructivism philosophy, students actively participate in laboratory practice, group problem-solving, and introspective talks. It works with actual data from viscometers and rheometers. This learner-centered, inquiry-based learning strategy encourages significant knowledge development and long-term information retention in the subject of food science and technology. The results of this study provide strong evidence that the integration of active learning strategies in the teaching of Food Rheology can significantly increase student achievement. A comparison of pretest and posttest results revealed substantial progress, with all students in the experimental group improving their scores, and the effect size was very large. These results indicate both statistical and practical significance for undergraduate students in Food Science and Technology.

3.4 Score Distribution and Gains

An analysis of the distribution of scores significantly reinforces the magnitude of the development that has occurred. It was found that the pretest score was in the range of

4 to 11, while the posttest score was in the range of 14 and 20, reflecting a clear higher change in all groups of learners. The results for gain scores (posttests) - pretest varied from 6 to 13, with a mean of 9.44 and a standard deviation (SD) of 1.63. Importantly, the results of the analysis showed that 100% of all students had higher scores, with no stable or regressive scores. This consistent positive trend reflects the effectiveness of active learning interventions, which can impact comprehensive learning development at all skill levels. In other words, such interventions are equitable in creating benefits for all learners without disparities in academic achievement.

4 Conclusion

This study demonstrated that implementing active learning strategies in a Food Rheology course for undergraduate Food Science and Technology students led to substantial and statistically significant improvements in learning achievement. The pretest-posttest study found that every student in the cohort increased their performance, with a very substantial effect size (Cohen's $d = 5.25$). These findings confirm the effectiveness of student-centered instructional strategies in improving both knowledge acquisition and higher-order thinking skills. Based on constructivist and experiential learning theories, the intervention successfully engaged students in applying theoretical knowledge to actual laboratory circumstances, resulting in a greater comprehension and long-term retention of rheological concepts.

Acknowledgments. The author would like to thank Suan Sunandha Rajabhat University, Bangkok, Thailand for providing financial support, equipment, and research facilities.

References

1. Groman, D.B., McNulty, M.A.: Food chemistry demonstrations: a simple approach to actively engaging high school students in the chemistry of foods. Food Chem. **157**, 1–568 (2014)
2. Hollis, F.H., Eren, F.: Implementation of real-world experiential learning in a food science course using a food industry-integrated approach. J. Sci. Educ. **15**(4), 109–119 (2016)
3. Zhao, L., Li, Z., Wang, Z., Caswell, B., Ouyang, J., Karniadakis, G.E.: Active learning of constitutive relation from mesoscopic dynamics for macroscopic modeling of non-Newtonian flows. J. Comput. Phys. **348**, 674–694 (2017)
4. Zhao, L., Li, Z., Wang, Z., Caswell, B., Ouyang, J., Karniadakis, G.E.: Active- and transfer-learning applied to microscale-macroscale coupling to simulate viscoelastic flows. J. Comput. Phys. **404**, 109–124 (2020)
5. Jeong, S., Kim, H., Lee, S.: Rheology-based classification of foods for the elderly by machine learning analysis. Appl. Sci. **11**(5), 2262 (2021)
6. Sangsuwon, C., Chopvitayakun, S., Aukkanit, N., Sirichokworrakit, S., Kularbphettong, K.: The effectiveness of the project-based learning in the perfume chemistry course. In: International Conference on Computational Science and Computational Intelligence, pp. 117–128. Springer Nature, Switzerland, Cham (2024)
7. Chutrtong, J., Narumon, B., Waradoon, C.: Service learning by STEM activity in secondary school at Prachin Buri Province, Thailand. Int. J. Inform. Educ. Technol. **9**(8), 580–583 (2019)
8. Kolb, D.A.: Experiential Learning: Experience as the Source of Learning and Development. Prentice Hall, Englewood Cliffs, NJ (1984)

Project-Based Active Learning in Searching Algorithms: Assessing Learning Gains via Pretest–Posttest and Correlation Analysis

Kunyanuth Kularbphettong[1(✉)], Nareenart Raksuntorn[1], Nuntaporn Aukkanit[1], Supatchalee Sirichokworrakit[1], and Pattarapan Roonrakwit[2]

[1] Suan Sunandha Rajabhat University, Bangkok 10300, Thailand
kunyanuth.ku@ssru.ac.th
[2] Silpakorn University, Bangkok 11120, Thailand

Abstract. The study presents the development and assessment of a project-based active learning approach for instructing search algorithms in an undergraduate course on algorithm analysis and design. The pedagogical framework emphasized problem-solving, coding exercises, and comparative evaluation of linear, binary, and interpolation search methodologies to include students in cooperative, practical projects. A total of 85 students participated in pretest and posttest evaluations, each scored out of 20 points, in addition to a laboratory assignment valued at 10 points to evaluate their learning outcomes. The active learning technique markedly improved students' comprehension of searching algorithms, evidenced by the statistically substantial rise in post-test scores relative to pretest scores, as indicated by paired t-test analysis. Positive relationships were seen among the pretest, posttest, and laboratory assignment scores, indicating that practical project-based activities improved theoretical understanding and promoted deeper comprehension. The findings demonstrate the successful incorporation of project-based active learning in algorithm education to enhance academic achievement and promote critical thinking and collaboration. The study offers evidence-based recommendations for educators seeking to improve computer science education using new teaching methods.

Keywords: Project-Based Learning · Active Learning · Searching Algorithms · Pretest–Posttest · Correlation Analysis

1 Introduction

Searching algorithms are one of the fundamental topics in computer science and algorithm design courses. They provide the basis for solving problems with numerous computational tasks, ranging from information systems to database indexing and data retrieval. Students often struggle to comprehend the theoretical principles and practical efficacy of diverse search strategies, such as linear, binary, and interpolation search, despite their importance. Traditional lecture-based learning typically leads to memorizing concepts rather than true comprehension, which affects students' ability to choose or apply the appropriate algorithms in real-world scenarios.

H. R. Arabnia et al. (Eds.): CSCI 2025, CCIS 2974, pp. 255–268, 2027.
https://doi.org/10.1007/978-3-032-28475-4_18

The pedagogical approach of active learning is increasingly recognized for enhancing comprehension, critical thinking, and student involvement [1, 2]. Project-based learning (PBL) and active learning mutually reinforce one another through a focus on engagement, collaboration, and experiential learning. PBL enables students to concentrate on valuable projects that necessitate cooperation, coding practice, and experimentation—skills essential for computer science graduates. Moreover, in contrast to conventional lecturing, active learning improves examination outcomes and lowers failure rates, as evidenced by a significant meta-analysis of undergraduate STEM courses [3–6]. Consequently, incorporating project-based active learning into algorithm instruction enhances higher-order thinking skills and bridges theoretical knowledge with practical application.

Nevertheless, recent research in computer science education have revealed certain deficiencies. Despite the extensive application of active learning in teaching programming fundamentals, software engineering, and basic data structures, there is a scarcity of research concentrated on algorithm-specific topics, such as searching strategies. Most teaching methods primarily rely on lectures or individual coding exercises, often neglecting a thorough evaluation of algorithmic efficiency and real-world practical application. Though project-based learning (PBL) has been examined in specific computer science settings, empirical research rigorously evaluating its impact on algorithm education is inadequate [7]. A quasi-experimental study conducted at Ibb University, Yemen, demonstrated that Problem-Based Learning (PBL) significantly improved students' achievement and self-efficacy in learning the C# programming language compared to traditional methods [8]. Research with teacher educators in Ghana indicated that technology-integrated project-based learning facilitated student-centered and collaborative learning, while also highlighting challenges such as insufficient resources, limited instructional time, and varying levels of instructor proficiency [9]. These findings underscore the potential and practical constraints of applying PBL, indicating the need for more empirical study to evaluate its effectiveness in algorithm-centric education [10–13].

This study aims to address these deficiencies by developing and evaluating a project-based active learning approach tailored for searching algorithms in an undergraduate course on algorithm analysis and design. The instructional design engages students in collaborative, hands-on projects that focus on problem-solving, coding practice, and the comparative analysis of search algorithms. The research utilizes a pretest-posttest design including 85 students, accompanied by a laboratory assignment, to evaluate its efficacy. The paired t-test is employed to evaluate learning gains, whereas correlation analysis examines the relationships among pretest, posttest, and laboratory performance.

The experiment was conducted in the first semester of the 2025 academic year at Suan Sunandha Rajabhat University (SSRU), Thailand, with second-year undergraduate students enrolled in the Algorithm Analysis and Design course. This context highlights the significance of the suggested teaching model in real classroom settings within higher education.

This research presents three key contributions. It introduces an innovative approach by integrating project-based active learning into the pedagogy of searching algorithms, surpassing traditional lecture-based methods. Furthermore, it provides empirical evidence of enhanced learning outcomes through statistical analysis, demonstrating the method's effectiveness. Third, it offers insights into the interplay of foundational knowledge, practical application, and learning outcomes, so augmenting both pedagogy and educational research in computer science. This study provides evidence-based recommendations for instructors seeking innovative teaching practices that enhance algorithm learning, cultivate critical thinking, and strengthen collaborative skills.

2 The Research Hypothesis

- H1: Students' posttest results will significantly surpass their pretest scores, indicating that the project-based active learning process improves conceptual understanding of searching algorithms.
- H2: A positive correlation exists between pretest and posttest scores, suggesting that students with stronger prior knowledge are likely to achieve better learning outcomes after the intervention.
- H3: A positive correlation occurs between lab assignment scores and posttest scores, suggesting that experiential, project-based activities improve theoretical comprehension and result in superior learning outcomes.
- H4: The implementation of project-based active learning encourages personal development and collaborative skills, hence improving overall learning outcomes in computer science education.

3 Research Questions

- RQ1: Does the project-based active learning methodology significantly improve students' understanding of searching algorithms, as evidenced by the difference between pretest and posttest scores?
- RQ2: What is the relationship between students' pretest and posttest scores, and in what manner does prior knowledge influence subsequent learning outcomes?
- RQ3: What is the relationship between laboratory assignment scores and posttest performance, and to what extent do hands-on, project-based activities facilitate conceptual learning?
- RQ4: In what ways may the integration of project-based active learning enhance personal development and collaboration skills among computer science students?

4 Research Methodology

4.1 Research Design

This study employed a quasi-experimental design using a one-group pretest–posttest model to evaluate the effectiveness of a project-based active learning approach in teaching searching algorithms. The teaching strategy was implemented in the Algorithm Analysis and Design course during the first semester of the 2025 academic year at Suan Sunandha Rajabhat University (SSRU), Thailand. Eighty-five second-year undergraduate students that were enrolled in the course made up the participants.

The instructional approach integrated active learning techniques (e.g., think–pair–share, coding exercises, group discussions) with project-based learning (PBL), allowing students to work in teams on real projects that analyzed linear, binary, and interpolation search algorithms. Given the restricted parameters of this classroom and institutional setting, the study utilized a case study methodology to clarify the relationship between context and practice, thereby providing an in-depth understanding of the benefits of project-based active learning in algorithm education [14, 15].

4.2 Data Collection Instruments

Data were collected via three principal instruments:

1. **Pretest and Posttest**: Each assessment consisted of theoretical and practical questions related to search algorithms, each scoring 20 points. The pretest was carried out before the intervention, and the posttest was given after the conclusion of the project-based activities.
2. **Laboratory Assignment (Practical Exercise)**: Students worked in groups to implement searching algorithms and assess their effectiveness using actual or simulated datasets, receiving 10 points for their contributions.

   The laboratory assignment assessed coding accuracy, analytical abilities, and teamwork.
3. **Learning Logs and Observations**: Students' responses and classroom observations were taken into consideration to qualitatively supplement the quantitative data.

4.3 Design Procedure

This study's design methodology was divided into five main stages:

Step 1: Preparation: In the planning phase, the learning objectives for the searching algorithms module, including linear, binary, and interpolation search strategies, were clearly defined to align with course outcomes. Assessment instruments were created, consisting of a pretest and a posttest (each worth 20 points) to evaluate both conceptual and practical knowledge, along with a laboratory assignment (valued at 10 points) that concentrated on code implementation, experimental testing, and performance analysis of algorithms. Students were organized into small groups of four to five individuals to promote collaboration and active participation during project-based learning activities.

Step 2: Implementation of the Pretest: At the beginning of the semester, a pretest was given to all participating students to provide a baseline assessment of their prior knowledge on searching algorithms. The pretest results offer an initial evaluation of students' strengths and weaknesses, acting as a standard for measuring progress after the instructional intervention. Scores were systematically recorded for future statistical analysis.

Step 3: Performance of Project-Based Active Learning: The intervention was carried out through project-based active learning activities integrated into the Algorithm Analysis and Design course. Students were initially introduced to searching algorithms through brief lectures and guided experiments. They subsequently collaborated in groups to implement linear, binary, and interpolation search algorithms. Each group conducted trials using both unsorted and sorted datasets, comprising a dataset of 85 students, and recorded algorithm performance regarding comparisons, execution time, and accuracy. Active learning methodologies, such as think–pair–share, coding challenges, and peer conversations, were employed to promote critical thinking, facilitate information exchange, and increase engagement with the material.

Step 4: Post-Assessment and Laboratory Evaluation: Upon concluding project-based activities, students participated in a posttest to assess their conceptual understanding following the intervention. The laboratory assignment was evaluated based on coding precision, problem-solving effectiveness, and the exhibition of teamwork within groups. Additionally, students were required to give succinct reflections on their learning experiences, offering qualitative insights into the effects of project-based active learning on their understanding and collaborative skills.

Step 5: Data Analysis: The final section involved the analysis of the collected data using both descriptive and inferential statistics. Descriptive statistics, encompassing mean, standard deviation, minimum, and maximum values, were calculated for pretest, posttest, and lab assignment scores to provide an overview of student performance. Inferential statistics were employed using a paired t-test to see whether a statistically significant improvement existed between pretest and posttest results. Furthermore, correlation analysis (Pearson's r) was conducted to examine the relationships among pretest, posttest, and laboratory results. The results from these analyses were employed to evaluate the effectiveness of the project-based active learning approach in enhancing both conceptual understanding and practical application in search algorithms.

4.4 Data Analysis

This study's data analysis employed both descriptive and inferential statistical approaches, precisely aligned with the research hypotheses (H1–H4) and research questions (RQ1–RQ4). This dual approach enabled a comprehensive evaluation of students' academic achievement as well as their personal and collaborative development.

Descriptive Statistics
Descriptive statistics, including the mean, standard deviation (SD), minimum, and maximum values, were calculated for pretest, posttest, and laboratory assignment scores. The data provided an overview of students' performance distribution, highlighted key learning trends, and established a foundation for future inferential analysis. The analyses immediately enabled the exploration of RQ1, concerning changes in students' understanding before and during the intervention, and RQ3, which investigates performance patterns associated with laboratory activities.

Inferential Statistics
Two primary inferential techniques were employed:

- Paired Sample t-test

A paired sample t-test was conducted to compare pretest and posttest outcomes, evaluating whether the project-based active learning intervention produced statistically significant improvements. This study specifically examined H1, which claimed that posttest scores would significantly exceed pretest results, and RQ1, which explored the extent to which project-based active learning enhanced comprehension of searching algorithms.

- Correlation Analysis (Pearson's)

  Pearson's correlation analysis was conducted to examine the relationships among pretest, posttest, and lab assignment scores. Figure 4 illustrates the results, summarized as follows:

  Correlation analysis was applied to explore the interrelationships among pretest, posttest, and laboratory assignment scores. The choice between Pearson's was based on data distribution and underlying assumptions. Two key relationships were examined:

  - Pretest–Posttest Correlation: To evaluate the extent to which prior knowledge predicts subsequent performance, hence assessing H2 and addressing RQ2.
  - Lab Assignment–Posttest Correlation: To assess if practical, project-based laboratory activities improved theoretical comprehension, hence evaluating H3 and addressing RQ3.

- Qualitative Analysis

  In addition to quantitative studies, qualitative data from students' reflective learning journals and classroom observations were examined through content analysis. This enabled the identification of recurring themes relevant to personal development (e.g., critical thinking, self-directed learning) and collaborative engagement (e.g., collaboration, peer interaction). The results corroborated H4, which asserted that project-based active learning fosters self-development and teamwork, and RQ4, which investigated how these pedagogical approaches enhance personal and social skills in computer science education.

  The integration of descriptive, inferential, and qualitative analyses enables a thorough evaluation of the intervention's effectiveness, providing insights into cognitive learning outcomes and broader developmental aspects associated with project-based active learning.

5 Results

5.1 Descriptive Statistics

The descriptive analysis of the 85 participating students revealed an obvious pattern of learning progression in pretest, posttest, and laboratory assignment scores. Table 1 reveals that the mean pretest score was 7.76 (SD = 1.56, Min = 4, Max = 13) out of 20, demonstrating students' insufficient foundational understanding of searching algorithms. This result underscores the first challenges students face in comprehending both theoretical and practical components prior to the commencement of formal instruction.

Table 1. Descriptive statistics of students' Pretest, Posttest, and Lab scores (N = 85).

Measure	Pretest	Posttest	Lab
Mean	7.76	17.04	8.0
SD	1.56	1.23	0.84
Min	4	14	7
Max	13	20	9

Following the implementation of Project-Based Active Learning (PBAL), the posttest mean increased dramatically to 17.04 (SD = 1.23, Min = 14, Max = 20), reflecting an enhancement of almost nine points, or about 120%. The substantial enhancement, in addition to the lower standard deviation compared to the pretest, illustrates that PBAL not only boosted overall performance but also diminished score variability, signifying that less proficient students benefited considerably from the intervention. The findings of the laboratory assignment highlight the practical effectiveness of PBAL. The average laboratory score was 8.02 out of 10 (SD = 0.84, Min = 7, Max = 9), indicating that most students successfully completed the project-based tasks with consistently high performance levels. The restricted score distribution improves the dependability of PBAL in cultivating practical proficiency in search approaches.

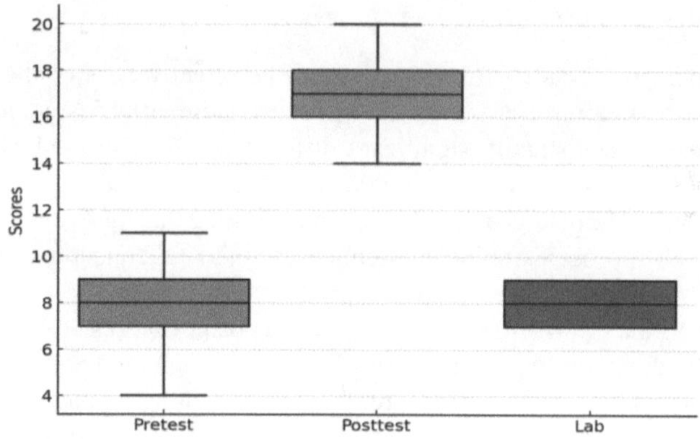

Fig. 1. Distribution of Pretest, Posttest, and Lab Scores

Figure 1 illustrates the distribution of pretest, posttest, and laboratory scores. The boxplots demonstrate that pretest scores exhibited considerable variability in the lower range, while posttest scores were tightly clustered in the upper range, suggesting a uniform improvement across the cohort. Similarly, laboratory scores were restricted to a narrow range, underscoring the reliability of project-based learning outcomes. Figure 2 presents the mean scores with standard deviations, emphasizing the notable performance difference between pretest and posttest results, and demonstrating the consistently high level of laboratory achievement.

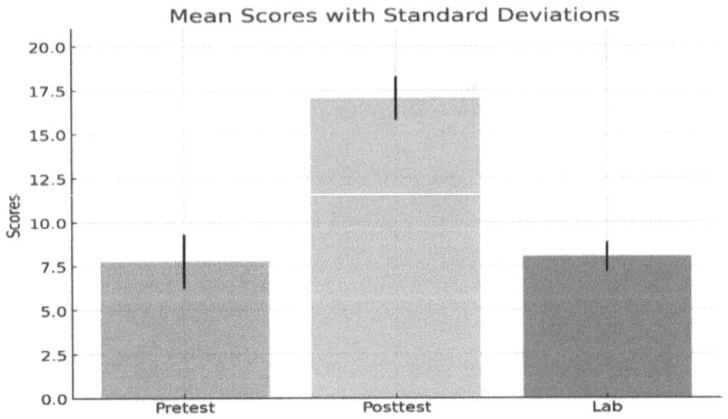

Fig. 2. Mean scores of Pretest, Posttest, and Lab with standard deviations (N = 85).

The results support constructivist learning theory and experiential learning models, which propose that learners acquire deeper knowledge through active engagement, practical experimentation, and collaborative problem-solving. The enhancement in both conceptual understanding (posttest) and practical application (lab) provides strong evidence that PBAL is an effective instructional framework for teaching searching algorithms in computer science education.

5.2 Paired t-test Results

A paired sample t-test was performed to assess the effectiveness of the instructional intervention by comparing the pretest and posttest scores of 85 students. The analysis demonstrated a statistically significant difference, $t\ (84) = -54.21$, $p < 0.001$, indicating that students' scores post-intervention were significantly higher than pre-intervention scores. The results indicate H1, demonstrating that the Project-Based Active Learning (PBAL) approach significantly enhances students' conceptual understanding of searching algorithms.

Table 2 shows the mean difference in pretest and posttest scores as 9.27 points (SD = 1.58), with individual improvements varying from a minimum of 5 points to a maximum of 13 points and the consistent upward trend observed among all participants indicates that the PBAL model was advantageous for both high-achieving students and those with less robust prior knowledge.

Table 2. Paired t-test results comparing Pretest and Posttest scores (N = 85).

	Mean	Mean Diff.	t (df)	*p* value	Interpretation
Pretest	7.76	9.27	−54.21(84)	< 0.001	Significant improvement
Posttest	17.04				

Figure 3 illustrates the distribution of pretest and posttest scores through a violin plot, highlighting the mean (black diamond) and median (red circle). The figure demonstrates the broad distribution of pretest scores in the lower range, in contrast to the closely grouped posttest scores in the higher range, indicating a notable improvement and decreased variability after the PBAL intervention.

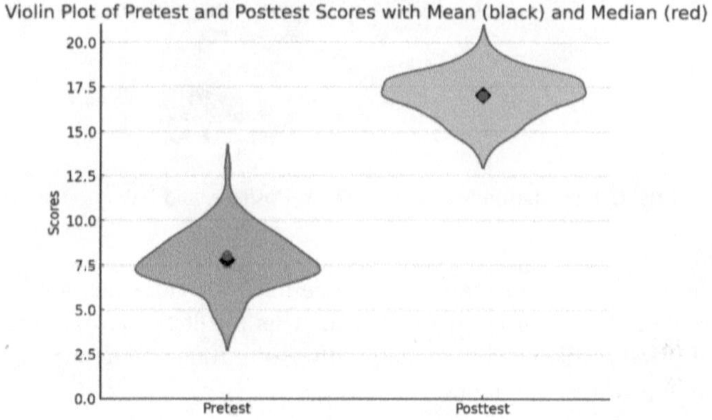

Fig. 3. Violin plot of Pretest and Posttest scores with mean and median.

The results demonstrate the importance of incorporating PBAL into computer science education, revealing an average increase of over nine points, which is statistically significant and pertinent to educational outcomes. This finding aligns with existing literature on constructivist and experiential learning frameworks, indicating that learners attain deeper and more lasting knowledge via collaborative, hands-on activities. Engaging students in authentic projects enabled PBAL to promote the internalization of algorithmic concepts, resulting in notable enhancements in conceptual understanding beyond simple rote memorization.

5.3 Correlation Analysis Result

Pearson's correlation analysis was conducted to examine the relationships among pretest, posttest, and lab assignment scores. Figure 4 illustrates the results, summarized as follows:

- The pretest-posttest analysis demonstrated a moderate positive correlation ($r = 0.378$), suggesting that students with greater baseline knowledge typically achieved higher posttest scores, thus corroborating hypothesis H2. The moderate strength suggests that PBAL reduced the dependence on prior knowledge, enabling students who were less proficient to make significant progress.
- H3 is partially supported by the posttest-lab correlation ($r = 0.229$), revealing a weak positive association and suggests that students who scored well on lab tasks tended to have a stronger conceptual comprehension and the weak relationship indicates that applied practice and theoretical mastery are interrelated yet distinct components of learning.

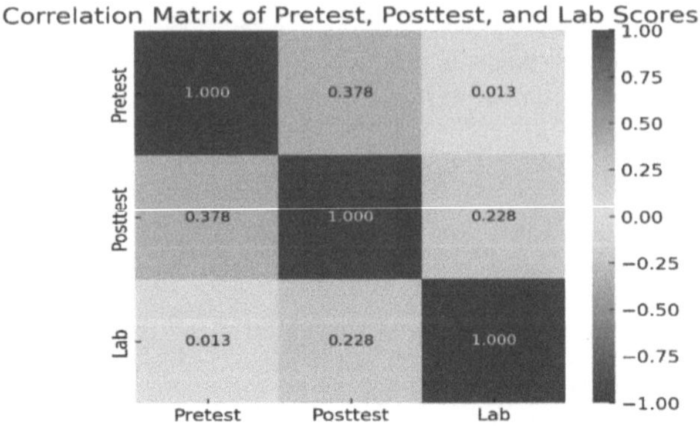

Fig. 4. Correlation Matrix of Pretest, Posttest, and Lab Scores.

- Pretest–Lab ($r = 0.013$, insignificant correlation): Previous theoretical knowledge was not very predictive of lab performance. This result emphasizes the scaffolding function of PBAL, which helps students with poorer foundations succeed in practical tasks through teamwork and guided practice.

All of the data show how theoretical assessments and real-world project-based activities work in tandem to improve algorithmic understanding.

5.4 Qualitative Findings

The examination of reflective learning logs and classroom observations corroborated H4, which posited that PBAL fosters personal development and collaborative skills. Three main themes were identified:

1. Self-directed learning: Students reported a heightened sense of accountability for their education, encompassing independent exploration of algorithmic concepts and examination of alternative coding strategies.
2. Classroom observations revealed that students often participated in peer-to-peer debates, during which they evaluated and enhanced algorithmic strategies, demonstrating the advancement of analytical reasoning skills.
3. Students indicated improved communication skills, heightened confidence in group interactions, and a deeper appreciation for collaborative problem-solving, highlighting the influence of PBAL on enhancing teamwork.

The qualitative findings substantiate H4, demonstrating that PBAL improved academic performance and developed essential soft skills. The findings align with constructivist and experiential learning theories, emphasizing that substantial knowledge is generated through active participation, reflection, and collaboration.

6 Discussion

This research assessed the efficacy of a Project-Based Active Learning (PBAL) method in instructing searching algorithms and examined its influence on conceptual comprehension and practical performance and the results demonstrated significant statistical evidence of improvement and provided insights into the relationship among prior knowledge, practical application, and collaborative learning.

1. **Learning Gains and Hypothesis H1 (Research Question 1)**
   The notable enhancement observed from pretest to posttest ($M = 7.76$ to $M = 17.04$, $p < 0.001$) corroborates H1 and responds to RQ1, indicating that PBAL significantly improved students' conceptual understanding of searching algorithms. This finding aligns with previous studies indicating that active methodologies enhance academic performance and engagement in higher education [16]. García-Holgado [17] demonstrated that active methodologies implemented through virtual campuses improved student performance in software engineering. Morais [18] noted heightened engagement in project-based activities. Problem-based learning (PBL) enhances performance and student perceptions in computer science engineering [19]. This study expands existing research to encompass searching algorithms, an essential but inadequately examined aspect of computer science education.
2. **Prior Knowledge and Hypothesis H2 (RQ2)**
   The moderate correlation between pretest and posttest scores ($r = 0.378$) supports H2 and addresses RQ2, suggesting that students with higher foundational understanding generally attained superior outcomes. The relationship's moderate strength implies that by assisting pupils with less prior knowledge, PBAL helped to close performance gaps. This aligns with constructivist learning theory, highlighting the importance of scaffolding to support learners in the Zone of Proximal Development. Morais [18] reported comparable findings, demonstrating that project-based methodologies enhanced engagement among students from diverse backgrounds. Chávez [19] demonstrated that project-based learning (PBL) resulted in equitable improvements in student learning outcomes.
3. **Applied Practice and Hypothesis H3 (RQ3)**
   The observed weak positive correlation between posttest and lab scores ($r = 0.229$) offers partial support for H3 and addresses RQ3. The relationship between student performance in lab assignments and posttest results is positive but modest. This suggests that applied practice and conceptual mastery are complementary but separate aspects of learning. This corroborates the findings of Hmelo-Silver [20] and aligns with those of Aires [21], who demonstrated that problem-based and project-based methodologies in programming improve collaboration and practical competence, despite variations in conceptual test performance and the significance of combining agile practices with active methodologies, suggesting that structured, iterative project work improves problem-solving skills and promotes adaptability, which are crucial in algorithm education [22].
4. **Collaboration, Equity, and Hypothesis H4 (RQ4)**
   The minimal correlation between pretest and lab scores ($r = 0.013$) indirectly supports H4, indicating that prior theoretical knowledge had a limited impact on applied task performance. This demonstrates that PBAL offered equitable assistance,

facilitating the success of students with weaker foundations in collaborative, practical environments. The consistently high laboratory performance (M = 8.02, SD = 0.84) supports this interpretation. The results correspond with the work [16], which highlighted that both educators and learners view active methodologies as promoting engagement and collective accountability. In a similar vein, [18] indicated that project-based learning in software engineering enhanced teamwork and collaborative practices. Additionally, [22] showed that the integration of agile and active learning methodologies improved collaboration and engagement in introductory programming for high school students, highlighting the wider relevance of PBAL in various educational settings.

7 Conclusion

7.1 Conclusion

This research assessed the efficacy of a Project-Based Active Learning (PBAL) method in instructing second-year undergraduate students on searching algorithms at Suan Sunandha Rajabhat University in the first semester of the 2025 academic year. The results indicated that PBAL significantly improved students' conceptual understanding, as shown by the notable increase in pretest to posttest scores, and supported consistently high performance in project-based laboratory assignments. H1 was validated by the paired t-test, while H2 and H3 were supported by the correlation analysis, which also indirectly supported H4. These findings demonstrated that PBAL enhanced collaboration and equity in outcomes in addition to fostering individual learning gains and the findings collectively address all four research questions and offer strong empirical evidence for the effectiveness of PBAL in algorithm education.

7.2 Theoretical Implications

The study supports constructivist learning theory, indicating that students attain a deeper understanding through collaborative, hands-on activities. The results correspond with experiential learning theory, highlighting the importance of combining conceptual understanding with active experimentation. This study extends the application of PBAL to algorithm-specific content, particularly searching algorithms, thereby contributing new insights to the field of computer science education, which has primarily concentrated on programming or general data structures in previous research.

7.3 Practical Implications

The findings underscore the importance of incorporating PBAL into algorithm courses. Instructors may implement PBAL to diminish dependence on rote memorization while promoting skills in problem-solving, coding, and critical thinking. This approach promotes equity by facilitating the success of students with limited prior knowledge in project-based tasks through collaboration and structured guidance. The utilization of various assessment measures, including pretests, posttests, and laboratory assignments, highlights the significance of evaluating both conceptual and applied aspects of learning, thereby providing a more thorough understanding of student progress. These insights are significant for educators and curriculum designers aiming to enhance computer science education and adequately prepare students for real-world challenges.

7.4 Future Directions

This study provides substantial evidence supporting the effectiveness of PBAL. Future research could improve these findings by incorporating qualitative methods, such as interviews or reflective journals, to gain deeper insights into students' collaborative learning processes. Comparative studies incorporating control groups or addressing diverse algorithmic topics would improve the generalizability of the results. Investigating the long-term retention of knowledge acquired through PBAL will provide significant insights into sustaining learning outcomes beyond a single course.

Acknowledgements. This research was funded by Suan Sunandha Rajabhat University.

References

1. Calderon Ribeiro, M.I., Mestrinho Passos, O.: A study on the active methodologies applied to teaching and learning process in the computing area. IEEE Access. **8**, 219083–219097 (2020). https://doi.org/10.1109/ACCESS.2020.3042747
2. Korbuakaew, T., Robroo, I., Korbuakaew, S.: The development of online active learning management with application for pre-service teachers. Int. J. Inf. Educ. Technol. **12**(1), 77–81 (2022)
3. Freeman, S., et al.: Active learning increases student performance in science, engineering, and mathematics. Proc. Natl. Acad. Sci. USA. **111**(23), 8410–8415 (2014). https://doi.org/10.1073/pnas.1319030111
4. Yong, S.T., Tiong, K.M.: A blended learning approach. Int. J. Inf. Commun. Technol. Educ. **18**, 1–16 (2022)
5. Chutrtong, J., Boonman, N., Chutrtong, W.: Service learning by STEM activity in secondary school at Prachin Buri Province, Thailand. Int. J. Inf. Educ. Technol. **9**(8), 580–583 (2019)
6. Holmes, V.L., Hwang, Y.: Exploring the effects of project-based learning in secondary mathematics education. J. Educ. Res. **109**(5), 449–463 (2016)
7. Sánchez-García, R., Reyes-de-Cózar, S.: Enhancing project-based learning: a framework for optimizing structural design and implementation—a systematic review with a sustainable focus. Sustainability. **17**(11), 4978 (2025). https://doi.org/10.3390/su17114978
8. Ahmed Qasem, A.: The effectiveness of project-based learning in development of C# programming language skills among education technology students. Discov. Educ. **4**, 223 (2025). https://doi.org/10.1007/s44217-025-00633-0
9. Aidoo, B.: Teacher educators experience adopting problem-based learning in science education. Educ. Sci. **13**, 1113 (2023). https://doi.org/10.3390/educsci13111113
10. Syahril, Purwantono, Wulansari, R.E., Nabawi, R.A., Safitri, D., Kiong, T.T.: The effectiveness of project-based learning on 4Cs skills of vocational students in higher education. J Tech Educ Train. **14**(3), 29–37 (2022). https://publisher.uthm.edu.my/ojs/index.php/JTET/article/view/10322
11. Ngereja, B., Hussein, B., Andersen, B.: Does project-based learning (PBL) promote student learning? A performance evaluation. Educ. Sci. **10**(11), 330 (2020). https://doi.org/10.3390/educsci10110330

12. Lazić, B.D., Knežević, J.B., Maričić, S.M.: The influence of project-based learning on student achievement in elementary mathematics education. S. Afr. J. Educ. **41**, 1909 (2021)
13. Yusuf, A.R., Marji, M., Sutadji, E., Sugandi, M.: Integration of STEM project-based learning into 21st century learning and innovation skills (4Cs) in vocational education using SEM model analysis. Hacet. Univ. J. Educ. **38**, 454–469 (2023)
14. Kularbphettong, K., Raksuntorn, N., Roonrakwit, P., Boonseng, C.: Enhancing student's learning by integrating the concept of project-based and challenge-based learning. In: Arabnia, H.R., Deligiannidis, L., Shenavarmasouleh, F., Amirian, S., Ghareh Mohammadi, F. (eds.) Computational Science and Computational Intelligence. CSCI 2024. Communications in Computer and Information Science, vol. 2504. Springer, Cham (2025). https://doi.org/10.1007/978-3-031-94943-2_17
15. Bryman, A.: Methods and methodology. Qual. Res. Organ. Manag. Int. J. **2**, 159–168 (2008)
16. Crisol-Moya, E., Romero-López, M.A., Caurcel-Cara, M.J.: Active methodologies in higher education: perception and opinion as evaluated by professors and their students in the teaching-learning process. Front. Psychol. **11**, 565113 (2020). https://doi.org/10.3389/fpsyg.2020.565113
17. García-Holgado, A., Vázquez-Ingelmo, A., García-Peñalvo, F.J., Conde, M.R.: Improvement of learning outcomes in software engineering: active methodologies supported through the virtual campus. IEEE Rev. Iberoam. Tecnol. Aprendiz. **16**, 143–153 (2021)
18. Morais, P., Ferreira, M.J., Veloso, B.: Improving student engagement with project-based learning: a case study in software engineering. IEEE Rev. Iberoam. Tecnol. Aprendiz. **16**, 21–28 (2021)
19. Amaya Chávez, D., Gámiz-Sánchez, V.M., Cañas Vargas, A.: Problem-based learning: effects on academic performance and perceptions of engineering students in computer sciences. J. Technol. Sci. Educ. **10**, 306–328 (2020). https://doi.org/10.3926/jotse.979
20. Hmelo-Silver, C.E.: Problem-based learning: what and how do students learn? Educ. Psychol. Rev. **16**(3), 235–266 (2004). https://doi.org/10.1023/B:EDPR.0000034022.16470.f3
21. Aires, J.P., Aires, S.B.K., Pereira, M.J., Alves, L.M.: Using the methodology problem-based learning to teaching programming to freshman students. Int. J. Inf. Educ. Technol. **13**(5), 448–455 (2023). https://doi.org/10.18178/ijiet.2023.13.5.1842
22. Moreno, B.: Combining agile and active methodologies in teaching introduction to programming to high school students. In: Proc Workshop Challenges Comput Appl Educ, vol. 8, p. 45. Universidad de Alicante (2019)

A Curriculum Framework for Algorithms and Data Structures: Enhancing Skill Development and Critical Thinking

Dewan Tanvir Ahmed(✉)

Department of Computer Science, University of North Carolina at Charlotte, Charlotte, NC, USA
dahmed@charlotte.edu

Abstract. This paper investigates the design and implementation of a graduate-level Algorithms and Data Structures curriculum and proposes a revised version that enhances both theoretical understanding and practical relevance. The study emphasizes the importance of balancing algorithmic foundations with applied problem-solving, ensuring that students not only master formal concepts but also develop skills that prepare them for real-world computational challenges. The curriculum incorporates several key improvements. First, the integration of active learning enriches the traditional lecture-heavy approach by introducing coding challenges, in-class exercises, and collaborative activities to improve retention and confidence. Second, a stronger focus on real-world applications contextualizes abstract concepts by linking algorithms to domains such as task ordering, networking, scheduling, and biological data analysis. Third, a diversified assessment strategy—covering class tests, projects, participation, and exams—ensures that student performance is evaluated comprehensively across different skill sets. The curriculum also maintains progressive skill development by gradually increasing the complexity of the content. Finally, it introduces practical algorithms and trade-off analysis to strengthen critical thinking and professional decision-making. Student learning outcomes reveal promising results, with most learners achieving proficiency in algorithm design, trade-off evaluation, and applied problem-solving. Nonetheless, the data highlight the need for an additional framework to help struggling students transition from competency to mastery levels.

Keywords: Curriculum · Engagement · Pedagogy · Feedback · Class Activities

1 Introduction

From a design perspective, an effective curriculum reflects institutional goals. A solid design incorporates the progression of concepts, diverse instructional strategies, and appropriate assessments. As industries evolve and technologies

H. R. Arabnia et al. (Eds.): CSCI 2025, CCIS 2974, pp. 269–279, 2027.
https://doi.org/10.1007/978-3-032-28475-4_19

advance, the curriculum must adapt, integrating new tools and approaches while maintaining academic standards. A key part of quality education is keeping course material updated and aligned with the learning outcomes set by the department or institution. The goal is to ensure that students gain the right skills to stay competitive in the job market. This alignment not only meets accreditation standards but also benefits students, faculty, and the institution.

Aligning teaching with assessment provides a clear link between what students are taught and how their progress is evaluated. This not only maintains academic standards and accountability but also improves the overall learning experience, promotes consistency across programs, and enhances the reputation of the institution.

This paper investigates the graduate-level Algorithms and Data Structures curriculum presented in [1] and outlines a modified version of it. The course framework balances theoretical depth with practical application, giving students both algorithmic foundations and real-world problem-solving opportunities. Assessments are varied and intentionally tied to learning outcomes, ranging from coding assignments and analytical exercises to collaborative projects. This ensures that students are evaluated in ways that reflect their growth. By combining accreditation requirements with best practices in curriculum design, the course provides a robust, future-focused learning experience while serving as a model for ongoing refinement and enhancement [2,3].

The key improvements made in the structure and content are summarized below:

- Active Learning: Weekly in-class problem solving, mini coding assignments, and real-world problem-solving.
- Applied Focus: Real-world examples across various domains, including text search, optimization, and best practices.
- Balanced Assessment: A mixture of homework, class tests, activities, class participation, projects, and examination.
- Skill Development: Greater emphasis on implementation, experimentation, and critical thinking.

In this paper, Student Learning Outcomes (SLOs) are evaluated based on the presented curriculum. The course has four SLOs:

- SLO1: Data Structure Mastery - Students will design, implement, and select advanced data structures, justifying their choices for specific applications.
- SLO2: Algorithm Proficiency - Students will analyze computational complexity, compare algorithms, and choose appropriate methods for problem-solving.
- SLO3: Algorithm and Data Structure Design - Students will create efficient solutions for complex challenges while evaluating trade-offs between options.
- SLO4: Problem-Solving Skills - Students will apply algorithms and data structures to real-world problems and assess the efficiency and correctness of their solutions.

The rest of the paper is organized as follows. Section 2 describes the course design and course delivery strategies adopted. Section 3 presents observations from the experiments and recommendations for instructors. Finally, the paper concludes in Sect. 4.

2 Course Design and Improvement

This paper examines an Algorithms and Data Structures course to understand the impact of curriculum design, teaching methodologies, and assessment criteria on student learning outcomes. First, I briefly present the course curriculum and structure to provide context for their impact on students' academic performance and skill development.

The course is designed for a 15-week semester and aims to build a strong foundation in algorithms and data structures while ensuring students develop both theoretical understanding and practical problem-solving skills.

- In Week 1, students are introduced to algorithms, data structures, and the foundations of algorithm analysis. They begin with experimental evaluation, counting primitive operations, and learning asymptotic notation in an intuitive way. Simple case studies, such as insertion sort and its loop invariant, are analyzed for best-case and worst-case performance.
- Week 2 focuses on fundamental data structures, including stacks, queues, linked lists, and dynamic arrays with amortized analysis. Basic tree structures and traversals methods are introduced, along with dictionaries, binary search, and binary search trees. Students also study hash tables, hashing functions, and collision resolution strategies such as chaining, open addressing, double hashing and universal hashing.
- In Week 3, the course covers sorting algorithms in detail. Students explore divide-and-conquer through merge sort and quicksort, including pivot strategies, runtime analysis, and recurrence relations. The lower bound for comparison based sorting and linear-time sorts, such as counting, radix, and bucket sort, are also examined. The first in-class test is held.
- Weeks 4 and 5 are devoted to recurrence solving methods and dynamic programming. Students study approaches such as recursion trees and the Master Theorem, and apply dynamic programming to classic problems like matrix chain multiplication, 0/1 knapsack, coin change, and longest common subsequence, etc.
- Week 6 introduces balanced search trees. Students learn AVL trees, including rotations, insertions, deletions, and height balancing, along with red-black trees as used in practice. Self-adjusting data structures, such as splay trees and self-organizing lists, are also studied.
- During Weeks 7 and 8, the course moves into graph theory. Students learn graph representations such as adjacency lists and matrices and perform DFS and BFS with runtime analysis. Applications like connectivity, cycle detection, and topological sorting are emphasized. Strongly connected components, transitive closure, and reachability problems are also explored. The second in-class test is held.

- Week 9 includes a midterm exam followed by applied case studies in graph algorithms for routing and dependency analysis.
- In Weeks 10 and 11, students advance to shortest path problems and spanning trees. Dijkstra's, Bellman-Ford, and DAG shortest path algorithms are studied, along with Prim's and Kruskal's algorithms for minimum spanning trees. Disjoint set data structures with path compression and union-by-rank are also introduced. Students then explore greedy algorithms such as fractional knapsack, Huffman coding, and scheduling problems.
- Week 12 begins with randomized algorithms, including skip lists and their analysis. String matching algorithms such as the naive method and Boyer-Moore-Horspool are introduced. The third in-class test is conducted.
- Week 13 continues string algorithms with Boyer-Moore, KMP, Rabin-Karp, and DFA-based matching. Students also study tries, suffix trees, and applications to text and sequence matching.
- Week 14 presents advanced topics, including network flows. Students learn the Max-Flow Min-Cut theorem and implement Ford-Fulkerson and Edmonds-Karp algorithms.
- Finally, in Week 15, the course concludes with complexity class and approximation algorithms. Students are introduced to NP-completeness, reductions, and approximation techniques for hard problems such as vertex cover, set cover, and traveling salesman. Trade-offs between optimality and efficiency are emphasized.

The revised curriculum maintains a rigorous theoretical foundation while enriching the course with active learning, applied examples, varied assessments, modern relevance, and a stronger focus on trade-off analysis. This creates a more engaging and skill-focused learning experience that better prepares students for both academic and professional success [4–8]. The key improvements in the curriculum are given below:

Integration of Active Learning - The original curriculum was primarily lecture-based. The revised version introduces mini coding assignments and in-class problem-solving sessions, allowing students to apply concepts immediately after learning them. This approach helps improve retention, boost confidence, and bridge the gap between theoretical concepts and coding practice.

Heavy Focus on Applied and Real-World Problems - While the original structure offered strong theoretical coverage, it placed limited emphasis on practical applications. The improved version explicitly connects topics to real-world problems: graph algorithms are tied to social networks and routing; string matching to DNA sequence alignment; optimization to scheduling and resource allocation; and network flow to traffic management. This approach demonstrates the practical utility of algorithms and motivates students toward deeper learning.

Incorporate a Balanced Assessment Strategy - The revised assessment strategy incorporates a balance of class tests, homework assignments, active participation, coding projects, and a final examination. This multifaceted approach provides multiple avenues for students to demonstrate their mastery of the course material.

Maintain Progressive Skill Development - Following the original design sequence, the content is presented in order of increasing difficulty. The structure facilitates learning through progressive complexity: early weeks focus on foundational skills, case studies, and introductory assignments, while later weeks address advanced material that builds upon prior knowledge. This scaffolded approach prevents students from becoming overwhelmed and systematically strengthens their problem-solving abilities.

Connect Industry and Future Relevance - While the original curriculum focused on traditional topics such as sorting, dynamic programming, and graph algorithms, the revised version incorporates foundational exposure to practical algorithms with applications in machine learning, networking, and data mining. This ensures that students are better prepared for real-world computing challenges and emphasizes the professional relevance of algorithmic thinking in modern careers.

Promote Critical Thinking - While the original curriculum emphasized correctness and efficiency, the new version incorporates rigorous trade-off analysis across multiple algorithmic approaches. Students now evaluate greedy versus dynamic programming strategies, randomized versus deterministic methods, and exact versus approximation algorithms. This focus strengthens critical thinking and prepares students to select the most suitable solutions within complex professional contexts.

3 Observation and Remarks

While teaching a course, multiple approaches can be applied to enhance student engagement and learning outcomes. These approaches may include traditional lectures, whiteboard explanations, animations, YouTube or recorded lecture videos, in-class problem-solving sessions, active learning strategies, and collaborative group projects [6,7]. Instructors also have the option to adapt or introduce other techniques that align with their teaching style and along with the needs of students. To better understand student learning experience, an optional anonymous survey was conducted. A total of 54 students from the Algorithms and Data Structures course participated and shared their responses.

Q1: Upon completing the course, would you be able to: a) design and implement advanced data structures, such as AVL trees, Skip List, graphs, and hash tables, b) select appropriate data structures for specific application scenarios and justify their choices.

The survey results for SLO 1 (Data Structure Mastery), based on 54 responses, indicate that the majority of students demonstrated strong proficiency in designing, implementing, and justifying the use of advanced data structures. A combined 79.6% of students rated themselves as either "Good" (44.4%) or "Excellent" (35.2%). Students in these categories reported the ability to appropriately select and apply complex data structures—including AVL trees, skip lists, graphs, and hash tables—while justifying their selections based on efficiency

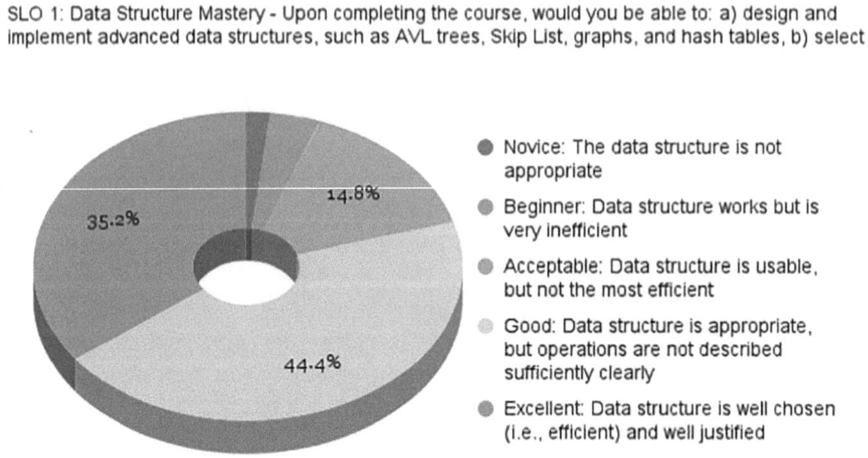

Fig. 1. Data Structure Mastery in terms of design and implementation

and suitability. This outcome demonstrates the effectiveness of the pedagogy in equipping students with both technical mastery and critical decision-making skills (Fig. 1).

Meanwhile, 14.8% of students rated themselves as "Acceptable," suggesting they could identify and implement data structures but struggled with providing strong justifications or optimizing their efficiency. This group reflects an intermediate level of achievement and highlights the need for more practice in evaluating trade-offs and articulating reasoning behind data structure selection. Only a small proportion of students assessed themselves as "Beginner" or "Novice," indicating limited ability to design or justify the use of data structures. These cases suggest that while the course was largely successful, a minority of students would benefit from additional scaffolding, targeted exercises, and feedback to solidify their understanding.

Overall, the result demonstrates that the majority of students achieved the intended learning outcome for SLO 1. Most students were able to design and implement advanced data structures effectively, while also making appropriate and well-justified choices, though continued instructional support is needed to lift students from the acceptable range into higher mastery levels.

Q2: At the end of the course, would be able to: a) analyze the time and space complexities of algorithms effectively, b) compare different algorithms, and can make appropriate choices for various problem-solving situations.

The survey also assessed whether, by the end of the course, students were able to: (a) effectively analyze the time and space complexities of algorithms, and (b) compare various algorithms to make appropriate choices for specific problem-solving scenarios (Fig. 2).

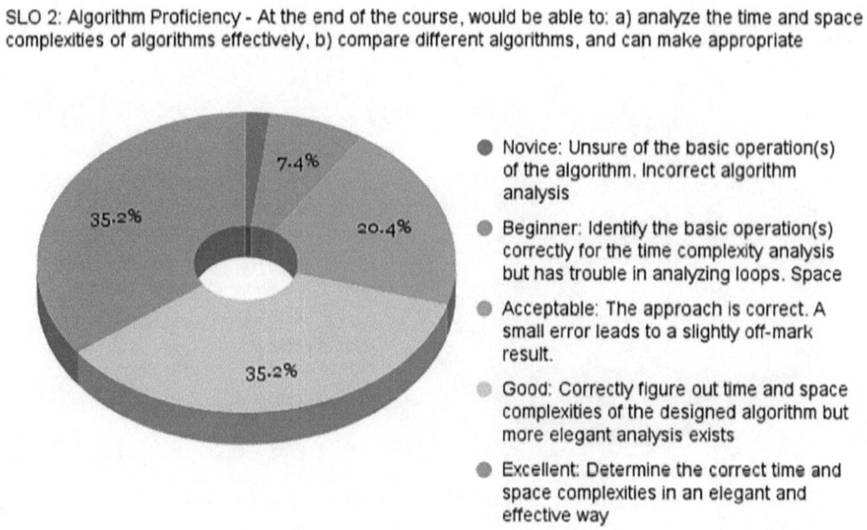

Fig. 2. Algorithmic Proficiency in terms of complexity analysis and appropriate algorithm selection

Based on 54 responses, the results show that 35.2% of students rated themselves at the "Good" level, indicating confidence in analyzing complexities and comparing algorithms, though with some room for improvement. Another 35.2% rated themselves as "Excellent," showing strong mastery and the ability to determine complexities efficiently while justifying their choices clearly. On the other hand, 20.4% of students placed themselves in the "Acceptable" category, showing partial understanding but lacking depth and consistency. About 9.2% identified as "Beginner" or "Novice", being able to recognize basic complexities but struggling with deeper comparisons. Overall, the data suggests strong achievement, with a majority of 70.4% (Good + Excellent) demonstrating proficiency or higher. However, around 9.2% of students remain in the Beginner or Novice categories, which signals a need for reinforcement of foundational skills. The Acceptable group (20.4%) also highlights students who are close to proficiency but require more structured practice with trade-offs and justifications.

In conclusion, the results reflect positive outcomes, with most students achieving proficiency in algorithm analysis and comparison. However, targeted interventions such as foundational reviews, additional problem-solving exercises, and scaffold comparisons could help lower-performing students' progress while pushing more learners from "Good" to "Excellent."

Q3: Upon completion of the course, would you be able to: a) design efficient algorithms and data structures to address complex computational challenges, b) analyze the trade-offs between different algorithmic and data structure options.

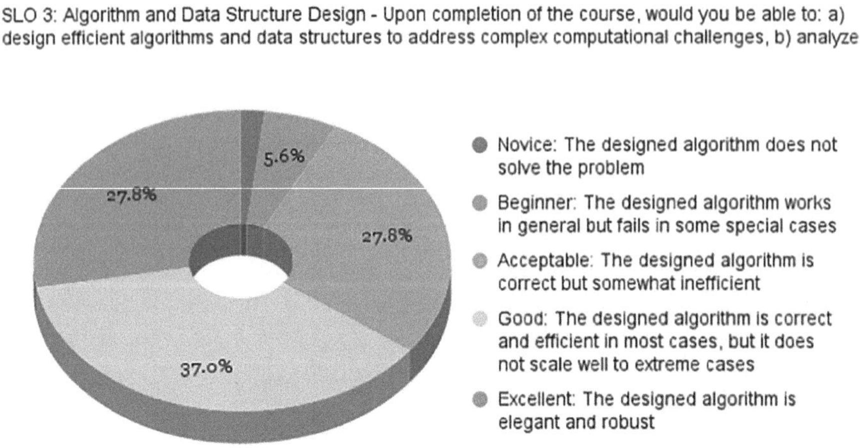

Fig. 3. Algorithm and Data structure design skills

This question evaluated whether students, upon completion of the course, can (a) design efficient algorithms and data structures to address complex computational challenges and (b) analyze trade-offs among different algorithmic and data structure options (Fig. 3).

The responses suggest that a strong majority of students developed meaningful proficiency in these advanced skills. Most students, 64.8%, reported being either "Good" or "Excellent", reflecting their ability to not only design appropriate solutions but also to evaluate competing approaches critically. This indicates that the course was effective in preparing students to balance efficiency, correctness, and practicality when faced with complex computational problems. A smaller proportion, 27.8%, of students rated themselves as "Acceptable", showing they can design and analyze algorithms but may still lack consistency or confidence when dealing with highly complex challenges. Only a limited percentage identified as "Beginner" or "Novice", suggesting that while some students struggled with foundational aspects of design and trade-off analysis, these were not widespread issues.

Overall, the data reveal encouraging results: the majority of learners attained strong competency in algorithm and data structure design, as well as in evaluating trade-offs, which are essential for real-world computational problem-solving. However, the presence of a small group of students in the lower categories highlights the need for additional scaffolding, more hands-on practice with trade-off case studies, and targeted support to ensure that all students can progress toward higher proficiency levels.

Q4: By the end of the course, would you be able to: a) apply algorithms and data structures to solve real-world problems, demonstrate advanced problem-solving abilities, b) evaluate the efficiency and correctness of algorithmic solutions.

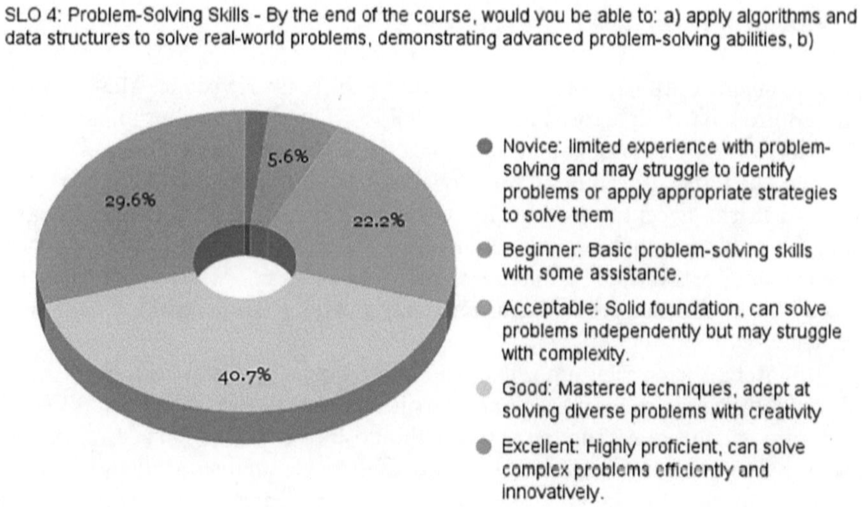

Fig. 4. Problem solving skills

The survey results for SLO 4 (Problem-Solving Skills), based on 54 student responses, indicate that most students achieved strong levels of proficiency. A combined 70.3% of respondents rated themselves as either “Good” (40.7%) or “Excellent” (29.6%), suggesting that most students were able to effectively apply algorithms and data structures to real-world problems while also evaluating the efficiency and correctness of their solutions. These outcomes reflect the success of the course pedagogy in building advanced problem-solving abilities and encouraging both accuracy and creativity in algorithmic thinking. At the same time, 22.2% of students assessed themselves at the “Acceptable” level, showing they had a solid foundation and could solve problems independently but struggled with efficiency or handling more complex tasks. This group highlights an area for growth, where additional practice, scaffolded instruction, and targeted feedback could help move students toward higher levels of mastery. Only a small percentage of students identified as “Beginner” (5.6%) or “Novice” (1.9%). These students likely experienced significant challenges with problem-solving and would benefit from early intervention, additional resources, or peer support to strengthen their skills (Fig. 4).

The data show that 70.3% of students achieved proficiency or higher in problem-solving skills, confirming the effectiveness of the teaching pedagogy in meeting SLO 4. However, approximately 30% of students remain in the Acceptable to Beginner range, highlighting the need for additional structured support to help them apply algorithms efficiently to real-world problems.

4 Conclusion

The assessment data support the effectiveness of these revisions. Most students achieved proficiency in algorithm design, trade-off analysis, and problem-solving, validating the curriculum's alignment with Student Learning Outcomes (SLOs). However, the data also point to some challenges: approximately 30% of students remained in the acceptable-to-beginner range. This suggests the need for additional mechanisms, such as guided tutorials, targeted problem-solving sessions, enhanced support and accommodations. These steps could help struggling students build confidence while simultaneously pushing advanced learners toward higher levels of mastery.

Ultimately, the curriculum reflects an important step toward a more dynamic and impactful approach to teaching algorithms and data structures. While the results are encouraging, the curriculum should be viewed as part of an ongoing cycle of refinement. By continuing to collect feedback, analyze student outcomes, and adapt teaching strategies, this framework can evolve to meet the changing demands of both academia and industry. In doing so, it sets a model for future curriculum design that not only meets accreditation standards but also prepares students to thrive as innovative problem-solvers in a rapidly advancing digital world.

Disclosure of Interests. The author has no competing interests to declare that are relevant to the content of this article.

References

1. Ahmed, D.T.: Creating a holistic graduate-level curriculum for algorithm and data structures with targeted educational objectives. In: ICERI Proceedings, pp. 10034–10040 (2024)
2. Bennedsen, J., Caspersen, M.E.: Failure rates in introductory programming. SIGCSE Bull. **39**(2), 32–36 (2007). https://doi.org/10.1145/1272848.1272879
3. Porter, L., Guzdial, M., McDowell, C., Simon, B.: Success in introductory programming: what works? Commun. ACM **56**(8), 34–36 (2013). https://doi.org/10.1145/2492007.2492020
4. Ahmed, D.: Discovering effective learning methods and impact of team-based programming projects in graduate level courses. In: Proceedings of the 2021 International Conference on Computational Science and Computational Intelligence (CSCI), pp. 1099–1102 (2021). https://doi.org/10.1109/CSCI54926.2021.00232
5. Jin, K.H.: Students' engagement in collaborative active learning - online vs. face-to-face. In: Proceedings of the 53rd ACM Technical Symposium on Computer Science Education V. 2 (SIGCSE 2022), Providence, RI, USA, p. 1135 (2022). https://doi.org/10.1145/3478432.3499107
6. Ahmed, D.T.: Resources for encouraging academically challenged students and their impact. In: 2023 International Conference on Computational Science and Computational Intelligence (CSCI), pp. 1750–1754. IEEE Computer Society, Los Alamitos, CA, USA (2023). https://doi.org/10.1109/CSCI62032.2023.00288

7. Ahmed, D.T.: Preferred course modality and effective teaching methods for graduate level courses. In: Proceedings of the 53rd ACM Technical Symposium on Computer Science Education V. 2 (SIGCSE 2022), p. 1083. Association for Computing Machinery, New York, NY, USA (2022). https://doi.org/10.1145/3478432.3499058
8. Jong, M., Shang, J., Lee, F., Lee, J.: Harnessing computer games in education. Int. J. Distance Educ. Technol. (IJWLTT) **3**, 54–61 (2008)

Improving Capstone Team Outcomes Through Dynamic Skill Matching and Preference Alignment

Brandon Pardi[1], Garret Castro[2], Michael Pisman[2(✉)], Avash Adhikari[2], and Santosh Chandrasekhar[2]

[1] University of California, Davis, Davis, CA 95616, USA
[2] University of California, Merced, Merced, CA 95343, USA
mpisman@ucmerced.edu

Abstract. Team-based projects are a cornerstone of engineering and computing courses, but unstructured team formation often leads to poor project outcomes due to misaligned student interests and inadequate skill coverage. This paper introduces a novel, three-stage methodology for creating effective student teams by integrating student preferences with project skill requirements.

In the first stage, students complete a survey to report their project interests and self-assessed skills. Next, a Large Language Model (LLM) analyzes project descriptions to extract the necessary skills for each project's success. Finally, a dynamic assignment algorithm matches students to projects, simultaneously maximizing skill coverage and preference alignment. The algorithm iteratively prioritizes projects with unfulfilled skill needs to optimize team balance.

Preliminary evaluations show our approach produces teams with higher skill coverage and better preference satisfaction compared to random or manual assignment approaches. Our approach also overcomes limitations of widely-used tools like CATME Team-Maker, which do not explicitly account for project skill fulfillment. Our findings point toward an effective and customizable strategy for improving student motivation and learning outcomes in project-based courses.

Keywords: Capstone Projects · Automated Team Formation · Project-Based Learning · Computer Science Education · Greedy Algorithm · Skill-based Matching · Preference Optimization · Large Language Models(LLMs) · Stable Matching

1 Introduction

Capstone and senior design courses in engineering and computing education are typically organized around project-based learning, providing students with opportunities to collaboratively apply their acquired knowledge and skills to authentic and professionally relevant problems. These courses often involve

H. R. Arabnia et al. (Eds.): CSCI 2025, CCIS 2974, pp. 280–294, 2027.
https://doi.org/10.1007/978-3-032-28475-4_20

teams working on real-world projects that are proposed by industry or academic sponsors. The success of such project-based courses strongly relies on effective team formation to ensure that student teams possess both the necessary skills to successfully complete projects and a strong interest alignment with their assigned projects to maintain student motivation and agency. However, without a carefully designed assignment process, students often end up assigned to projects that either do not match their interests or lack critical skill coverage, which can result in reduced motivation, lack of agency, and poor project outcomes [16].

In practice, team formation methods can be broadly classified into student-led and instructor-led approaches. While student-led team formation can lead to increased agency, it often produces suboptimal results in scenarios where projects have predefined skill requirements and objectives [22]. More specifically, student-led approaches typically fail to ensure that students are matched with projects that are aligned with their skill sets, thus negatively impacting project success. On the other hand, instructor-led team formation methods can be more effective in aligning student skills and preferences with explicit project requirements. However, when done manually, an instructor-led process is both labor-intensive and susceptible to bias, particularly when involving complex and nuanced metrics such as detailed skill sets and individual student preferences [23]. These challenges have led to a growing interest in data-driven or automated methods to streamline and improve team formation.

Existing automated tools, such as CATME Team-Maker [19], address several common challenges of team formation by grouping students based on demographic factors and personality traits. However, CATME lacks explicit mechanisms for detailed skill-to-project matching and does not dynamically integrate student preferences into the assignment process. This necessitates manual intervention post-assignment, adding significant cognitive load on instructors and potentially reducing the effectiveness of teams [14]. Other methods employing optimization techniques, heuristic algorithms, or machine learning approaches, while capable of forming balanced teams, similarly fail to simultaneously address detailed project-specific skill coverage and student preferences dynamically. Furthermore, approaches relying purely on clustering methods or stable matching algorithms typically focus only on optimizing student satisfaction without adequately considering the detailed skill needs of each project. Recent advances in natural language processing, particularly the emergence of large language models (LLMs), offer promising solutions for automating skill extraction from project descriptions to facilitate more accurate team formation.

In this paper, we propose a novel, data-driven team formation algorithm that dynamically balances detailed project-specific skill requirements with individual student preferences. Our approach comprises of three stages: (1) student surveys to capture detailed and granular skill sets and project preferences; (2) LLM-driven extraction of essential project skills from sponsor-provided descriptions; and (3) a dynamic matching algorithm that iteratively prioritizes currently unfulfilled skills and integrates student preferences during team formation. To systematically evaluate our approach, this study addresses the following research questions:

RQ1: Does dynamic skill weighting increase project skill coverage compared to static weighting approaches?
RQ2: Does our algorithm improve student preference satisfaction compared to manual and random assignment baselines?
RQ3: Does the combined approach of skill-preference matching outperform existing team formation methods in terms of overall team formation quality?

We validate the effectiveness of our approach using real data from capstone courses at University of California, Merced (UC Merced), collected during the Fall 2023 and Spring 2024 semesters. Analysis of our results indicates that our algorithm significantly improves skill coverage compared to manual and random assignments, while maintaining strong alignment with student preferences, demonstrating its potential to enhance student motivation and agency, and improve project outcomes in capstone courses.

Contributions. To summarize, the main contributions of our work are as follows:

- A novel three-stage methodology for automated team formation that combines student skill surveys, LLM-based project skill extraction, and dynamic preference-aware matching.
- An LLM-driven approach for automatically extracting essential technical skills from project descriptions, enabling more accurate skill-to-project alignment.
- A dynamic matching algorithm that iteratively prioritizes unfulfilled project skill requirements while simultaneously optimizing student preference satisfaction.
- Empirical validation on real capstone course data demonstrating significant improvements in skill coverage and preference alignment compared to manual and random assignment methods.

2 Related Work

The problem of forming effective student teams in capstone and senior project courses has been widely studied, leading to a variety of algorithmic approaches ranging from rule-based heuristics to optimization and machine learning techniques. While existing tools provide structured methods for team assignment, they often focus on student attributes within teams. Moreover, these tools lack the ability to simultaneously perform fine-grained skill matching and dynamically integrate student preferences, both of which are crucial for successful team formation in capstone settings.

Recent work has further highlighted that many published team-formation algorithms narrowly target specific formulations (e.g., a fixed objective mix, a single preference model, or rigid constraints), limiting generalizability across diverse courses. This fragmentation makes it difficult for instructors to compare methods or adapt them to evolving pedagogical goals [16].

Additionally, recent advances in natural language processing (LLMs) enable automated extraction of project-required skills from textual descriptions [7,8,18]. This section reviews prior work on tools and algorithms, the role of LLMs, and key research gaps our approach addresses.

CATME Team-Maker and Related Methods: CATME Team-Maker [19] provides structured surveys and semi-automated grouping to balance traits (e.g., leadership, availability). However, it does not explicitly match student skills to project requirements, nor does it dynamically weight scarce skills as assignments progress, often requiring post-processing by instructors [14].

In a multi-instructor study, most CATME users reported performing manual "sanity checks" and re-running configurations, citing concerns about opaque scoring, criterion weighting, default biases, and limited transparency when using sensitive attributes [14]. Meulbroek et al. [21] propose integrating Gale–Shapley [11] to bring student preferences into capstone assignment, but the implementation optimizes student satisfaction only and does not provide skill-to-project matching, shifting burden to instructors. In contrast, we target both preference alignment and explicit project skill coverage.

Other Criterion-Based Heuristic Matching Methods: FASTT [5] uses a multi-round selection mechanism and fairness criteria to balance teams, but does not account for project preferences. Genetic-algorithm families [1,4] optimize homogeneity/diversity proxies across teams; while useful for balance, they typically do not integrate granular project skill coverage with student project choices.

Machine Learning Methods: Clustering approaches (e.g., k-means over project choices) can structure cohorts [2] but do not guarantee per-project skill coverage. Cognitive and diagnostic models (e.g., SDINA, Bayesian roles) [3,20] estimate proficiency and roles from assessment and peer data, improving team heterogeneity and dynamics, yet they require rich longitudinal data and typically do not incorporate student project preferences explicitly. ML–assisted heuristics such as MCTS pipelines introduce predictive signals (availability, proficiency growth) [6], but often rely on large activity datasets [10,17] and are rarely validated in situated capstone deployments.

Optimization and Integer Linear Programming (ILP) Methods: Optimization based methods (e.g., ILP) can provide guarantees for specific objectives (interest similarity, balance) [9,26]. Closer to capstone needs, Kessler et al. [16] propose a hierarchical ILP-based method to coordinate multiple pedagogical goals, and Hammond et al. [13] incorporate student project preferences alongside skills and instructor constraints. While expressive, ILP approaches can be slow at scale and solver-dependent. In contrast, our greedy matching algorithm runs in $O(n^3)$, scales with cohort size, and maintains both project-specific skill fulfillment and project-choice alignment during assignment. Moreover, our algorithm is designed to minimize assignment bias and ensure equitable distribution of student expertise and interests across all projects.

Complexity Considerations and Hierarchical Constraints: Natural formulations that jointly enforce capacity, multi-skill coverage, and preference constraints are NP-hard, requiring careful objective design and practical compromises for course-scale deployments [16]. This motivates adaptive or hybrid strategies like ours that quickly approximate quality solutions while exposing tunable levers (e.g., preference weighting) to instructors.

In summary, existing methods tend to capture broad characteristics or preferences but rarely optimize both project-specific skill fulfillment and project choice simultaneously with dynamic weighting. To the best of our knowledge, our approach is one of the first to utilize LLM-derived skills and instructor constraints to combine both "bottom-up" and "top-down" methods to form teams. Our algorithm optimizes project preference and skill coverage while maintaining scalability which positions this work among the first capstone-focused methods to use both agency and fine-grained skill fulfillment in a single, scalable pipeline at classroom scale.

3 Methodology

Here we introduce our proposed three-step team formation process that begins with conducting a student survey to capture student skills and project preferences, followed by extraction of project skills using a LLM and finally, the matching system to form the teams while optimizing skill coverage and student preference. The general overview of our system is shown in Fig. 1.

3.1 Student Survey

To effectively match students to projects, information on each student's technical skill sets and project preferences is collected through a survey at the beginning of the course. Prior to completing the surveys, all students are given access to the summary descriptions of all projects, which include the background and problem statement for each project as submitted by the sponsors. In the skill survey, students self-report and score their skills based on their experience from a predetermined list. The list of skills are selected either because projects in prior semesters had required them or to anticipate future projects needing them. Our approach can be easily adapted to work with any set of skills or student *traits* as long as they can be identified as requirements for the projects for the matching system to consider in its calculations (Sect. 3.3). In the preference survey, students provide their preference score for all projects on a discrete scale of one to five, with five indicating the highest preference score.

3.2 Skill Extraction (via LLM)

While use of LLMs for extracting specific skills in educational projects is limited, there are results that showcase their ability to find skills from text associated with educational content [18], and there are several empirical studies that show LLMs are an effective method for extracting skills from unstructured text in the

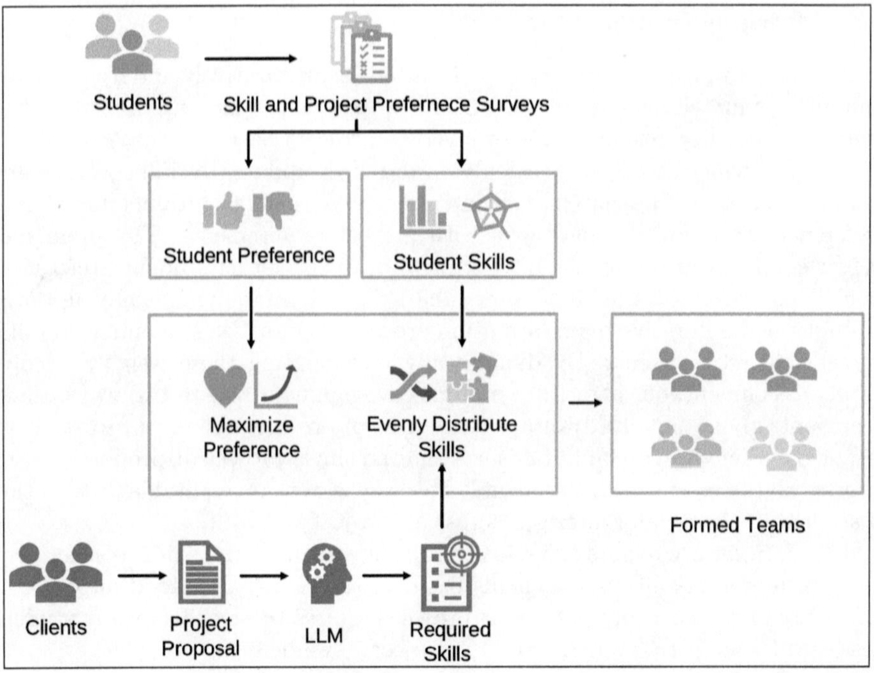

Fig. 1. Flow diagram giving a generalized overview of the team assignment algorithm.

professional industry [7]. Furthermore, LLMs have also been shown to outperform previous methods such as Named Entity Recognition (NER) for extracting skills specifically from unstructured textual data [8]. Our current implementation uses a preexisting fine-tuned instruct version of Llama 3.2 [12,24] applied to a novel use case of LLM skill extraction, aiding capstone project team formation by selecting skills required for project completion.

To identify the skills required for each project, the project descriptions are parsed and sent to the LLM along with the same skill list that is given to the students. It is then asked to rate the necessity of each skill to complete the project on a scale from 0 to 5. A 0 in this case is given when a skill doesn't help for any plausible approach to the project, and a 5 is given when a skill is realistically indispensable for a project. Many project often have multiple possibilities of completion (e.g. one could use Pytorch or Tensorflow to achieve the same results) so the prompt includes such caveats, and instructed to consider multiple potential solutions. This workflow results in a structured JSON list of skills and their rated necessity, that is then manually reviewed and given minor or no corrections.

3.3 Matching System

Our student-to-project matching algorithm works by iteratively and dynamically computing a match score for every student-project pair, reflecting the strength of their compatibility. Students are assigned to projects based on these computed scores, after which scores are recalculated to reflect updates in skill coverage and team composition. The match score integrates two *weighted* parameters: student preferences and skill compatibility with project requirements. The preference weight can be adjusted by the instructor to prioritize either student preferences or skill coverage. At the same time, the skill weight is dynamically updated throughout the iterative team formation process to prioritize students who fulfill unmet skill requirements. By dynamically recalculating these weighted scores during assignment, our algorithm minimizes assignment bias in two ways. First, it prevents giving unfair advantages (in terms of project preference) to students assigned earlier. And second, it ensures that no single project disproportionately receives all the most suitable students. This way, we ensure equitable distribution of student expertise and interests across all projects.

Let P, S, and K denote the sets of projects, students, and skills, respectively. Here, K represents all possible skills listed in the survey described in Sect. 3.1, regardless of whether any particular project requires these skills or any student possesses them. For a student s and a project p, we define:

$PS_s(p)$: Student s's preference score for project p.
$K_p \subseteq K$: Set of all skills required for p.
$K'_p \subseteq K_p$: Set of currently unfulfilled skills for p.
$SS_s(k)$: Student s's self-scoring of skill k.
α: A tunable instructor-defined parameter called the "preference weight."
$\beta_p(k)$: A dynamic parameter called the "skill weight" of skill k for project p.

The instructor sets the value of α to control the influence of student preferences on team assignment. For each skill $k \in K_p$, $\beta_p(k)$ is initialized to 1. Otherwise, if $|K'_p| > 0$ (there are still some unfulfilled skills for a project), $\beta_p(k)$ is computed as $|K_p|/|K'_p|$. A skill is considered *unfulfilled* if none of the students currently assigned to that project have self-rated their proficiency in that skill at a level of 3 or higher (on the survey's 1–5 scale).

Based on this notation, we define an auxiliary function to our student-to-project algorithm. The "match score" $MS_{s,p}$ between student s and project p is computed as follows:

$$MS_{s,p} = \alpha * PS_s(p) + \sum_{k \in K_p} \beta_p(k) * SS_s(k) \quad (1)$$

Notice that as the algorithm progresses, and more students are assigned to the project, the value of $\beta_p(k)$ for unfulfilled skills will increase leading to a higher match score for students that satisfy those skills. This dynamic adjustment of the skill-weight enables better skill coverage for the projects.

ASSIGNMENT ALGORITHM: Students are iteratively assigned using a greedy process (Algorithm 1). For each round, the highest $MS_{s,p}$ among unassigned students and available projects is selected; ties are broken by preference score, then randomly. Scores are recomputed after each assignment until all students are placed. Projects with higher average preference scores are given more capacity, up to a fixed team-size limit (five in our case). This approach ensures fairness and maximizes skill coverage.

Algorithm 1. Greedy Student-to-Project Assignment with Dynamic Skill Weighting

Require: List of students S, list of projects P, preference weight α
Ensure: Student-to-project assignments
while some $s \in S$ is unassigned **do**
  max_score $\leftarrow -\infty$ best_pair $\leftarrow$ None
  for unassigned student $s \in S$ **do**
    for project $p \in P$ with space **do**
      Let K_p be required skills and K'_p be the currently unfulfilled skills
      Initialize all $\beta_p(k) \leftarrow 1$ for $k \in K_p$
      if $|K'_p| > 0$ **then**
        for skill $k \in K'_p$ **do**
          $\beta_p(k) \leftarrow |K_p|/|K'_p|$
      Compute match score: $MS_{s,p} \leftarrow \alpha * PS_s(p) + \sum_{k \in K_p} \beta_p(k) * SS_s(k)$
      if $MS_{s,p} >$ max_score **then**
        max_score $\leftarrow MS_{s,p}$, best_pair $\leftarrow (s,p)$, max_pref $\leftarrow PS_s(p)$
      else if $MS_{s,p} =$ max_score **then**
        if $PS_s(p) >$ max_pref **then**
          best_pair $\leftarrow (s,p)$, max_pref $\leftarrow PS_s(p)$
        else if $PS_s(p) =$ max_pref **then**
          Randomly decide whether to update best_pair
  Assign student s to project p from best_pair and mark s as assigned and update p's assigned list

ON SUPPORT FOR COURSES WITH LABS: The capstone course at UC Merced includes both lecture and lab sessions. Outside of their self-organized meetings, student teams attend these lab sessions to meet and plan their tasks for the week. Although projects can receive students from any lab, we enforce a constraint that all students assigned to a given project must come from the same lab. This constraint ensures that teams have overlapping availability during lab hours, facilitating consistent collaboration. To support this requirement, we introduce a supplementary algorithm that runs prior to Algorithm 1, effectively partitioning the set of projects across all the labs before student-to-project assignments are made.

We refer to the lab associated with the students assigned to a project as the project's *assigned lab*. The supplementary algorithm starts by computing the number of teams needed in each lab by dividing the number of students with an instructor-defined base team size, rounded up. To assign projects to labs, we formulate the problem as a version of the college admissions problem, solvable via the Gale-Shapley algorithm [11], which seeks a stable matching between projects and labs, i.e., a matching without blocking pairs.

The preference list for each lab is constructed by sorting the projects in descending order based on the average project preference score from that lab's students. Conversely, the preference list for each project is constructed by sorting the labs in descending order of the average *lab skill frequency* of each of the project's required skills. A skill's *lab skill frequency* is defined as the percentage of students in that lab who self-report having that skill. Once preference lists are constructed, we match the labs to the projects using the Gale-Shapley algorithm.

While the Gale-Shapley algorithm ensures stability in lab-to-project assignments, it does not account for skill diversity within student teams. As such, once projects are assigned to labs using this method, team formation proceeds using our main algorithm (Algorithm 1) within each lab, where skill-based considerations are fully integrated.

4 Results and Discussion

To evaluate the effectiveness of our approach, we utilize real-world data gathered from the computer science capstone program at UC Merced. This capstone program is offered every semester as a required culminating experience for senior-level computer science students. In this program, external sponsors submit project ideas, and student teams are subsequently formed to complete these projects over a semester. For team assignments in all semesters, we aimed to maintain reasonable team sizes of three to five students, guided by existing research on optimal team sizes [25,27,28] and the typical scope of capstone projects at UC Merced.

The data analyzed was collected from the Fall 2023 and Spring 2024 semesters. In Fall 2023, the dataset included 16 projects and 68 students, with each project-team pairing being unique, i.e., no project was assigned to more than one student team. In Spring 2024, due to increased enrollment and insufficient availability of unique projects, multiple student teams were assigned to some of the projects. Specifically, Spring 2024 had 22 unique projects, with 5 assigned to two teams each. Thus, the Spring 2024 dataset consisted of 27 projects involving 122 students.

4.1 Quantitative Findings

We evaluate our team formation algorithm using two metrics: *Percentage of Skills Fulfilled*, which measures the proportion of required project skills covered by assigned students, and *Average Preference Toward Assigned Project*, which captures the average student preference for their assigned project. Let $K_s \subseteq K$ be the set of skills possessed by student s (for which they assigned a score >1 in the skill survey). Also, let $S_p \subseteq S$ be the set of students assigned to project p. The two metrics can be mathematically defined as follows:

$$\text{Percentage of Skills Fulfilled} = \frac{\left| \left(\bigcup_{s \in S_p} K_s \right) \cap K_p \right|}{|K_p|}$$

$$\text{Average Preference Toward Assigned Project} = \frac{1}{|S_p|} \sum_{s \in S_p} PS_s(p)$$

To evaluate the effectiveness of our approach, we compare its performance across three scenarios. In the first scenario, students are assigned to projects using our algorithm, which balances skill coverage and preference alignment through a dynamic match scoring system described in Sect. 3.3. In the second scenario, the instructor manually forms teams using the same survey data, attempting to optimize both skill coverage and student preferences. The third scenario serves as a baseline, in which students are assigned to projects at random without considering any input data.

Table 1. Tables 1A and 1B: Performance metrics of various assignment methods. The values are the average performance across each project. The highest scores for each method are in bold.

(%)	Fall 2023	Spring 2024
Ours	**98.4**	**91.9**
Manual	90.4	89.6
Random	81.4	89.8

(A) Percentage of Skills Fulfilled

(%)	Fall 2023	Spring 2024
Ours	87.2	86.3
Manual	**88.7**	**87.6**
Random	58.2	64.0

(B) Average Preference Toward Assigned Project

Analysis of our results from Fig. 2 is summarized in Table 1. Table 1A presents the average *Percentage of Skills Fulfilled* for the Fall 2023 and Spring 2024 semesters. Here, our algorithm outperforms both the manual and random assignments across both semesters. Although the manual approach performed better than random assignment, it lagged behind the automated method in effectively meeting project skill requirements. Table 1B shows the average values of the *Average Preference Toward Assigned Project* computed across all projects for both semesters under consideration. Among the three assignment strategies, the manual approach achieved the highest average *Student Preference Toward Assigned Project* metric, with our algorithm performing competitively. Both the manual and algorithmic approaches substantially outperformed the random baseline.

These findings highlight a key trade-off inherent in manual assignment strategies. Assigning students to projects based on preferences is relatively straightforward, since preference scores are singular, easily comparable values. However,

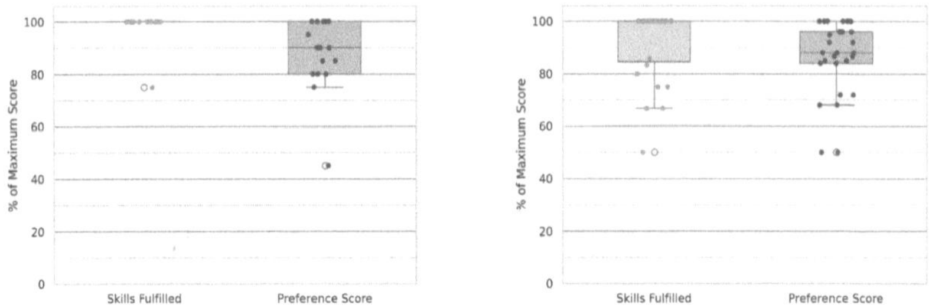

(a) Ours (Fall 2023 and Spring 2024)

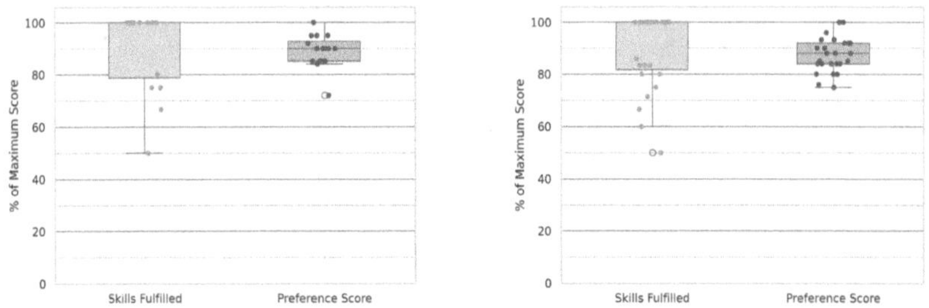

(b) Manual Assignment (Fall 2023 and Spring 2024)

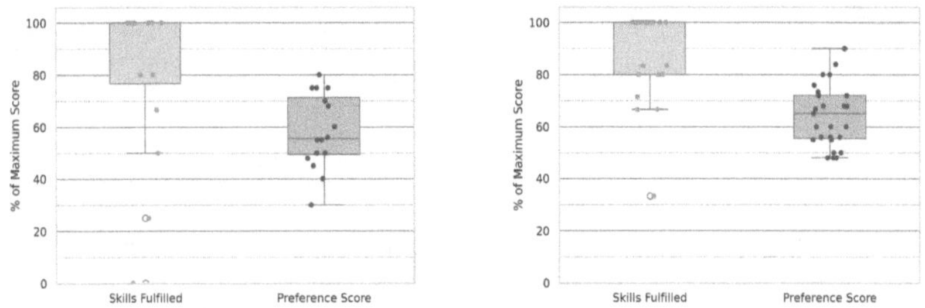

(c) Random Assignment (Fall 2023 and Spring 2024)

Fig. 2. Comparison of three team formation methods – (a) our algorithm, (b) manual, and (c) random – applied to Fall 2023 (left) and Spring 2024 (right). The "Skills Fulfilled" and "Preference Score" metrics in the plots correspond to the *Percentage of Skills Fulfilled* and *Average Preference Toward Assigned Project* described in text.

manually optimizing for skill coverage is far more complex. Instructors need to deal with wide variability in both student skill profiles and project requirements, requiring detailed inspection of dozens of individual ratings. This effort imposes a significant cognitive load on the instructor, thereby increasing the likelihood of oversights. On the other hand, our algorithm is specifically designed to address this challenge by simultaneously accounting for both preference and skill alignment through a dynamic, data-driven scoring mechanism. The results reflect the strength of our approach as it surpasses manual assignment in skill coverage while maintaining a competitive preference alignment. Finally, it is worth noting the practical cost of manual assignment. Creating teams manually required over

20 h of instructor effort to carefully review data and assign students. In contrast, the algorithm produces assignments of higher quality in a matter of milliseconds, demonstrating not only better balance but also high efficiency.

4.2 Qualitative Findings

We also collected a brief satisfaction survey from 22 previous Capstone students to understand how team formation experiences mapped to perceived fairness, learning, and engagement. This work was approved by our institution's Institutional Review Board (IRB). The survey was comprised of the following 7 questions:

Q1: Did the process of assigning teams/projects feel fair?
Q2: Overall, how happy are you with the project you got?
Q3: Did the team collectively have the needed skills?
Q4: Did you personally get to use or grow your skills in the work?
Q5: Was the project aligned with what you like?
Q6: Did you actually learn new tech or knowledge?
Q7: What matters more to you in assigning teams?

Figure 3a shows aggregated boxplots for Q1, Q2, Q3, Q4, Q5, and Q6 (Likert 1–5) and Fig. 3b visualizes multi-select priorities from Q7.

Three themes dominated: desire for an interesting topic, preference to work with specific teammates, and aligning work with one's skills or career direction. Qualitatively, students prioritizing an "interesting topic" sometimes reported lower satisfaction when reality did not match expectations, while those emphasizing "preferred teammates" were often more satisfied but also more likely to note uneven contributions. Together with the boxplots, these findings suggest that managing expectations about project scope, supporting light-weight accountability for contributions, and capturing skill/career signals early can improve both perceived fairness and overall satisfaction.

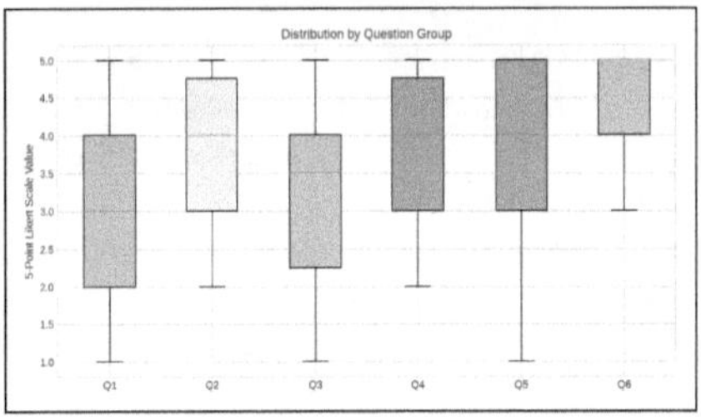

(a) Aggregated boxplots for Q1 (fair assignment), Q2 (project satisfaction), Q3 (team skill mix), Q4 (my skills used), Q5 (interest kept me engaged), and Q6 (learned new skills).

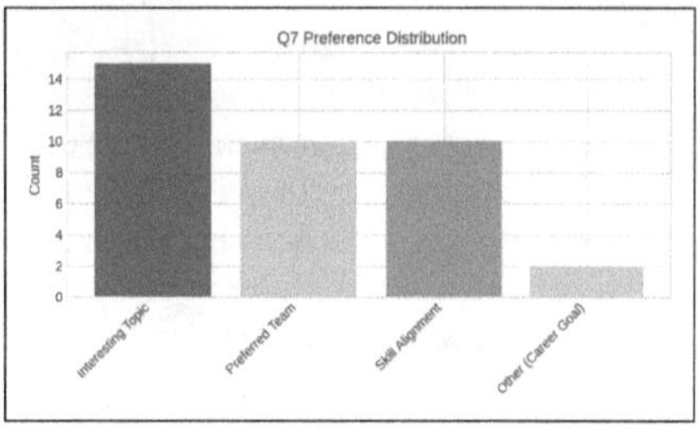

(b) Q7 multi-select priorities (counts): Interesting Topic, Skill Alignment, Preferred Team, Other (Career Goal).

Fig. 3. Qualitative survey results summarizing distributions across core satisfaction/learning items (a) and student priorities for team assignment (b).

5 Conclusions and Future Work

In this paper we presented a dynamic, data-driven algorithm for forming student teams in capstone and senior design courses that integrates student preferences with project skill requirements. By combining LLM-based skill extraction with dynamic weighting of skills and preferences, our approach addresses limitations of existing tools such as CATME. Evaluation on real course data showed that it improves skill coverage while maintaining high levels of student satisfaction, outperforming random assignment and reducing instructor workload compared to manual assignment.

Despite these advantages several limitations remain. Reliance on self-reported skills and preferences leaves the algorithm vulnerable to misrepresentation. The preference weight parameter α requires manual tuning, and the current method can favor projects with many required skills at the expense of those with rare but critical skills. All skills are currently treated equally, with fulfillment based on a single student's intermediate-level rating, and the LLM-generated project skills still require manual verification.

Our ongoing work is focused on addressing these issues through adaptive preference weighting, normalization strategies to handle skill rarity, refined fulfillment thresholds, and faculty-validated datasets for improving skill extraction. We are also investigating batch and multi-pass assignment strategies and stable matching variants such as the classified stable matching problem [15]. These refinements aim to enhance fairness, scalability, and educational impact, enabling instructors to form teams that are both technically capable and well-aligned with student interests.

References

1. Agarwal, J., Piatt, E., Imbrie, P.K.: Team formation in engineering classrooms using multi-criteria optimization with genetic algorithms. In: 2022 IEEE Frontiers in Education Conference (FIE), pp. 1–6. IEEE, Uppsala, Sweden (2022). https://doi.org/10.1109/FIE56618.2022.9962741
2. Akbar, S., Gehringer, E.F., Hu, Z.: Improving formation of student teams: a clustering approach. In: Proceedings of the 40th International Conference on Software Engineering: Companion Proceeedings, pp. 147–148. ACM, Gothenburg Sweden (2018). https://doi.org/10.1145/3183440.3195057
3. Alberola, J.M., Del Val, E., Sanchez-Anguix, V., Palomares, A., Dolores Teruel, M.: An artificial intelligence tool for heterogeneous team formation in the classroom. Knowl.-Based Syst. **101**, 1–14 (2016). https://doi.org/10.1016/j.knosys.2016.02.010
4. Bergey, P., King, M.: Team machine: a decision support system for team formation: team machine: a decision support system. Decis. Sci. J. Innov. Educ. **12**(2), 109–130 (2014). https://doi.org/10.1111/dsji.12027
5. Bulmer, J., Fritter, M., Gao, Y., Hui, B.: FASTT: team formation using fair division. In: Goutte, C., Zhu, X. (eds.) Canadian AI 2020. LNCS (LNAI), vol. 12109, pp. 92–104. Springer, Cham (2020). https://doi.org/10.1007/978-3-030-47358-7_9

6. Chen, Y., Zhang, L., Ding, Y., Guo, L., Bian, K.: Multi-task oriented team formation in online collaborative learning. Expert Syst. Appl. **259**, 125289 (2025). https://doi.org/10.1016/j.eswa.2024.125289
7. Decorte, J.J., Verlinden, S., Hautte, J.V., Deleu, J., Develder, C., Demeester, T.: Extreme multi-label skill extraction training using large language models, July 2023. https://doi.org/10.48550/arXiv.2307.10778
8. Deußer, T., Hillebrand, L., Bauckhage, C., Sifa, R.: Informed named entity recognition decoding for generative language models, August 2023. https://doi.org/10.48550/arXiv.2308.07791
9. Dias, T.G., Borges, J.: A new algorithm to create balanced teams promoting more diversity. Eur. J. Eng. Educ. **42**(6), 1365–1377 (2017). https://doi.org/10.1080/03043797.2017.1296411
10. Feng, W., Tang, J., Liu, T.X.: Understanding dropouts in MOOCs. In: Proceedings of the AAAI Conference on Artificial Intelligence, vol. 33, pp. 517–524 (2019)
11. Gale, D., Shapley, L.S.: College admissions and the stability of marriage. Am. Math. Mon. **69**(1), 9–15 (1962)
12. Grattafiori, A., et al.: The Llama 3 herd of models. arXiv preprint arXiv:2407.21783 (2024)
13. Hammond, M., Martinez, J., Herzog, J.B.: Work in progress: an optimization model for assigning students to multidisciplinary teams by considering preferences and skills. In: 2023 ASEE Annual Conference & Exposition. ASEE Conferences, Baltimore, Maryland, June 2023. https://doi.org/10.18260/1-2-44448. https://peer.asee.org/44448
14. Hastings, E.M., Ojha, V., Austriaco, B.V., Karahalios, K., Bailey, B.P.: Composing team compositions: an examination of instructors' current algorithmic team formation practices. Proc. ACM Hum.-Comput. Interact. **7**(CSCW2), 1–24 (2023). https://doi.org/10.1145/3610096
15. Huang, C.C.: Classified stable matching. In: Proceedings of the Twenty-First Annual ACM-SIAM Symposium on Discrete Algorithms, pp. 1235–1253. SIAM (2010)
16. Kessler, A., Scheiber, T., Schmitz, H., Lykourentzou, I.: A hierarchical integer linear programming approach for optimizing team formation in education. arXiv preprint (2025). https://doi.org/10.48550/arXiv.2506.02756
17. Kulshrestha, J., Oliveira, M., Karaçalık, O., Bonnay, D., Wagner, C.: Web routineness and limits of predictability: investigating demographic and behavioral differences using web tracking data. In: Proceedings of the International AAAI Conference on Web and Social Media, vol. 15, pp. 327–338 (2021)
18. Kwak, Y., Pardos, Z.A.: Bridging large language model disparities: skill tagging of multilingual educational content. Br. J. Edu. Technol. **55**(5), 2039–2057 (2024). https://doi.org/10.1111/bjet.13465
19. Layton, R.A., Loughry, M.L., Ohland, M.W., Ricco, G.D.: Design and validation of a web-based system for assigning members to teams using instructor-specified criteria. Adv. Eng. Educ. **2**(1), 1–28 (2010)
20. Liu, Y., et al.: Collaborative learning team formation: a cognitive modeling perspective. In: Navathe, S.B., Wu, W., Shekhar, S., Du, X., Wang, S.X., Xiong, H. (eds.) Database Systems for Advanced Applications, vol. 9643, pp. 383–400. Springer, Cham (2016). https://doi.org/10.1007/978-3-319-32049-6_24
21. Meulbroek, D., Ferguson, D., Ohland, M., Berry, F.: Forming more effective teams using CATME TeamMaker and the Gale-Shapley algorithm. In: 2019 IEEE Frontiers in Education Conference (FIE), pp. 1–5. IEEE, Covington, KY, USA, October 2019. https://doi.org/10.1109/FIE43999.2019.9028552

22. Oakley, B., Felder, R.M., Brent, R., Elhajj, I.: Turning student groups into effective teams. J. Stud. Centered Learn. **2**(1), 9–34 (2004)
23. Parker, R., Sangelkar, S., Swenson, M., Ford, J.D.: Launching for success: a review of team formation for capstone design. Int. J. Eng. Educ. **35**(6), 1926–1936 (2019)
24. QuantFactory: Meta-llama-3-8b-instruct-GGUF (2024). https://huggingface.co/QuantFactory/Meta-Llama-3-8B-Instruct-GGUF. Hugging Face model repository; quantized GGUF conversions for llama.cpp. Accessed 15 Sept 2025
25. Rising, L., Janoff, N.S.: The scrum software development process for small teams. IEEE Softw. **17**(4), 26–32 (2000)
26. Sadeghi, H., Kardan, A.A.: Toward effective group formation in computer-supported collaborative learning. Interact. Learn. Environ. **24**(3), 382–395 (2016). https://doi.org/10.1080/10494820.2013.851090
27. Sutherland, J., Sutherland, J.: Scrum: The Art of Doing Twice the Work in Half the Time. Crown Business, New York (2014)
28. Wheelan, S.A.: The effect of team size on the performance of continuous improvement teams: is seven really the magic number? Small Group Res. **40**(3), 247–262 (2009)

Blockchain-Based Framework for Effective Proctoring System in Open Distance Electronic Learning (ODeL) Online Examinations: Hardware and Software Control Perspective

Prince Yaw Owusu Amoako(✉), Ernest Mnkandla, and Emmanuel Freeman

Centre for Augmented Intelligence and Data Science (CAIDS), School of Computing, College of Science, Engineering and Technology, University of South Africa, Pretoria, South Africa
{amoakpyw,mnkane}@unisa.ac.za

Abstract. This research presents a theoretical framework for a blockchain-based proctoring system designed to enhance the integrity of Open Distance electronic Learning (ODeL) online examinations from both hardware and software control perspectives. Employing a theory-led design framework, the study conducts a comparative analysis of multiple detailed case studies to derive broad theoretical insights that address security, authenticity, and user trust challenges inherent in remote assessments. By leveraging blockchain's immutability and decentralized nature, the framework aims to create a transparent, tamper-proof system for real-time exam supervision and credential verification. Although practical implementation is forthcoming, this paper establishes a solid conceptual foundation that bridges existing technological gaps and points toward scalable, secure solutions for ODeL proctoring challenges.

Keywords: Proctoring system · Open Distance electronic Learning · Blockchain

1 Introduction

Open Distance electronic Learning (ODeL) has rapidly transformed the landscape of education worldwide by offering flexible, accessible learning opportunities that overcome traditional barriers such as geographic location, time constraints, and socioeconomic factors [1]. This mode of education enables learners especially non-traditional students, working adults, and those in remote areas to access higher education without the need for physical presence on campuses [2, 3]. ODeL fosters lifelong learning by utilizing digital platforms, e-resources, virtual libraries, and interactive tools that support self-paced and personalized learning [4]. The flexibility and scalability of ODeL have been critical in expanding educational access, particularly in emerging economies where social and infrastructural challenges traditionally limit opportunities [5, 6].

With the surge in ODeL enrolment and reliance on digital platforms, the demand for secure and reliable online examination systems has escalated significantly to preserve academic integrity and ensure fair assessments [7]. The need for robust proctoring

H. R. Arabnia et al. (Eds.): CSCI 2025, CCIS 2974, pp. 295–311, 2027.
https://doi.org/10.1007/978-3-032-28475-4_21

solutions in ODeL is driven by concerns about cheating, identity fraud, and unauthorized access during online exams [8]. Modern secure online exams incorporate advanced technologies such as biometric verification, artificial intelligence (AI)-powered behavior monitoring, encrypted data transmission, and lockdown browsers, which collectively create trusted digital assessment environments [9]. These systems not only enhance trust in the validity of online assessments but also support scalability and accessibility without compromising security [10, 11]. Consequently, institutions offering ODeL seek proctoring frameworks that are transparent, user-friendly, and resilient to technological limitations to meet the expectations of both learners and accrediting bodies [9, 12].

The expansion of remote proctoring has introduced significant challenges centered on maintaining academic integrity, protecting user privacy, and ensuring technical reliability [13]. One primary concern revolves around safeguarding exam integrity, as remote settings inherently increase the risk of cheating through unauthorized assistance, identity fraud, and use of prohibited materials [14]. Despite advances in AI-based monitoring and biometric verification, the ability to fully deter dishonest practices remains limited, prompting ongoing debates on the effectiveness of these technologies to replicate traditional in-person supervision [15, 16].

Privacy concerns also loom large, as remote proctoring often requires continuous video and audio recording, raising issues about data security, consent, and potential misuse of personal information [17]. High-profile cases and regulatory scrutiny underscore the importance of transparent data policies and ethical safeguards to maintain learner trust [18]. Moreover, technical reliability is a persistent challenge, with connectivity issues, software malfunctions, and hardware limitations sometimes disrupting exam experiences and undermining fairness [19, 20]. Recent surveys reveal that approximately 30% of remote exam takers report technical difficulties severe enough to impact their performance, highlighting the need for resilient, user-friendly proctoring solutions [21, 22]. These trends underscore the complex balancing act educational institutions face in adopting remote proctoring that is secure, respectful of privacy, and technically dependable.

Blockchain technology is increasingly recognized as a promising solution for enhancing security, transparency, and trust in online proctoring systems [23]. Fundamentally, blockchain provides a decentralized and immutable ledger that securely records every transaction or event, ensuring that exam data, proctoring logs, and authentication records cannot be altered or tampered with [24]. When integrated into online exam environments, blockchain enables secure and verifiable tracking of exam activities, timestamps interactions, and creates transparent audit trails accessible to authorized stakeholders. This immutable record-keeping can greatly reduce disputes regarding exam integrity and bolster confidence among institutions, instructors, and learners [25].

Moreover, smart contracts, self-executing codes stored on the blockchain, automate enforcement of exam rules such as time limits, access restrictions, or authentication protocols, making the proctoring process more reliable and less susceptible to human error or manipulation [9]. By combining biometric data from hardware controls with blockchain's cryptographically secured storage, the framework ensures the candidate's identity is securely verified and that exam data is protected during transmission and storage. Blockchain's decentralization also mitigates risks associated with centralized server failures or cyberattacks, improving system resilience [24]. Thus, blockchain technology offers a robust foundation for designing future-proof online proctoring systems that uphold academic integrity while respecting privacy and operational efficiency. Hence, this research aims to develop a theoretical framework for a blockchain-based proctoring system focusing on hardware and software control aspects.

2 Literature Review

2.1 ODeL Proctoring Technologies

The current state of online proctoring technologies in Open Distance electronic Learning (ODeL) reflects significant advancements aimed at preserving academic integrity while facilitating remote assessments. These technologies typically combine biometric authentication (such as facial recognition and fingerprint scans), real-time AI-driven behavior analysis, screen recording, and browser lockdown features to monitor exam sessions remotely [3]. The use of AI algorithms enables the detection of suspicious activities including eye movement, unusual background noise, and prohibited materials, allowing institutions to flag potential cheating incidents efficiently [7, 26]. Additionally, cloud-based platforms have improved scalability, making it possible to conduct high-stakes exams with large student populations across widespread geographic locations [27].

Despite these advancements, several limitations persist in ODeL contexts. Technical challenges such as unstable internet connections, hardware incompatibility, and software glitches frequently disrupt exam processes, affecting user experience and assessment fairness [28]. Privacy concerns remain paramount, as continuous video and audio monitoring raise ethical questions related to data security, consent, and the intrusive nature of surveillance [29, 30]. There are also issues with accuracy and bias in AI detection systems, which may result in false positives or negatives, undermining trust in the proctoring outcomes [31]. Moreover, the dependency on sophisticated technologies may exclude learners in under-resourced areas lacking access to required devices or stable connectivity, posing equity challenges for ODeL programs [32]. These limitations necessitate ongoing research and innovation to design adaptable, reliable, and ethically sound proctoring solutions suitable for diverse ODeL environments.

2.2 Overview of Blockchain Technology

Blockchain technology is fundamentally a decentralized, distributed ledger system that securely records and stores transactions across a network of multiple computers rather than relying on a central authority [33]. This decentralization enhances transparency and trust because each participant in the network holds a synchronized copy of the ledger, and no single entity controls or can alter the data unilaterally [34]. The blockchain is composed of blocks which include data records, timestamps, and cryptographic hashes linking each block to its predecessor, forming an immutable chain that prevents modification or tampering of past transactions [35]. This immutable and transparent nature ensures accountability and verifiability, making blockchain a powerful tool for applications requiring high security and trustworthiness.

Key features of blockchain include smart contracts and cryptography [33]. Smart contracts are self-executing codes stored and run on the blockchain that automatically enforce and validate contractual agreements based on predefined conditions, reducing the need for intermediaries and increasing efficiency [36]. Cryptography, specifically public key cryptography, secures the blockchain by enabling unique digital signatures for participants and ensuring that data added to the ledger is authentic and confidential [33, 37]. Together, these features create a trusted environment where data integrity is guaranteed, processes are automated, and all actions are transparently recorded. These fundamentals position blockchain technology as a promising foundation for enhancing security, transparency, and trust in various domains including online proctoring systems for education.

2.3 Blockchain Applications in Educational Assessment and Proctoring

Blockchain technology has found growing applications in educational assessment and proctoring, addressing challenges of transparency, security, and trust in digital learning environments [38]. Several institutions and ed-tech companies have implemented blockchain to create tamper-proof records of academic credentials, examination results, and certifications, thereby combating fraud and enhancing credibility [39, 40]. For instance, blockchain-based systems enable immutable storage of exam results and digital badges, allowing learners and employers to verify qualifications easily without relying on intermediaries [41]. Such applications not only protect the integrity of assessment data but also streamline administrative processes by decentralizing record-keeping.

In the context of online proctoring, blockchain's decentralized ledger can securely log exam activities, timestamps, and candidate verification events, creating an auditable trail that resists tampering [42]. Some pioneering frameworks integrate blockchain with biometric authentication and AI monitoring to ensure candidate identity and exam integrity simultaneously [43, 44]. Smart contracts have been employed to automate exam rules enforcement—such as time limits and conditional unlocking of exams—enhancing reliability and fairness [25]. However, practical adoption remains limited due to technological complexity, scalability concerns, and integration challenges with existing Learning Management Systems (LMS). Nonetheless, ongoing research and pilot projects indicate substantial potential for blockchain to revolutionize education assessment and proctoring by fostering trust, security, and transparency in increasingly digital academic landscapes [35, 39, 40].

2.4 Existing Key Blockchain-Based Platform's Architecture and Specific Implementation Outcomes

The existing blockchain-based platforms present significant architectural composition providing diverse specific implementation outcomes. This section reveals the architecture and the specific outcomes of the key blockchain-based platforms.

EduCTX

Architecture: EduCTX is a permissioned blockchain platform designed specifically for higher education credit management. The platform represents credits as ECTX tokens, each corresponding to one European Credit Transfer and Accumulation System (ECTS) credit, managed on a distributed peer-to-peer (P2P) network where the nodes are member Higher Education Institutions (HEIs). EduCTX uses a delegated proof-of-stake (DPoS) consensus mechanism, promoting a democratic, non-profit governance structure where HEIs vote for nodes authorized to validate and seal blocks. Students hold blockchain wallets where their credits accumulate, and smart contracts facilitate credit transfer and verification. Privacy is upheld through multi-signature addresses enabling students to receive but not transfer tokens, ensuring data security and ownership control [45].

Implementation Outcomes: EduCTX has been piloted with positive results demonstrating streamlined credit transfer across universities without traditional administrative delays, enhancing transparency and trust. Verification by employers or institutions is efficient via blockchain APIs without risking data misuse. The system reduces fraud by making transcripts immutable and universally verifiable, improving international credit recognition and student mobility [45]. The platform's scalable architecture positions it for broader application beyond ECTS credits to diplomas and certificates.

Blockcerts

Architecture: Blockcerts is an open standard for issuing and verifying blockchain-based academic certificates. It utilizes public blockchains like Bitcoin or Ethereum to anchor certificate data cryptographically. The system issues digitally signed certificates to learners that can be stored in digital wallets. Verification occurs independently by relying parties through public ledger data, eliminating dependence on central databases. Certificates include cryptographic proofs that validate authenticity and prevent unauthorized alterations [41].

Implementation Outcomes: Blockcerts is widely adopted globally for diploma and credential management in universities and online courses, successfully reducing verification times from weeks to seconds. It empowers learners with lifelong access to their credentials and supports employers in conducting instant, reliable certificate validation [41]. However, Blockcerts mainly addresses credential issuance rather than the real-time nuances of exam proctoring.

Further comparative analysis of the mojor features of the EduCTX and ThinkExam blockchain-based platforms as the base case for evaluation to propose an effective blockchain-based proctoring is presented in this section.

Architecture: Built on top of Blockcerts standards, Learning Machine offers user-friendly digital wallets for storing and sharing blockchain-secured academic credentials. It integrates seamless blockchain verification with educational institution administrative systems, automating certificate issuance workflows. Its architecture supports web and mobile interfaces, allowing learners to control whom they share verifiable credentials with [46].

Implementation Outcomes: Learning Machine has demonstrated efficacy in enhancing learner data ownership and simplifying administrative processes across diverse education systems [46]. The platform boosts trust in digital certificates while supporting modular integrations, but it remains focused on record-keeping rather than directly interfacing with proctoring processes.

ThinkExam Blockchain Solutions
Architecture: ThinkExam's blockchain-based proctoring integrates exam authentication, secure paperless exam delivery, and immutable logging of exam sessions on a blockchain ledger. It combines biometric authentication hardware with blockchain-stored exam metadata and AI-powered monitoring data. Smart contracts automate exam time enforcement and result processing. The platform leverages a permissioned blockchain to control access and protect exam confidentiality while ensuring traceable activity logs that stakeholders can audit [23].

Implementation Outcomes: The platform has improved exam integrity by preventing unauthorized access and enabling robust dispute resolution through immutable exam logs. Institutions report higher confidence in online exam fairness and streamlined exam management [23]. However, the solution is relatively new, and wider adoption is ongoing.

Ethereum-based Smart Contract Solutions
Architecture: Several experimental platforms utilize Ethereum's smart contracts for automating exam rule enforcement, such as restricting exam duration, managing candidate eligibility, and automating grading workflows. These decentralized applications store transaction hashes and results on the blockchain, providing transparency and auditability. Integration with hardware (such as biometric verifiers and secure browsers) is attempted through off-chain or hybrid solutions [47].

Implementation Outcomes: While offering high programmability and security, practical use faces scalability and transaction cost challenges due to "gas fees" and large data throughput limitations. Interest remains high for future optimizations through Layer-2 solutions and private blockchains.

3 Methodology

In this research, a comparative case study analysis combined with theoretical evaluation is employed. Inspired by [48] this research adopts a theory-led design framework where multiple detailed case studies are analyzed comparatively to extract broader theoretical insights. The methodology involves building a research design based on preliminary theory, selecting units of analysis (cases), and establishing logical links between data and theoretical propositions.

This approach involves:

1. Comparative Case Study Analysis:
   Identify and analyze existing blockchain implementations or related proctoring systems in education or other sectors. Examine their architectural designs, adoption challenges, strengths, and weaknesses through detailed document analysis, published case studies, and reports. Draw parallels and contrasts with the proposed framework to infer potential performance and feasibility.
2. Theoretical Evaluation and Framework Mapping:
   Conduct an in-depth theoretical analysis of the blockchain protocols, security features, and hardware/software control mechanisms proposed. Map these against established standards, best practices, and known vulnerabilities in blockchain and digital proctoring literature. Evaluate how each component theoretically addresses problem requirements such as data integrity, authentication, and user privacy.

3. Gap and Risk Analysis:
   Identify gaps between the proposed framework and existing solutions or standards and assess potential risks or limitations through logical reasoning and scenario-based critique. This can include analyzing scalability concerns, latency issues, regulatory compliance, and user accessibility challenges.
   This method allows rigorous validation through documented evidence and solid theoretical reasoning, providing a strong foundation for the framework's design while avoiding the resource intensity of simulations or coordinating expert panels.

4 Proposed Theorithecal Framework

Table 1 presents a comparison of key blockchain-based platforms and applications in educational assessment and proctoring.

These platforms illustrate blockchain's ability to ensure immutability, transparency, and security in recording educational achievements and exam-related data. While many focus on credential verification, increasing effort is being placed on incorporating blockchain-based solutions for online proctoring to improve exam integrity and trustworthiness.

4.1 Comparative Case Study Analysis: EduCTX vs. ThinkExam in Blockchain-Based Proctoring for ODeL

Considering the EduCTX and ThinkExam as the most significantly blockchain-based related system in ODeL proctoring, insightful comparison are presented as follows:

- Blockchain Integration: EduCTX leverages blockchain's decentralized ledger to provide tamper-proof academic record management, ensuring traceability and trustworthiness, whereas ThinkExam lacks blockchain, relying on traditional centralized systems. This makes EduCTX more robust for post-exam credential verification but less focused on live exam integrity controls.
- Proctoring Features: ThinkExam's advanced software and hardware proctoring functionalities address immediate cheating prevention and real-time monitoring needs, essential for ODeL online examinations. EduCTX's framework does not directly incorporate these features but could benefit from integrating proctoring controls.
- Scalability and Focus: EduCTX aims for scalability across institutions and international credit portability, beneficial for broader educational ecosystems. ThinkExam targets exam security and cheating deterrence at the point of assessment but may face trust issues due to its centralized nature.
- Potential Synergy: Combining EduCTX's blockchain features with ThinkExam's robust proctoring controls could yield a comprehensive system offering both secure, transparent academic credentialing and effective, tamper-resistant online exam supervision.

This comparative analysis reveals that while EduCTX excels in blockchain-based credential verification, ThinkExam leads in practical proctoring technologies. For ODeL proctoring systems, incorporating blockchain with advanced hardware/software controls, as proposed in this research, would address critical gaps in both security and operational effectiveness observed in these platforms.

Table 1. Comparison of key blockchain-based platforms and applications in educational assessment and proctoring.

Platform or Application	Main Features	Use Case in Education	Strengths	Limitations
EduCTX (UNESCO-backed)	Blockchain-based higher education credit system, immutable records, global verification	Storing, verifying academic credits and diplomas	Trusted by institutions globally, decentralized credential verification	Focus mainly on credit transfer, less on proctoring
Blockcerts	Open standard for blockchain credentials, verifiable digital diplomas and certificates	Issuance and verification of academic certificates	Open-source, widely adopted, portable	Limited integration with exam proctoring workflows
Learning Machine	Digital wallets for credentials using Blockcerts standard, blockchain verification	Secure storage and sharing of digital credentials	User-friendly wallets, supports lifelong learning	Primarily credential management, not exam monitoring
MobiChain	Mobile blockchain platform for possible educational records and assessment	Real-time, tamper-proof tracking of assessments	Mobility and accessibility focused	Requires mobile infrastructure, limited proctoring features
Etherium-based Smart Contract Platforms	Custom smart contract powered systems for exam management, rule automation	Automated exam rule enforcement and result handling	Highly customizable and programmable	Higher complexity, gas fees, scalability challenges
OpenCerts	Government-backed blockchain credential issuance (e.g., Singapore)	Secure diploma issuance and verification	Government supported, enhances trust	Limited application beyond credential issuance

(continued)

Table 1. (*continued*)

Platform or Application	Main Features	Use Case in Education	Strengths	Limitations
ThinkExam Blockchain Solutions	Secure exam authentication, automated proctoring access, tamper-proof exam logs	Ensuring exam integrity, proctoring digital exams	Integrates proctoring data on blockchain	Emerging tech, adoption still growing
Blockchain EduChain	Integrated credentialing, learning analytics, and proctoring data logging	Holistic educational ecosystem management	Combines academic records and monitoring	Complexity and adoption barriers

4.2 Feature-Based Analysis of EduCTX and ThinkExam Blockchain-Based Platforms

Further comparative analysis of the mojor features of the EduCTX and ThinkExam blockchain-based platforms as the base case for evaluation to propose an effective blockchain-based proctoring is presented in this section.

EduCTX Features

EduCTX operates as a permissioned blockchain network where participating HEIs serve as validating nodes in a Delegated Proof of Stake (DPoS) consensus. Each HEI retains a copy of the distributed ledger containing cryptographically secured academic credits issued to students. The credits are tokenized as ECTX tokens equivalent to ECTS credits, stored in student-controlled blockchain wallets.

Key Components:

- Node Network: HEIs run validator nodes responsible for block creation and validation.
- Student Wallets: Learners receive tokens representing earned credits in wallets controlled via private keys to ensure ownership and security.
- Smart Contracts: Automate credit issuance, transfer requests, and verification processes. They also manage multi-signature authentication mechanisms protecting student privacy.
- APIs and Interfaces: Web portals and institutional software connect with blockchain APIs to display verified data and facilitate credit requests.

Security and Privacy:

- Multi-signature wallets require both student and institution consent for credit transfers.
- Immutable ledger prevents unauthorized modifications without consensus.
- Distributed nature reduces risks of central data breaches.

ThinkExam Blockchain Proctoring Features

ThinkExam integrates blockchain for secure exam data logging with AI-driven remote proctoring and biometric authentication hardware. The architecture includes a permissioned blockchain network to manage access, tokenized exam sessions, and smart contract-driven automations.

Key Components:

- Blockchain Ledger: Stores encrypted metadata about exam sessions, timestamps of exam starts, stops, and suspicious events detected by AI.
- Biometric Authentication System: Uses face recognition or fingerprints to confirm candidate identity prior to blockchain recording.
- AI Proctoring Module: Runs in parallel, analyzing video and audio streams, flagging suspicious behaviors logged to blockchain.
- Smart Contracts: Automatically enforce exam rules such as timed access, auto-submit triggers, and grading condition checks.
- Exam Portal Interface: Provides students and administrators real-time feedback and post-exam audit data linked to the blockchain.

Security and Privacy:

- Exam data is encrypted and stored on the blockchain with access control via permissioned consensus.
- Biometric data is used for identity verification but stored off-chain to ensure privacy compliance.
- AI-flagged incidents are logged immutably for transparent review.

4.3 Architectural Analysis of EduCTX and ThinkExam Blockchain-Based Platforms

Based on the feature-based analysis of the two platforms under consideration, architectural analysis are presented. Figure 1 shows the derived EduCTX Blockchain Architecture.

The EduCTX Blockchain Architecture presented in Fig. 1 operates as follows:

1. Distributed Node Network (HEIs)
    - Represents multiple Higher Education Institutions interconnected as validating nodes in a permissioned blockchain network.
    - Each node stores a full copy of the immutable ledger containing all issued academic credits.
2. Blockchain Ledger
    - A decentralized ledger storing ECTX tokens (digital representations of credits).
    - Transactions (credit issuance, transfer requests) are cryptographically secured and timestamped on the ledger.
3. Smart Contract Layer
    - Automates validation of credit issuance and verification processes.

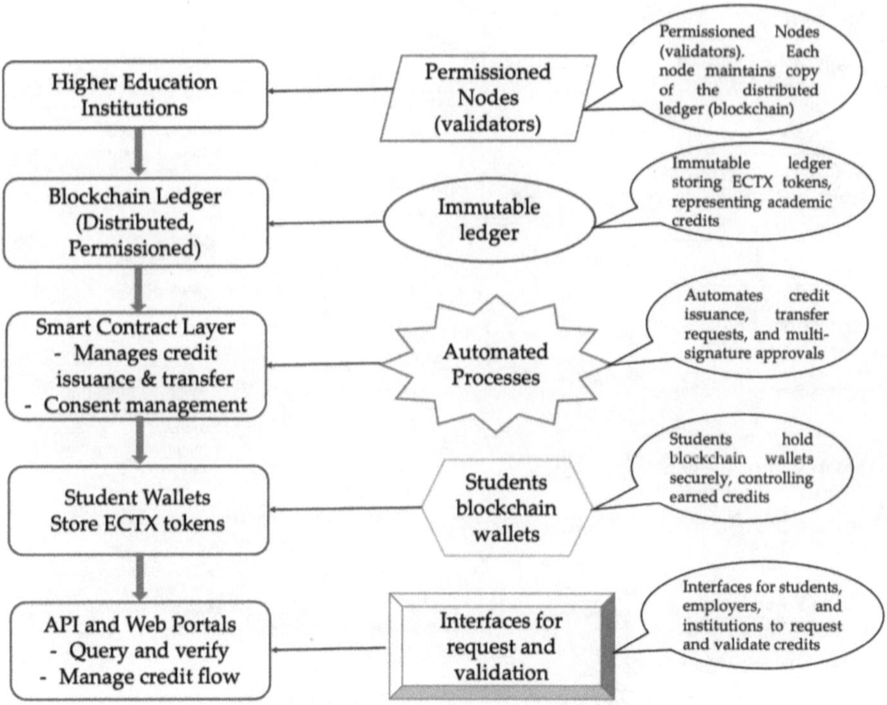

Fig. 1. EduCTX Blockchain Architecture.

- Handles multi-signature approvals ensuring students and institutions control credit transfers jointly.

4. Student Wallets

- Secure digital wallets held by students to accumulate and manage credits as tokens on the blockchain.
- Students control private keys that safeguard ownership.

5. API/Web Portals

- Interfaces for students, employers, and institutions to request, verify, or view credits without blockchain knowledge.
- Connects with the blockchain back-end to ensure data integrity.

EducTX architecture ensures transparency, decentralization, privacy, and streamlined credit transfers between institutions globally.

Figure 2 shows the derived architecture of the ThinkExam Blockchain Proctoring.

In reference to Fig. 2, the ThinkExam Blockchain Proctoring Architecture operates as follows:

1. Candidate Biometric Authentication

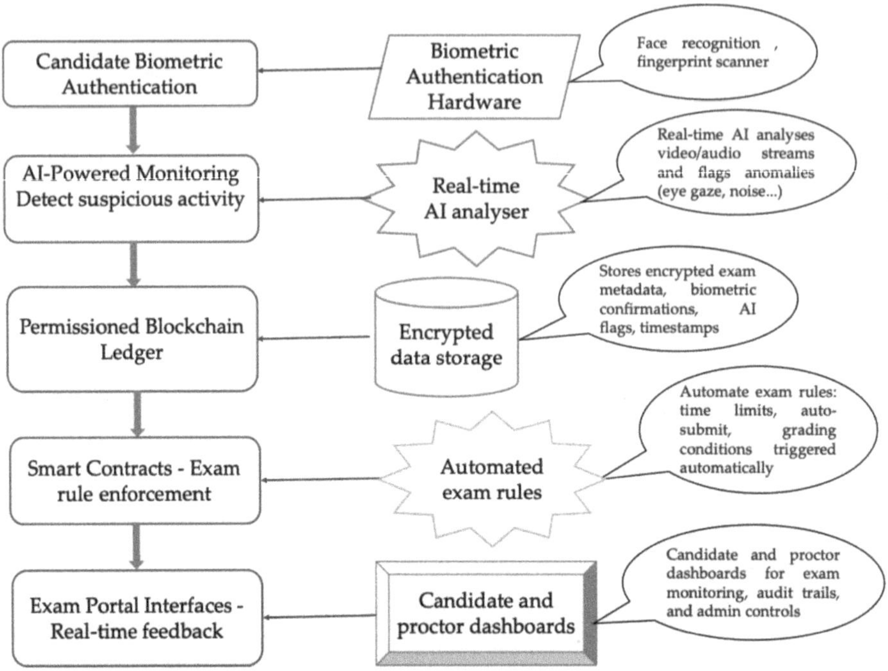

Fig. 2. ThinkExam Blockchain Proctoring Architecture

- Biometric devices (facial recognition, fingerprint scanners) verify candidate identity before admission to exams.
- This data is validated and linked with blockchain records without storing sensitive biometric data on-chain, ensuring privacy.

2. AI-Powered Monitoring Module

- Real-time video/audio analysis detects suspicious exam behaviors (e.g., gaze deviation, background noise).
- Flags generated are encrypted and logged.

3. Permissioned Blockchain Ledger

- Stores encrypted logs of exam metadata, timestamps, candidate authentication events, and AI flags.
- Accessible to authorized exam administrators and auditors for transparent review.

4. Smart Contracts

- Enforce exam rules such as time limits, submission deadlines, and conditional triggers (e.g., auto-submit on timeout).
- Automate grading if integrated with assessment engines.

5. Exam Portal Interface

- Web or mobile platforms for candidates and proctors to access exams and monitoring results in real-time.
- Administrators use dashboards linked to blockchain logs for audit and appeals.

ThinkExam's platform combines hardware validation, AI behavioral analysis, and blockchain's immutability to offer a comprehensive, secure, and audit-ready online proctoring solution.

To address the critical gaps in the two platforms a framework mapping is presented as follows:

- EduCTX addresses the post-exam academic record integrity and verification challenge with blockchain's distributed ledger technology, making it suitable for credential management but not directly tackling real-time exam monitoring or proctoring.
- ThinkExam fills the gap in operational exam security with dynamic proctoring mechanisms but lacks a decentralized trust layer and permanent auditing capabilities that blockchain provides.
- A combined framework could integrate EduCTX's blockchain-based immutability and consensus mechanisms with ThinkExam's hardware/software proctoring tools, offering end-to-end security—from exam supervision to secure certification.

The integration can theoretically enhance data transparency, reduce fraud, and ensure privacy while maintaining effective real-time controls tailored for Open Distance eLearning (ODeL) examinations.

4.4 Effective Blockchain-Based Proctoring and Credentialing Architecture with Hardware Control

Figure 3 presents the proposed theoretical framework for blockchain-based for effective proctoring in ODeL online examination.

The proposed framework operate to achieve effective proctoring in ODeL as follows:

- Blockchain Ledger: A unified permissioned ledger hosts both academic credit tokens (from EduCTX) and tamper-proof logs of exam sessions including timestamps, biometric authentications, and AI-generated flag events (ThinkExam). This immutable ledger ensures transparent verification of both academic achievements and exam integrity.
- Smart Contracts: These automate educational processes such as credit issuance and multi-signature consent management from EduCTX, and simultaneously enforce exam rules (time limits, auto-submit triggers) and identity verification protocols inspired by ThinkExam.
- Student Wallets: Students control their academic credentials via blockchain wallets secured by multi-signature mechanisms to enhance privacy and ownership. These wallets also link closely to the blockchain-based exam log system for holistic record-keeping.
- Exam Portal Interfaces: Candidates and proctors interact with real-time interfaces that provide exam access, monitoring, and incident alerts. Administrators and external verifiers can audit records seamlessly through authorized access.

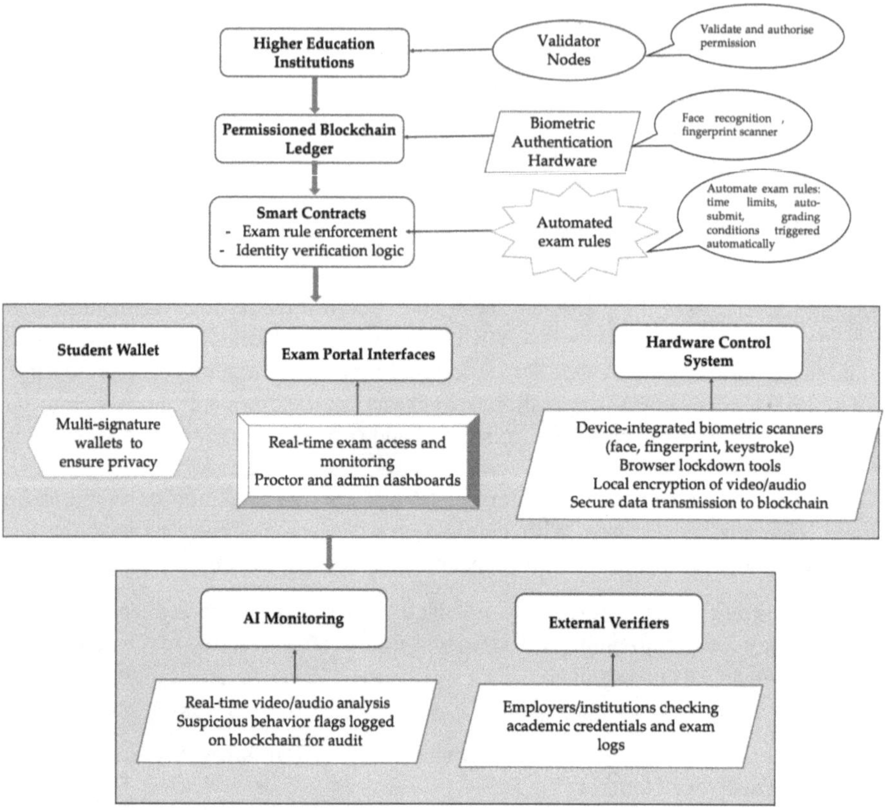

Fig. 3. Theoretical framework for blockchain-based for effective proctoring

- Hardware Control Mechanism: The student's device integrates biometric authentication hardware (e.g., facial recognition, fingerprint scanner, keystroke dynamics) that secures candidate identity verification locally before exam sessions. The system includes browser lockdown to prevent unauthorized access or application switching. Audio/video streams are encrypted locally on the device before uploading to the blockchain network to safeguard privacy and data integrity.
- AI Monitoring: A parallel AI-powered module analyses biometric feeds in real-time, detecting suspicious behaviors and flagging potential security breaches. These flags are immediately recorded immutably on the blockchain, providing a transparent audit trail.
- External Verifiers: Employers, universities, or accreditation bodies verify academic credentials and exam logs by querying the blockchain network, enhancing trust and reducing fraud.

This wholistic architecture offers a comprehensive, secure, scalable solution that combines blockchain's decentralization and immutability with advanced proctoring technologies and secure hardware controls.

5 Conclusion

In conclusion, this ongoing research lays a strong theoretical foundation for a blockchain-based proctoring system tailored for Open Distance electronic Learning (ODeL) online examinations, focusing on both hardware and software control perspectives. Utilizing a theory-led design framework, the study undertakes a comparative analysis of multiple detailed case studies to extract broad theoretical insights. These insights inform the design of a secure, transparent, and efficient proctoring framework that leverages blockchain's immutability and decentralized trust to enhance exam integrity in remote learning environments. While practical implementation remains a future step, the current paper establishes critical conceptual groundwork that bridges existing proctoring challenges with opportunities afforded by emerging blockchain technologies.

This theoretical framework addresses essential components such as identity verification, data security, and real-time monitoring controls, setting the stage for subsequent development and testing phases. By synthesizing empirical findings from diverse case studies with blockchain principles, the research promises to advance innovative solutions to persistent issues in online exam supervision. Future work will focus on translating this framework into implementable systems, including prototype development, integration with educational platforms, and evaluation of system performance and user experience. Ultimately, this research aims to contribute significant value to ODeL institutions by providing a scalable, trustworthy, and technologically sophisticated proctoring approach aligned with the evolving demands of digital education.

Acknowledgments. We gratefully acknowledge the organizers and program committee of the *Computational Science and Computational Intelligence* conference for providing an excellent forum for sharing and discussing this work.

Disclosure of Interests. The authors declare that they have no known competing financial interests or personal relationships that could have appeared to influence the work reported in this paper.

References

1. Gunter, A.: The geography of distance education: spatial disparities, accessibility, and impact across place. S. Afr. Geogr. J. **107**, 1–19 (2025)
2. Demuyakor, J.: Coronavirus (COVID-19) and online learning in higher institutions of education: a survey of the perceptions of Ghanaian international students in China. Online J. Commun. Media Technol. **10**(3) (2020)
3. Mwansa, P., Chanda Chansa, T., Loveness, N.: Assessing the effect of open distance and E-learning (ODeL) in enhancing women's access to higher education in Zambia. World J. Adv. Res. Rev. **25**(1), 1805–1816 (2025)
4. Jung, I.: Personalized education for all: the future of open universities. Open Praxis. **16**(1), 24–36 (2024)
5. UNESCO: Education for all global monitoring report 2015: education for all 2000–2015: achievements and challenges (2015)

6. Ashraf, N., Ali, A.: Flexibility and freedom: a comprehensive study of distance education in 21st century at higher education in Pakistan. J. Soc. Sci. Rev. **5**(1), 390–402 (2025)
7. Holden, O.L., Norris, M.E., Kuhlmeier, V.A.: Academic integrity in online assessment: a research review. Front. Educ. **6** (2021)
8. Jalilzadeh, K., Rashtchi, M., Mirzapour, F.: Cheating in online assessment: a qualitative study on reasons and coping strategies focusing on EFL teachers' perceptions. Lang. Test. Asia. **14**(1) (2024)
9. Heinrich, E.: A systematic-narrative review of online proctoring systems and a case for open standards. Open Praxis. **17**(3), 485–499 (2025)
10. Nigam, A., et al.: A systematic review on AI-based proctoring systems: past, present and future. Educ. Inf. Technol. **26**(5), 6421–6445 (2021)
11. Sakhipov, A., Omirzak, I., Fedenko, A.: Beyond face recognition: a multi-layered approach to academic integrity in online exams. Electron. J. e-Learn. **23**(1), 81–95 (2025)
12. Chugh, R., et al.: Implementing educational technology in higher education institutions: a review of technologies, stakeholder perceptions, frameworks and metrics. Educ. Inf. Technol. **28**(12), 16403–16429 (2023)
13. Hilliger, I., et al.: Trustworthy remote assessments: a typology of pedagogical and technological strategies. J. Comput. Assist. Learn. **38**(6), 1507–1520 (2022)
14. Gamage, K.A.A., et al.: Contract cheating in higher education: impacts on academic standards and quality. J. Appl. Learn. Teach. **6**(2), 1–13 (2023)
15. Perkins, M., Gezgin, U.B., Roe, J.: Reducing plagiarism through academic misconduct education. Int. J. Educ. Integr. **16**(1), 1–15 (2020)
16. Coghlan, S., Miller, T., Paterson, J.: Good proctor or "big brother"? Ethics of online exam supervision technologies. Philos. Technol. **34**(4), 1581–1606 (2021)
17. Terpstra, A., De Rooij, A., Schouten, A.: Online proctoring: privacy invasion or study alleviation? In: Proceedings of the 2023 CHI Conference on Human Factors in Computing Systems, pp. 1–20 (2023)
18. Gold Nmesoma, O., et al.: Ethical considerations in data collection and analysis: a review: investigating ethical practices and challenges in modern data collection and analysis. Int. J. Appl. Res. Soc. Sci. **6**(1), 1–22 (2024)
19. Hillier, M., Kumar, N., Wijenayake, N.: e-Examinations: the impact of technology problems on student experience. In: Open Conference on Computers in Education (OCCE), pp. 35–45, Mumbai (2020)
20. Celeste, R.J., Osias, N.: Challenges and implementation of technology integration: basis for enhanced instructional program. Am. J. Arts Hum. Sci. **3**(2), 106–130 (2024)
21. Terblanche, E.A.J., van Rooyen, A.A., Enwereji, P.C.: Auditing students' perceptions of online assessments and e-proctoring systems. Discov. Educ. **3**(1) (2024)
22. Makgakga, T.P.: Take-home examination proctoring technologies: undergraduate mathematics education students' perceptions in an open distance learning environment. Interdiscip. J. Educ. Res. **7**(1), a19 (2025)
23. Ramani, R., et al.: CryptoProctor in elevating online exams through blockchain technology. In: 2024 International Conference on Computing and Data Science (ICCDS), pp. 1–5 (2024)
24. Sattar, M.R.I., et al.: An advanced and secure framework for conducting online examination using blockchain method. Cyber Secur. Appl. **1** (2023)
25. ThinkExam: The role of blockchain technology in securing online exams and credentials (2025). https://thinkexam.com/blog/the-role-of-blockchain-technology-in-securing-online-exams-and-credentials/
26. Sevnarayan, K., Maphoto, K.B.: Exploring the dark side of online distance learning: cheating behaviours, contributing factors, and strategies to enhance the integrity of online assessment. J. Acad. Ethics. **22**(1), 51–70 (2024)

27. Govea, J., et al.: Optimization and scalability of educational platforms: integration of artificial intelligence and cloud computing. Computers. **12**(11) (2023)
28. Henderson, M., et al.: Online examinations: factors that impact student experience and perceptions of academic performance. Australas. J. Educ. Technol. **40**(4), 1–17 (2024)
29. Mutimukwe, C., et al.: Privacy in online proctoring systems in higher education: stakeholders' perceptions, awareness and responsibility. J. Comput. High. Educ. **38**, 732–761 (2025)
30. Barrett, L.: Rejecting test surveillance in higher education. SSRN Electron. J., 1–90 (2021)
31. Jacobs, L., Mncube, S.: Proctoring as a human substitution for online summative assessments in a comprehensive open distance e-learning institution: opportunities and obstacles. Indep. J. Teach. Learn. **18**(2), 93–108 (2023)
32. Mpungose, C.B.: E-learning platforms at ODeL institutions of higher education: creating a real-world learning. Educ. Inf. Technol. **30**(12), 17207–17223 (2025)
33. Anwar, F., et al.: A comprehensive insight into Blockchain technology: past development, present impact and future considerations. Int. J. Adv. Comput. Sci. Appl. **13**(11) (2022)
34. Shamim, A., et al.: Blockchain technology for secure transactions: a decentralized approach to data integrity and trust. Crit. Rev. Soc. Sci. Stud. **3**(2), 828–845 (2025)
35. Atlam, H.F., et al.: Blockchain forensics: a systematic literature review of techniques, applications, challenges, and future directions. Electronics. **13**(17) (2024)
36. Kaka, G.E., et al.: Smart contracts: a new form of contract in modern day. Imam Ja'afar Al-Sadiq Univ. J. Leg. Stud. **4**(2) (2025)
37. Luo, X., et al.: A survey on the application of blockchain in cryptographic protocols. Cybersecurity. **7**(1) (2024)
38. Dong, S., et al.: Blockchain technology and application: an overview. PeerJ Comput. Sci. **9**, 1–50 (2023)
39. Azad, R.U., et al.: Blockchain applications in education: the future of learning. In: 2023 26th International Conference on Computer and Information Technology (ICCIT), pp. 1–6 (2023)
40. Delgado-von-Eitzen, C., Anido-Rifón, L., Fernández-Iglesias, M.J.: Blockchain applications in education: a systematic literature review. Appl. Sci. **11**(24), 1–24 (2021)
41. Sharples, M., Domingue, J.: The blockchain and kudos: a distributed system for educational record, reputation and reward. In: Verbert, K., Sharples, M., Klobučar, T. (eds.) Adaptive and Adaptable Learning. EC-TEL 2016. Lecture Notes in Computer Science. Springer, Cham (2016)
42. Alam, S., et al.: A Blockchain-based framework for secure educational credentials. Turk. J. Comput. Math. Educ. **12**(10), 5157–5167 (2021)
43. Salem, S.H.G., et al.: Blockchain-based biometric identity management. Clust. Comput. **27**(3), 3741–3752 (2023)
44. Kumar, U.H., Sridhar, S., Murthy, T.N.: A decentralized approach to tamper-proof online examination results. Int. J. Novel Res. Dev. **10**(7), 101–106 (2025)
45. Turkanovic, M., et al.: EduCTX: a blockchain-based higher education credit platform. IEEE Access. **6**, 5112–5127 (2018)
46. Cuya, K.C., Palaoag, T.D.: Blockchain in higher education: advancing security, verification, and trust in academic credentials. Nanotechnol. Percept. **20**(23), 373–386 (2024)
47. Patil, S.R., et al.: Smart contract enabled online examination system based on blockchain. Int. J. Adv. Res. Sci. Commun. Technol. **3**(9), 386–397 (2023)
48. Rini, J.: WP2 Conceptual framework for comparative multiple case study analysis. The RECIPES project, pp. 1–38 (2019)

Methods for Summarization of the Context Graph in Elaboration Processes of Research Documents

Sho Onami[1(✉)], Ryo Onuma[1], Hiroki Nakayama[1], Hiroaki Kaminaga[1], Youzou Miyadera[2], and Shoichi Nakamura[1]

[1] Fukushima University, 1, Kanayagawa, Fukushima 960-1296, Japan
{onami,kami}@cs.sss.fukushima-u.ac.jp, {onuma,hnakayama, nakamura}@sss.fukushima-u.ac.jp

[2] Tokyo Gakugei University, 4-1-1, Nukuikita, Koganei, Tokyo 184-8501, Japan
miyadera@u-gakugei.ac.jp

Abstract. Researchers repeatedly create and elaborate on research documents while simultaneously performing multiple tasks such as surveys, presentation of work results, and discussions. To pursue research effectively, sufficient understanding the elaboration contexts—which consists of the artifacts, referenced documents, and the history of the work—is essential. However, it becomes increasingly difficult as the research period extends. We have aimed to develop methods for automatically generating the elaboration context graph on the basis of analyzing the work history data in elaboration processes. This graph visually represents the documents which were created or referenced in the work, screenshots of the PC desktop embodying the work scenes, and the relationships among them. This paper details methods for summarizing the elaboration context graph. We also describe experiments conducted using a prototype and discuss the effectiveness of the summarization method based on their results.

Keywords: Elaboration Context Graph · Graph Summarization · Understanding the Work Histories · Elaboration Contexts of Research Documents

1 Introduction

Research activities involve multiple tasks related to creating and editing various documents, such as writing papers and preparing presentation materials. These activities are not limited to a single document but progress alongside multiple concurrent tasks, including web searches for relevant information, discussions with other researchers, and consulting existing materials. Therefore, effectively carrying out this series of tasks requires a proper understanding of the elaboration contexts—the documents used or referenced, the artifacts, and the work circumstances and histories during the creation and elaboration of research documents—and the ability to reflect such understanding in subsequent work. However, elaborating research documents often spans long periods, during which the number of utilized and created documents grows rapidly. Managing and keeping track of these documents manually is challenging, and it is difficult to vividly recall the work scenes and circumstances without any cues.

H. R. Arabnia et al. (Eds.): CSCI 2025, CCIS 2974, pp. 312–327, 2027.
https://doi.org/10.1007/978-3-032-28475-4_22

To resolve these problems, there have been methods for document management [1–7], support systems for reviewing past work on a PC [8–13]. However, these methods are limited in the tasks they can target, making them insufficiently effective for understanding the deliverables and history of document creation and their elaboration work, which involve performing multiple tasks in parallel. On the other hand, there have been methods to support accumulation, understanding, and reuse of the artifacts and histories of intellectual work [14–17]. However, many of these methods require participants to register and maintain information in the system themselves, thereby imposing an additional burden.

Therefore, in this research, we accumulate historical data of document elaboration work performed by researchers on their PCs. Based on an analysis of this data, we develop methods for generating the elaboration context graph that visually represents the relationships among the documents involved in the elaboration process. However, when the work spans a long period, the amount of historical data grows and the resulting elaboration context graph becomes extremely large. To address this issue, we further develop methods for summarizing the graph to support the under-standing of the elaboration contexts.

Through these methods, we aim to provide new forms of support for understanding the elaboration context of research documents.

2 Problem and Support Direction

2.1 Elaboration Processes of Research Documents

In research activities, creating and editing documents while performing multiple tasks in parallel, and gradually accumulating artifacts, is indispensable. We refer to this process as the "elaboration process of research documents." In such tasks, it is essential to understand and utilize the elaboration contexts, which includes referenced documents, produced artifacts, and the work circumstances.

When these documents are managed manually, it is common to assign appropriate file names based on their content or document types and to organize them into folders according to the relevant research project. However, this approach makes it difficult to perfectly manage and track all documents, and valuable materials are often overlooked. Moreover, recalling the work circumstances and histories associated with each document is desirable, but file names and folder management alone are insufficient for this purpose.

Therefore, there is a strong need for effective methods that support the use of past elaboration contexts in subsequent work.

2.2 Related Work

There exist methods for document management [1–7]. Sambetbayeva et al. developed an intelligent electronic document management system (EDMS) model based on machine learning techniques [1], motivated by the increasing volume of information handled in electronic document management. Their approach incorporates machine learning methods—including text analysis and topic modeling—into key EDMS subprocesses such as document registration, classification, and retrieval, with the aim of improving the accuracy of tasks such as document attribute prediction, automated routing, and the detection of similar or duplicate documents. However, this work primarily targets improving the efficiency of document processing within EDMS, and does not address the understanding of complex, long-term work contexts, such as those involved in the iterative elaboration of research documents.

Moreover, there exist methods that support recalling past work on a PC [8–13]. Dingler et al. developed a method for creating activity summaries based on screenshots to support workers in reviewing their daily work on the computer [8]. This method utilizes eye-tracking data and reading detection algorithms to identify key activities throughout the day, allowing the creation of activity summaries that focus on the most important screenshots. However, this approach primarily supports reviewing a single day's activities and does not address long-term work tracking or the discovery of previously used documents.

Furthermore, there exist methods to support accumulation, understanding, and reuse of the artifacts and histories of intellectual work [14–17]. Kim et al. proposed a novel type of traceability for research projects involving software development, targeting relationships between documents, between documents and their elements, and among different versions of documents [14]. This method identifies which documents influence others through content-based analysis, thereby supporting both the tracking of different document versions and the understanding of relationships between documents. While this method has the potential to contribute to meticulous management of the relationships between documents and research outputs, it assumes that participants themselves register the documents they handle in the system, which may impose an additional burden on users

2.3 Problems

We focused on the following three issues to overcome in elaboration processes of research documents.

Problem 1: In the elaboration of research documents, the involvement of diverse processes and numerous related documents makes it difficult to manage and comprehend the overall work.

Problem 2: It is important to accurately review past work along with its associated circumstances and histories, but this is not easy.

Problem 3: When a task extends over a long period, the relationships among the documents and work situations involved in the elaboration process become increasingly complex, making it difficult to fully understand the elaboration contexts even if these elements are systematically organized.

2.4 Approach

To address these problems, we developed a method for supporting the understanding of work circumstances by organizing and visually presenting the documents involved in the elaboration processes of research documents (Fig. 1). We first constructed an environment that automatically accumulates the user's work history data. In addition to this, desktop screenshots representing the actual work scenes are also collected. Using the accumulated work history data, we then developed methods for generating an elaboration context graph that organizes the documents and screenshots involved in the user's work. Furthermore, we developed a graph summarization methods to facilitate the understanding of documents and work circumstances in large-scale elaboration context graphs. Based on these methods, we implemented a support system that assists users in understanding the elaboration contexts.

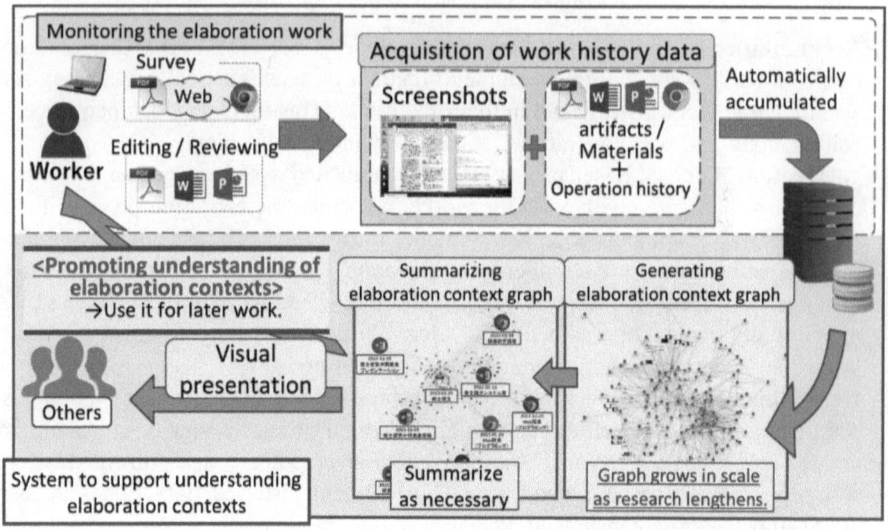

Fig. 1. Overview of support for understanding elaboration contexts.

3 Methods for Generating and Summarizing Elaboration Context Graph

3.1 Generating Elaboration Context Graph

The elaboration context graph visually represents the documents involved in the elaboration work, along with the circumstances surrounding their use, in an intuitive and comprehensible form. In this graph, documents that were edited or viewed during a series of elaboration tasks are represented as nodes. Screenshots of the PC desktop at the time of the work are also included as nodes, serving as cues that embody the actual work circumstances. Relationships among these nodes—such as simultaneous use during elaboration, citation links, and original/derived version relationships—are represented as edges. By tracing the documents connected to a specific document of interest, users can easily understand the work process across a sequence of tasks. Furthermore, users can intuitively identify past work scenes by using the screenshots as cues and understand the work circumstances together with the documents that were being used at those moments.

Here, we define the elaboration context graph. Elaboration context graph G is represented as $G = (V, E)$. V is a finite set of graph nodes, and v ($\in V$) is a node in the graph that represents a research document. E is a finite set of undirected connectors in the graph, and e ($\in E$) denotes a relationship defined between two pieces of research documents. Each node v stores attribute information such as its type (e.g., paper, presentation) and its name. Each edge e also stores attribute information, including the edge type (e.g., simultaneous use, copy-and-paste).

The elaboration context graph is generated in the following steps.

1) **Accumulation of work history data:** We developed a resident program that monitors users during they are handling documents on their PCs, automatically collecting and accumulating work history data in the background. This includes document access, editing, copy-and-paste operations, and active window switching.
2) **Generation of nodes:** Based on analysis of accumulated work history data, we extract documents that were involved in the worker's document elaboration process. First, we identify all documents accessed even once during the work session. We calculate the total active time for each document and adopt those active for a certain period or longer as nodes. Additionally, for screenshots, we adopt only those taken at the moment any document was active for a certain period or longer, and those taken immediately before the active window switched, as nodes.
3) **Generation of edges:** We analyze relationships between nodes and assign edges. The relationships between documents analyzed are primarily four types: "simultaneous use," "copy and paste," "file derivation/development," and "download from web pages." Edges are assigned for each by analyzing historical data. Edges are also assigned between screenshots and documents. At the time a screenshot is captured, documents displayed on the desktop are extracted, and edges are assigned between them.

3.2 Summarization of Elaboration Context Graph

When document elaboration work spans a long period, the number of documents involved becomes enormous, resulting in a large and complex graph. Therefore, it is essential to summarize the graph in accordance with the user's needs. In our method, document nodes are grouped by certain work periods and the most important documents are extracted from each group. Finally, the elaboration context graph is summarized by linking these groups with edges. This summarization enables users to more easily understand the elaboration contexts represented in the graph.

Summarization of elaboration context graph is defined as follows.

Definition 1: Given an elaboration context graph $G = (V, E)$, we call the mapping f: $V \rightarrow S$ the summary of G, where S is the set of summary nodes. For each summary node $s \in S$, the set of nodes in the original graph V that are mapped to s is defined as $F(s) = \{v \in V \mid f(v) = s\}$, and call $F(s)$ the set of original nodes of summary node s.

Definition 2: Given a graph $G(V, E)$ and a summary f, we call a graph $f(G) = G'(S, P, \sigma)$ that satisfies the following three constraints as a summary graph of G.

$$S = \{f(v) | \forall v \in V\} \tag{1}$$

$$P = \left\{(s,t) \,|\left(\cup_{i \in F(s), j \in F(t)}(i,j)\right) \in E,\ (i,j) \neq \varnothing\right\} \tag{2}$$

$$\sigma = (s,t) \in P \rightarrow \frac{\left|\left(\cup_{i \in F(s), j \in F(t)}(i,j)\right) \in E\right|}{|F(s)| \cdot |F(t)|} \tag{3}$$

Constraint (1) states that the summary node set is represented as a set of images of the original node set by f. Constraint (2) states that if there is at least one edge between the original node set pair $F(s)$ and $F(t)$ of summary nodes s and t, a summary edge is given between those summary nodes. Constraint (3) states that the weight of the summary edge is the density of the edge between $F(s)$ and $F(t)$.

This graph summarization method involves the following four steps (Fig. 2).

1) **Generating summary nodes:** The documents represented as nodes in the elaboration context graph are grouped according to their periods of use. Specifically, we estimate each document's usage duration by extracting the time windows in which it was active from the window history. We then group documents by analyzing overlaps in their usage periods. Each resulting group is treated as a node in the summary graph, referred to as a summary node.
2) **Selection of main node:** We quantify the work for each document using total active time and keystroke counts extracted from the work history data. Within each summary node, the document with the highest editing activity is identified and designated as the main node.
3) **Generating summary edges:** For each pair of summary nodes, we compute the degree of connectivity between the documents contained within them in the original graph. Specifically, we first determine the total possible number of edges between nodes in the two summary nodes. We then calculate the edge density by dividing the number of edges actually present in the original graph by this maximum possible number. Based on this measure, summary nodes are connected with weighted summary edges.
4) **Labeling summary nodes:** Summary nodes are labeled to allow users to quickly understand the work purpose each one represents. Specifically, the main task and its corresponding work period for each summary node are extracted by analyzing the file name of the main node, the folder where it is saved, and task information in the schedule management app. These extracted details are then assigned as the label for the summary node.

```
Algorithm GraphSummarization()
[Input]
・ G(V,E): Elaboration context graph
[Output]
・ G' = (S,P,σ): Summary Graph
[Global variable]
・ threshould_ED : Edge Density Threshold for Edge Selection
[Local variable]
・ DN : Set of document nodes in the elaboration context graph
・ SSN : Set of screenshot nodes for the elaboration context graph
・ SN [] : Set of summary nodes
・ SE [] : Set of summary edges
・ Rate_DN [] : Dict of workloads for node documents
・ DN_sortWR : Document nodes sorted by workload
[Method]
{
  //Retrieve nodes and edges from graph G.
  DN, SSN, E = acquisitionNodesAndEdges(G)
  // Summary Node Generation.
  for(∀dn∈ DN){
    //Calculate the workload for the document.
    Rate_DN[dn] = calculationWorkRate(dn);
  }
  //Sort document nodes in descending order by workload.
  DN_sortWorkRate =
      Sorting_DocumentNode(DN, Rate_DN);
  for(∀dn∈ DN_sortWorkRate){
    //Check whether the document node dn overlaps with the
      main node of each summary node. If so, retrieve that main
      node; otherwise, return NULL.
    MN = examine_UsingPeriod(SN, dn)
    if(MN){
      //Add dn to the summary node whose main node is MN.
      SN[MN].add(dn)
    }else{
      // Add a new summary node with dn as the main nodc.
      SN.add(dn)
    }
    // Get the set of screenshot nodes connected to "dn"
    Set_ssn = acquisition_screenshotNodes(dn, SSN)
    for(∀sn∈ Set_ssn) SN[dn].add(ssn);
  }
  for (∀mn∈ DN_sortWorkRate){
    SN[mn].label= labeling_SN (mn)
  }
  // Summary Edge Generation.
  for(∀sn1∈ SN){
    for(∀sn2∈ SN){
      // Calculate the total number of node combinations
        between the specified two groups of summary nodes.
      Number_possibleEdges =
          calculation_possibleEdgesNumber(sn1, sn2);
      // Get the number of edges connected in the actual input
        graph.
      ActualNumber_edges=
          acquisition_EdgesNumber(sn1, sn2);
      //Calculate edge density
      EdgeDensity =
          ActualNumber_edges / Number_possibleEdges
      // Assign summary edges weighted by edge density.
      SE.add((sn1, sn2));
      σ[(sn1, sn2)]= EdgeDensity;
    }
  }
  return G' = (SN, SE, σ)
}
```

Fig. 2. Graph Summarization Algorithm.

4 Support System for Understanding the Elaboration Contexts

On the basis of our methods, we implemented a prototype support system. The interface of the system is shown in Fig. 3. The system analyzes the accumulated work history data stored in the background, generates a detailed context graph, and presents it visually. When a user selects a document node in the graph, the contents of that document can be viewed directly within the system. Users can also understand the elaboration context by recalling the associated work scene using the screenshots as cues and by tracing and inspecting the documents used during that work.

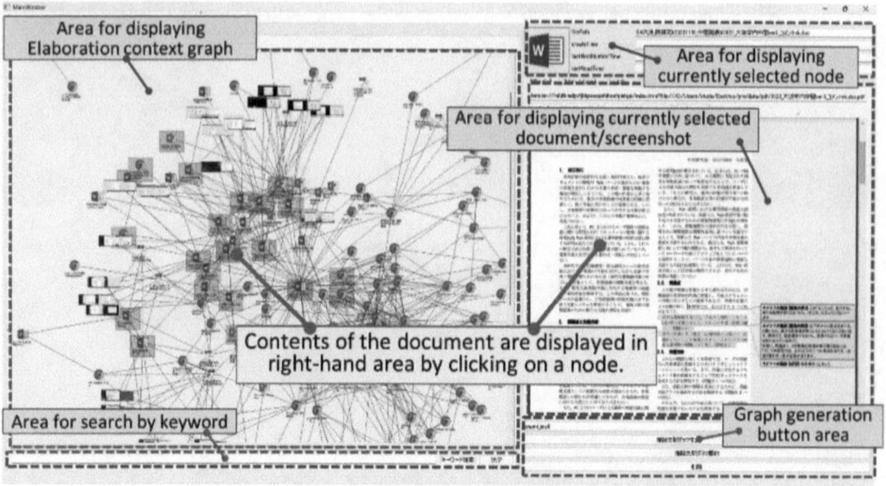

Fig. 3. Interface of support system.

5 Experiment and Discussion

5.1 Generation and Summarization of Elaboration Context Graphs of the Elaboration Work of Different Workers

Outline of Experiment. An experiment was conducted to verify the feasibility of generating and summarizing elaboration context graphs using actual elaboration-history data derived from research documents. Specifically, we asked one senior student and one master's student majoring in computer science (participants A and B) to perform research-document elaboration tasks. They were instructed to augment their base document—one of their own research documents, either a written paper or a set of presentation slides—while conducting web-based research on the specified topics. The target duration for the task was 90 min.

- Key related researches and their relevance to my research.
- Challenges and prospective ideas for advancing my research.

During these tasks, the prototype system captured and stored the task-history data. After the collaborators completed their work, we generated an elaboration-context graph and produced a summary graph based on an analysis of the collected history data.

Results and Discussion. Figures 4 and 5 show the elaboration context graphs generated on the basis of analyzing the work history data of participants A and B, respectively. First, we confirmed that the entire process—from acquiring the history data to generating the elaboration-context graphs—was successfully completed. The use of the elaboration-context graph makes it easy to grasp, for example, "the paper being written and the documents consulted alongside it at a glance," or "the work context when multiple documents are used, with screenshots serving as clues." It also highlights differences in participant behavior, such as spending considerable time editing a particular document versus devoting more time to web searches.

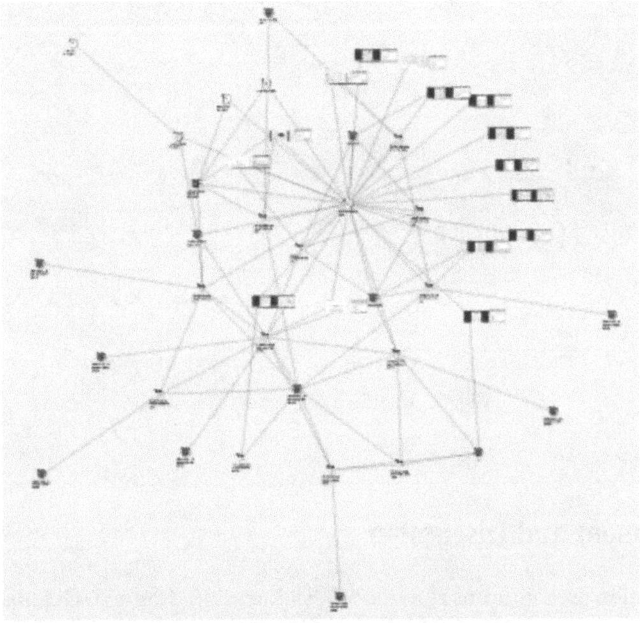

Fig. 4. Elaboration context graph of participant A.

Additionally, Figs. 6 and 7 summarize the elaboration-context graphs shown in Figs. 4 and 5. For both participants, the graphs were summarized as intended, allowing for easy comprehension of other documents referenced alongside the base document. Furthermore, clusters of tasks unrelated to editing the base document are represented as separate node clusters, enabling straightforward identification of the main tasks performed. Thus, even for relatively short work periods, the benefits of automatically generating and summarizing elaboration-context graphs can be confirmed.

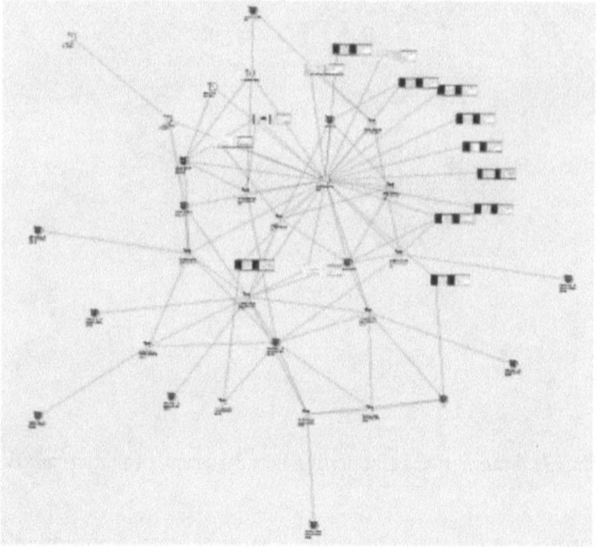

Fig. 5. Elaboration context graph of participant B.

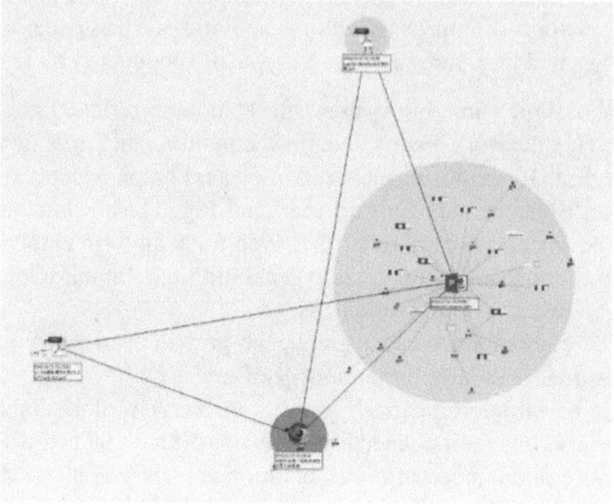

Fig. 6. Summarized elaboration context graph of participant A.

5.2 Generation and Summarization of Elaboration Context Graphs of Long-Term Elaboration Work

Outline of Experiment. An experiment was conducted to verify the feasibility and validity of generating and summarizing elaboration-context graphs for long-term elaboration tasks. In this study, we monitored the elaboration process of research documents on a PC used daily by a doctoral participant specializing in computer science over a one-year

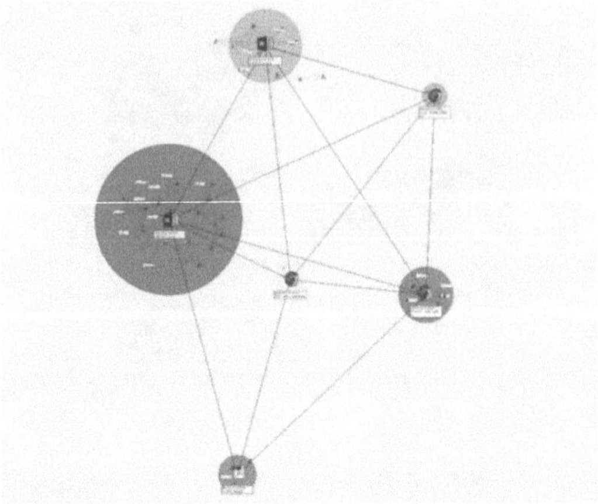

Fig. 7. Summarized elaboration context graph of participant B.

period using a prototype system, collecting historical data. Subsequently, elaboration-context graph generation and summarization were performed using two timespans: six months and one year. No specific tasks or constraints were imposed on the participant. Therefore, the experimental data consist of historical records from realistic research document elaboration work—comprising multiple activities such as paper writing, creating presentation slides, and literature review—as typically conducted by researchers.

Results and Discussion. First, Fig. 8 shows the elaboration context graph generated on the basis of analyzing the work history data for six months, and Fig. 9 shows its summary graph. Similarly, Fig. 10 shows the elaboration context graph generated on the basis of analyzing the work history data for one year, and Fig. 11 shows its summary graph. Through this experiment, we confirmed that even for long-term elaboration work, the entire process, from acquiring history data to generating and summarizing the elaboration context graph, can be carried out without issues.

Observing the elaboration context graph, the graph covering six months of work (Fig. 8) contains a substantially large number of nodes. While local observations can still be conducted as with shorter periods, gaining an overview of the elaboration context, such as the main tasks and their connections, remains difficult in its raw form. This issue becomes even more pronounced in the elaboration context graph covering one year of work (Fig. 10). Figure 12 shows an enlarged view of the graph shown in Fig. 10. It shows that this is not merely a visibility issue: the number of nodes, that is, intervening research documents, has increased further, and their relationships have become more complex. In both cases, examining the summarized elaboration context graphs (Figs. 9 and 11) makes it easy to grasp clusters of research documents used in the main tasks and the connections among them. Furthermore, an enlarged view of the summarized graph (Fig. 13) shows that each node group, or summary node, intuitively represents a set of documents involved in a single task which is easily comprehensible.

These results suggest that the methods for generating and summarizing elaboration context graphs are applicable to both short-term and long-term tasks and are useful for understanding the context of the work. Further investigation is important, for example, to identify challenges that may arise when generating and summarizing elaboration context graphs, depending on the research themes handled by participants and the content of the research documents being elaborated.

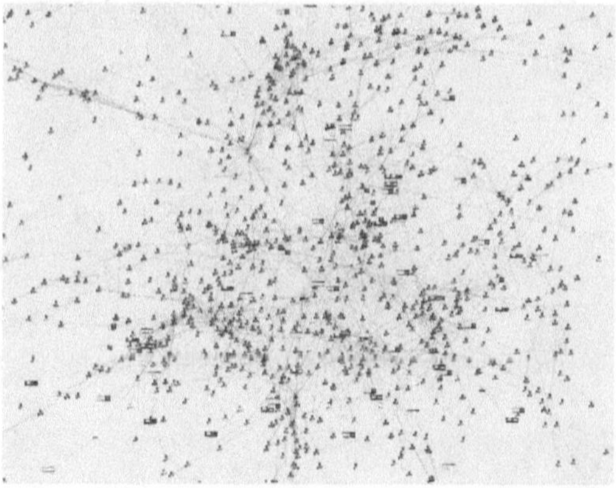

Fig. 8. Elaboration context graph generated on the basis of analyzing the six-month data.

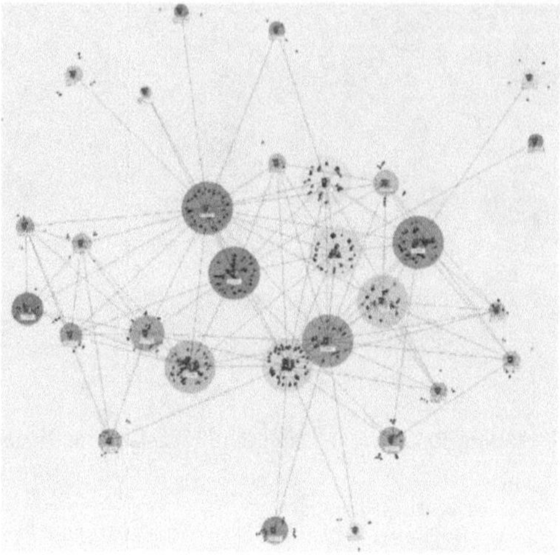

Fig. 9. Summarized elaboration context graph on the basis of analyzing the six-month data.

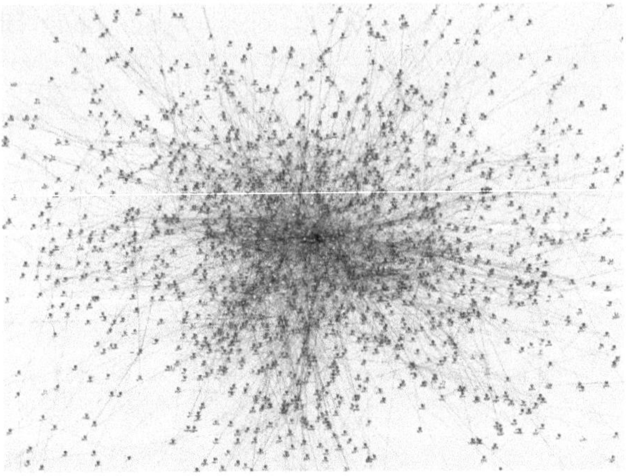

Fig. 10. Elaboration context graph generated on the basis of analyzing the one-year data.

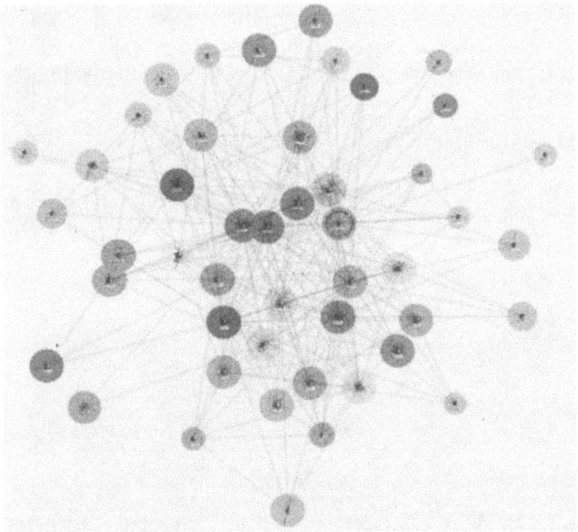

Fig. 11. Summarized elaboration context graph on the basis of analyzing the one-year data.

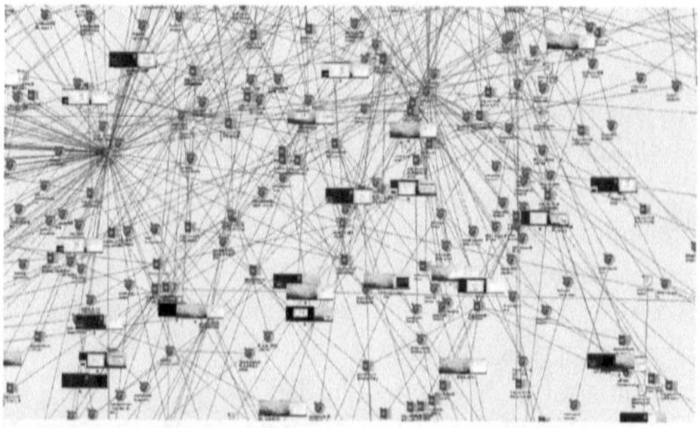

Fig. 12. Enlarged view of a part of the elaboration context graph shown in Fig. 10.

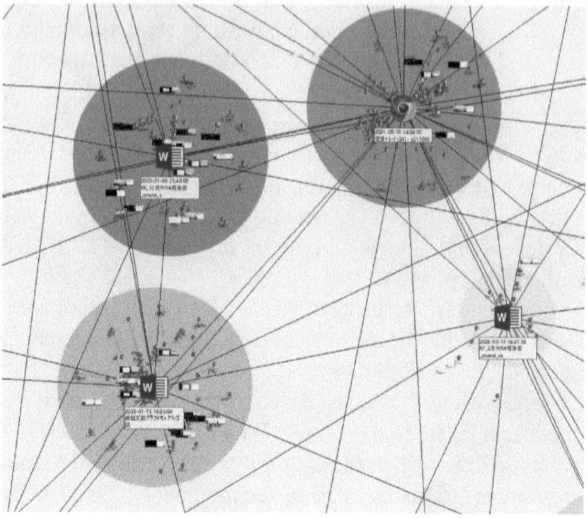

Fig. 13. Enlarged view of summary nodes in the summarized graph shown in Fig. 11.

6 Conclusion

In this research, we developed a method for generating elaboration context graphs and a graph summarization method to support the understanding of elaboration contexts. We also conducted experiments to evaluate and analyze the effectiveness of the proposed methods for generating and summarizing context graphs. Based on the results, we discussed the usefulness of the proposed approach.

Future work includes careful verification of the effectiveness of the proposed methods including the aspect of support effects to understanding of the elaboration contexts through repeated experiments applying the support system to actual research document elaboration work.

Acknowledgement. This work was supported in part by Japan Society for the Promotion of Science under Grant-in-Aid for Scientific Research (B) (No. 24K00446).

References

1. Sambetbayeva, M., Kuspanova, I., Yerimbetova, A., Serikbayeva, S., Bauyrzhanova, S.: Development of intelligent electronic document management system model based on machine learning methods. East.-Eur. J. Enterp. Technol. **1**(2), 68–76 (2022)
2. Oh, K.E.: A comprehensive investigation of researchers' shared file management practices in cloud storage. Hum.-Comput. Interact. **41**(1), 62–81 (2022)
3. Orlov, V.L., Kurako, E.A.: Electronic document management systems and distributed large-scale systems. In: Proceedings of the 10th International Conference on Management of Large-Scale System Development (MLSD), pp. 1–5 (2017)
4. Guo, F., Jahren, C.T., Turkan, Y.: Electronic document management systems for the transportation construction industry. Int. J. Constr. Educ. Res. **17**(1), 52–67 (2019)
5. Saito, N., Kanai, A., Tanimoto, S.: An organizational knowledge using method by extracting activity plan from documents in a shared folder. IPSJ J. **58**(1), 153–165 (2017)
6. Riba, P., Dutta, A., Dey, S., Lladós, J., Fornés, A.: Improving information retrieval in multi-writer scenario by exploiting the similarity graph of document terms. In: Proceedings of the 14th IAPR International Conference on Document Analysis and Recognition, pp. 475–480 (2017)
7. Cumpa, C.O., Anibal-Romero, B., Jiménez, H.J.B.: The impact of document management using good practices: a literature review. Int. J. Prof. Bus. Rev. **8**(11) (2023)
8. Dingler, T., Elagroudy, P., Matheis, G., Schmidt, A.: Reading-based screenshot summaries for supporting awareness of desktop activities. In: Proceedings of the 7th Augmented Human International Conference, pp. 1–5 (2016)
9. Hu, D., Lee, S.W.: ScreenTrack: using a visual history of a computer screen to retrieve documents and web pages. In: Proceedings of the 2020 CHI Conference on Human Factors in Computing Systems, pp. 1–13 (2020)
10. Jin, L., Feng, L., Liu, G., Wang, C.: Personal web revisitation by context and content keywords with relevance feedback. IEEE Trans. Knowl. Data Eng. **29**(7), 1508–1521 (2017)
11. Sappelli, M., Verberne, S., Kraaij, W.: Evaluation of context-aware recommendation systems for information re-finding. J. Assoc. Inf. Sci. Technol. **68**(4), 895–910 (2016)
12. Abela, C., Staff, C.: Behaviour mining for automatic task-keeping and visualisations for task-refinding. In: Proceedings of the 2016 ACM Conference on Human Information Interaction and Retrieval, pp. 23–32 (2016)
13. Huang, Z., Huang, J.: IRCHIVER: an information-centric personal web archive for revisiting past online sensemaking tasks. In: Proceedings of the 2025 ACM SIGIR Conference on Human Information Interaction and Retrieval, pp. 231–246 (2025)
14. Kim, S., Kim, H.Y., Kim, J.A., Cho, Y.: A study on traceability between documents of a software R&D project. In: Proceedings of International Conference on Multimedia and Ubiquitous Engineering, Lecture Notes in Electrical Engineering, vol. 354, pp. 203–210 (2015)
15. Limani, F., Latif, A., Borst, T., Tochtermann, K.: Scholarly resources structuring: use cases for digital libraries. In: Proceedings of Research Conference on Metadata and Semantics Research, Communications in Computer and Information Science, vol. 1057, pp. 248–259 (2019)

16. Au, C.H.: Wiki as a research support system—a trial in information systems research. In: Proceedings of the 2017 IEEE International Conference on Industrial Engineering and Engineering Management (IEEM), pp. 2271–2275 (2017)
17. Martin, L., Henrich, A.: RDFtex in-depth: knowledge exchange between LaTeX-based research publications and scientific knowledge graphs. Int. J. Digit. Libr. **25**, 517–535 (2024)

Methods for Estimating the Programming Logics in Latest Exercise from Source Codes Submitted to Previous Assignments

Ryo Onuma[1(✉)], Sho Onami[1], Hiroki Nakayama[1], Kaito Sato[1], Hiroaki Kaminaga[1], Youzou Miyadera[2], and Shoichi Nakamura[1]

[1] Fukushima University, 1, Kanayagawa, Fukushima 960-1296, Japan
{onuma,hnakayama,kami,nakamura}@sss.fukushima-u.ac.jp, {onami, kaito}@cs.sss.fukushima-u.ac.jp

[2] Tokyo Gakugei University, 4-1-1, Nukuikita, Koganei, Tokyo 184-8501, Japan
miyadera@u-gakugei.ac.jp

Abstract. The need for support is increasing as programming education becomes more widespread. Even for the same programming assignment, the approach to solving it (programming logic) varies among learners. Although effective instruction requires understanding each learner's programming logic, it is difficult to do so sufficiently within time constraints. It is possible to identify a learner's programming tendencies from the assignments they are currently working on in class. However, there are limitations to how instructors can observe learners' programming processes in detail. Moreover, learners typically have multiple past assignments in regular programming exercises. Since recent assignments alone do not sufficiently reveal learner tendencies, analysis of past assignments is necessary. However, logic analysis spanning multiple assignments requires additional effort.

In this research, we develop a model that automatically infers the logic of the latest assignment from the learner's most recent source code using machine learning. We aim to achieve comprehensive estimation of the learner's programming logic by incorporating logic tendencies across multiple assignments. Estimated results are presented to instructors, enabling them to identify groups of learners with similar programming logic. As a result, instructors can provide targeted instruction to each group efficiently, potentially reducing their teaching workload.

Keywords: Programming logics · Machine learning · Recent assignments · Support for programming exercise

1 Introduction

In recent years, opportunities for programming education have been increasing, and there is a growing need for supporting programming education. Mandatory programming education in Japanese elementary schools, junior high schools, and high schools has been implemented since the 2020 school year. This has had a significant impact on teachers

H. R. Arabnia et al. (Eds.): CSCI 2025, CCIS 2974, pp. 328–337, 2027.
https://doi.org/10.1007/978-3-032-28475-4_23

working in schools. Programming classes at universities are generally taught through "exercises" in which students think and create source code on their own. In exercises, learners often stumble and it is desirable for teachers to provide frequent guidance. In order for instructors to provide frequent guidance, they must understand each learner's logic and instruct accordingly. However, it is difficult for a small number of teachers to understand the logic of individual learners under time constraints.

In this research, we develop methods to identify groups of learners with similar programming logic by inputting learners' past exercise data into a machine learning model. Moreover, we develop a system that presents each learner's source code and logic to teachers in an easy-to-understand format. We aim to realize programming education support that corresponds to the logic of individual learners with the developed system.

2 Problems and Support Policy

2.1 Programming Logic

In programming exercises, learners work on programming tasks that are given to them. Even for the same task, the approach to solving it varies depending on the learner. This approach is referred to as "programming logic" in this paper. In order to provide effective guidance, it is necessary to accurately grasp the logic of each learner. However, it is difficult to grasp the programming logic under the actual time constraints. The trends in each student's learning in recent exercises can provide clues for predicting programming logic, but there are limits to what instructors can observe about learners' behavior.

Furthermore, because recent assignments may not accurately reflect students' programming styles, it is necessary to refer to the logic used in multiple exercises, which requires additional effort.

2.2 Related Research

There are reports on research supporting the assessment of learning progress in programming exercise classes [1–6]. Ichimura et al. [1] developed a method to enable the identification of common problems encountered during the coding process by analyzing learners' operation and error logs. Moreover, Satoh et al. [2] reported a study on the automatic detection of programming exercise participants' stumbling blocks. This research detected stumbling blocks by analyzing the differences between the final correct source code produced by learners and their source code at a given point in time and execution logs.

On the other hand, research on the classification of source code has been reported [7–18]. These studies are considered to be useful for understanding source code of the same lineage. However, these studies do not take into account understanding the learner's logic.

In summary, these studies support the identification of grammatical stumbling blocks in learning progress assessment. However, they do not consider stumbling blocks on the logic side. Therefore, it is necessary to support instructors to identify groups of learners following the same logic and provide corresponding support to each group.

2.3 Problems and Approach

We have summarized the problems that instructors encounter in understanding learners' programming logic during exercises as follows.

Problem 1: Instructors should teach according to each learner's programming logic, although it is difficult for a small number of instructors to grasp the programming logic of individual learners.

Problem 2: It is a difficult task for instructors to verify the source code of learners who have the same logic.

In this research, we design and define features by carefully examining past exercise data. We then develop a method for estimating the logic of learners using machine learning based on past exercise data (corresponding to Problem 1).

Furthermore, we develop a system that presents groups of learners and their source code for each logic in a form easily understood by instructors (corresponding to Problem 2). Figure 1 shows the overall flow of support for understanding learners' logic.

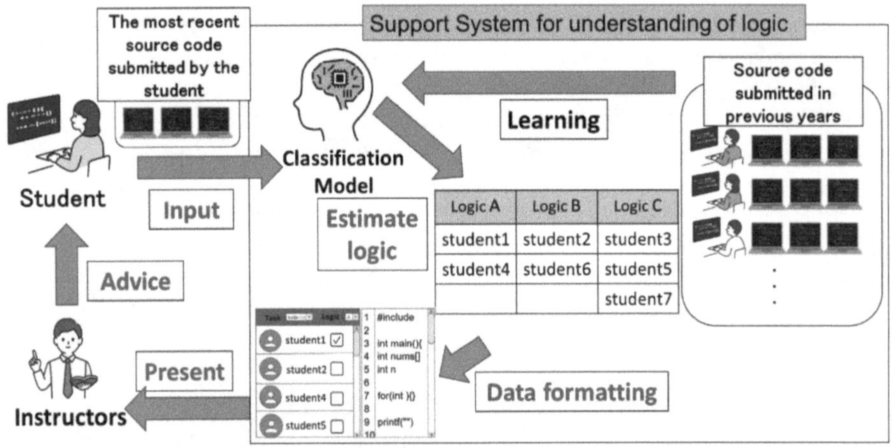

Fig. 1. Flow of support for understanding programming logic.

3 Methods for Estimating Programming Logic

3.1 An Overview of Methods for Estimating Programming Logic

Figure 2 shows an overview of logics estimation. First, we input the source code from the past three instances as features into the classification model. Trained classification model Classify the logic of the latest assignment using the vector of the most recent three submitted source codes as features. By analyzing past patterns, we can predict how learners will approach the latest challenges, enabling efficient instruction. In this research, we estimate programming logics using the following procedure.

1) Source code tokenization and preprocessing
2) Designing exercise data feature values
3) Designing machine learning models
4) Evaluating machine learning models

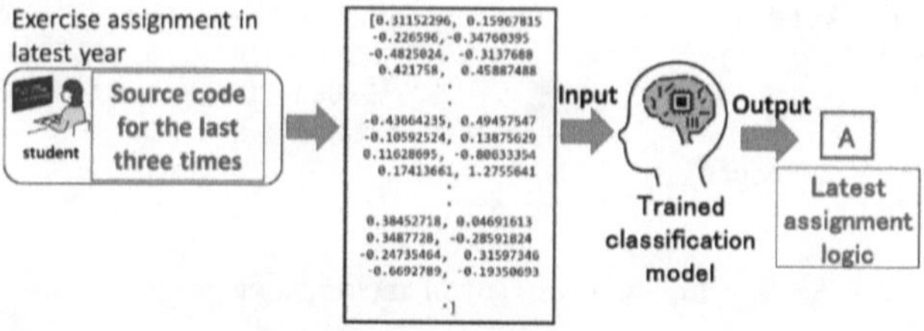

Fig. 2. An overview of logics estimation.

3.2 Source Code Tokenization and Preprocessing

We construct a dataset using C language source code, as C language is often used as the programming language taught in classes. We divide the source code into a list of the minimal lexical units of words by tokenization (Fig. 3). The tokenization process uses the Python regular expression library "re." These are the basis for extracting useful features for machine learning while retaining the semantic information of the source code.

This method captures the semantic relationships between words in vector space. Therefore, variable names, variable positions, and comments in source code affect machine learning. In this research, we aim to standardize variable names and the positions where variables are declared, and remove comments from the source code. By preprocessing unrelated to logic in this way, we can extract more features from the logic.

```
#include <stdio.h>
int main() {
      int x = 5;
      printf("Hello,world! %d
",x);
      return 0;
}
```

```
['include','<stdio.h>',
'int','main','(',')','
{','int', 'x', '=','5',
';', 'printf', '(',',',
'x', ')',';','return',
'0',';','}']
```

Fig. 3. Tokenization overview.

3.3 Feature Design for Exercise Data

We use the source code from the most recent three submissions to reflect the learner's current state of understanding. We design features by vectorizing text data for input into machine learning models. First, we convert the tokens generated from the source code into vectors. Therefore, we represent documents as relatively low-dimensional dense vectors. Specifically, in this research, we treat token sequences in source code as documents and perform distributed representation as 100-dimensional vectors (Fig. 4).

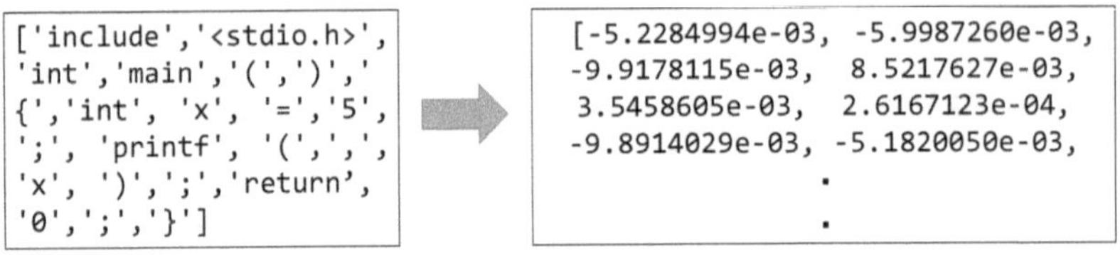

Fig. 4. Overview of vectorization.

Moreover, in this research, we combined the vectors of the most recent three source codes with equal dimensionality into a single 300-dimensional feature (Fig. 5). Based on this, we use the source code submitted for the most recent three assignments as features to estimate the logic of the latest assignment's source code.

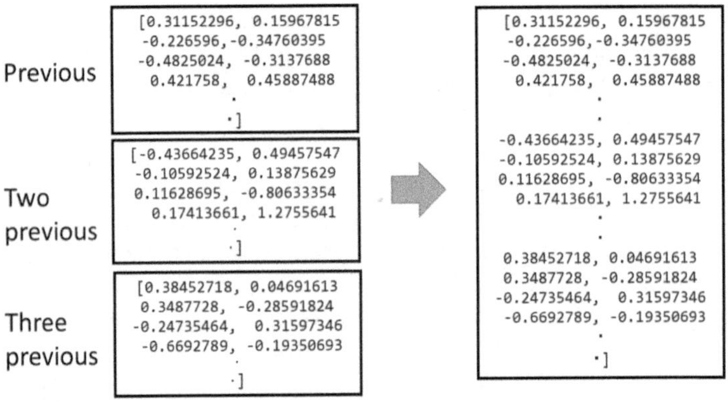

Fig. 5. Vector combination.

3.4 Design of Machine Learning Models

The aim of this research is to develop a method that can accept and classify learners' diverse logics. We develop a model that can predict the logic of a current learner's latest task based on the logic tendencies of past learners. We use scikit-learn which is a Python machine learning library for model development. scikit-learn provides numerous algorithms for supervised and unsupervised learning. It enables consistent processing from data preprocessing to model evaluation, allowing efficient creation of classification models.

In this research, we employ supervised learning-based classification model that requires relatively less data compared to deep learning and unsupervised learning. We use source code from previous years' exercise assignments as training data for the classification model. Furthermore, for supervised learning, we set the features to include the source code from the three most recent assignments of the target subject. This data is labeled with the logic type for the latest task and used to train the classification model. We classify the learner's test data by inputting it into the pre-trained model. Classification models include linear classifiers and nonlinear classifiers. In this research, we select nonlinear classifier to achieve classification into multiple logic types.

The following are examples of 5 representative nonlinear classifiers.

- k-Nearest Neighbor Algorithm
- Decision Tree Classifier
- Random Forest Classifier
- Neural Network
- Support Vector Machine

We design machine learning models using these algorithms and select the optimal algorithm by comparing them.

3.5 Evaluating Machine Learning Models

We evaluate machine learning models using one-fold cross-validation, a type of cross-validation technique. This method evaluates all samples in the dataset, sequentially selecting each sample one at a time as test data. At that time, the remaining samples are used as training data to train the machine learning model. After training is complete, classification is performed using the extracted test data. Based on the classification results, we verify the number of test data that were correctly classified. The average value of the performed evaluation is compared for each classifier to conduct the assessment.

4 Support System for Understanding Programming Logics

We develop a system that presents the source code of each learner within a group of learners who have the same logic. Figure 6 shows the system interface. The system presents specific logic types, enabling the instructor to understand each learner's logic type. By presenting learners grouped by logic and corresponding source code to instructors, we support logic-based instruction.

5 Experiment and Discussion

5.1 Experiment Overview

The aim of our experiment was to determine the success or failure of the machine learning model and to understand the issues involved. The training data includes the most recent three sets of source code for 120 students. We conduct leave-one-out cross-validation on the labeled source code and compare the classification accuracy of several algorithms to determine the optimal algorithm The experiment was conducted under two conditions. Experiment 1 assumed that the most recent task contained problems similar to the latest task. Experiment 2 assumed that the most recent task did not contain problems similar to the latest task. Specifically, Experiment 1 was conducted using the following tasks.

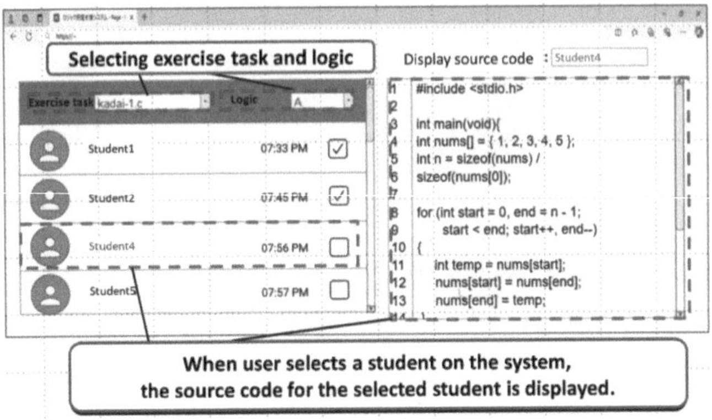

Fig. 6. Interface of the logic understanding support system.

Task 1: Create a program that determines whether an input value is a prime number.
Task 2: Create a program that outputs prime numbers up to 1000 in ascending order.
Task 3: Create a program to add two 3x3 matrices.
Task 4: Create a program that takes two 3x3 matrices as input and performs matrix multiplication.

Experiment 2 used the same set of tasks as Experiment 1, except that Task 3 was modified. The exercise task involves creating a program capable of performing addition, subtraction, multiplication, and division on two integers. This modification ensures that the task is not similar to the final assignment targeted for logic inference.

Task 3: Create a calculator program capable of performing addition, subtraction, multiplication, and division on two integers.

The possible logic for Task 4 in the latest assignment is as follows:

Logic A: Pointer management in two-dimensional arrays in Task 4
Logic B: Index management in two-dimensional arrays in Task 4
Logic C: Index management in one-dimensional arrays in Task 4

Finally, we calculated the accuracy rate and recall by confirming the matching status between the estimation results of our method and the correct data. The accuracy rate is an indicator showing the proportion of correctly classified predictions out of all predicted results, enabling an intuitive grasp of the overall performance of the method. Therefore, it is well-suited for evaluating how accurately the model performs classification overall. On the other hand, recall is a metric indicating the proportion of actual positive examples correctly identified as such. In this study, it is crucial not to overlook targets that should be detected; therefore, evaluation must focus on recall. Using recall allows assessment of how comprehensively the method captures targets without omission. We analyze how well classification models perform under different conditions through these experiments.

5.2 Results and Discussion

The results of Experiment 1 and Experiment 2 are shown in Tables 1 and 2, respectively. In Experiment 1, when problems similar to the latest exercise assignment were included in the most recent three assignments, we achieved high accuracy values. Moreover, Experiment 2, when no similar assignments were included, produced acceptable results, though its accuracy was slightly lower compared to Experiment 1. Therefore, these results suggests that the similarity between assignments is a significant factor whose impact on model performance cannot be ignored. On the other hand, it was found that the recall varies depending on the logic. In Experiment 1, k-nearest neighbor, random forest, neural network, and support vector machine classifiers achieved higher recall for Logic B than for the other logics. Similarly, Experiment 2 shows a high recall rate for Logic B.

In Experiment 2, the model learned syntactic differences between if statements and switch statements as distinct logical features, although these differences were not intended to represent programming logic in this study. As a result, this behavior affected the recall rates for Logic A and C, suggesting the need to treat such constructs as equivalent during training.

Table 1. Results of the Experiment 1.

Algorithm	Accuracy rate	Recall score of logic A	Recall score of logic B	Recall score of logic C
k-nearest neighbor algorithm	0.90	0.85	1.00	0.88
Decision tree classifier	0.90	0.95	0.92	0.84
Random forest classifier	0.98	0.98	0.98	1.00
Neural network	0.96	0.93	1.00	0.95
Support vector machine	0.98	0.95	1.00	0.98

Table 2. Results of the Experiment 2.

Algorithm	Accuracy rate	Recall score of logic A	Recall score of logic B	Recall score of logic C
k-nearest neighbor algorithm	0.60	0.52	0.94	0.57
Decision tree classifier	0.59	0.57	0.59	0.62

(continued)

Table 2. (*continued*)

Algorithm	Accuracy rate	Recall score of logic A	Recall score of logic B	Recall score of logic C
Random forest classifier	0.76	0.78	0.86	0.65
Neural network	0.83	0.79	0.94	0.80
Support vector machine	0.77	0.77	0.94	0.65

6 Conclusion

This paper outlines a logic inference method based on machine learning that leverages past exercise data. Moreover, it describes the outline of a logic understanding support system. In particular, we described a method for estimating logic from the most recent three source code versions.

In the future, we aim to repeatedly verify and improve the proposed method through experiments applying it to actual programming exercises. Moreover, we aim to construct optimal algorithm selection and hybrid methods by conducting detailed analyses of the impact of quantitative similarity evaluations and feature selection on model performance.

References

1. Ichimura, S., Kajinami, T., Hirano, H.: A trial to support understanding a programming exercise class. J. Inf. Process. Soc. Jpn. **54**(12), 2518–2527 (2013)
2. Satoh, A., Tanaka, T.: Prototype of programming exercise support system able to visualize coding status. In: 2022 12th International Congress on Advanced Applied Informatics (IIAI-AAI), pp. 256–259. IEEE (2022)
3. Igaki, H., Saito, S., Inoue, A., Nakamura, R., Kusumoto, S.: Programming process visualization for supporting students in programming exercise. IPSJ J. **54**(1), 330–339 (2013)
4. Oda, K., Kato, T., Kambayashi, Y.: Development and evaluation experiment of a classroom support system for programming education using tangibles educational materials. In: 2024 12th International Conference on Information and Education Technology (ICIET), pp. 67–71. IEEE (2024)
5. Abe, K., Fukawa, Y., Tanaka, T.: Dealing with stumbling in C language programming using visual programming environment. Int. J. Learn. Technol. Learn. Environ. **3**(1), 35–50 (2020)
6. Matayoshi, Y., Nakamura, S.: askTA: online practice classroom support system considering passivity. Comput. Softw. **39**(1), 1_55–1_71 (2022)
7. Uchiyama, T., Niimi, A.: Retrieving code fragments using Word2Vec considering the influence of nearby words peculiar to source code. In: IPSJ/SIGSE Software Engineering Symposium (SE2017), pp. 146–152 (2017)

8. Fujiwara, Y., Choi, E., Yoshida, N., Inoue, K.: Comparing deep learning-based approaches for source code classification. IEICE Trans. D. **J104-D**(8), 622–635 (2021)
9. Makihara, K., Matsushita, M., Inoue, K.: Source code evaluation method by machine learning using source code feature metrics. IEICE Tech. Rep. **119**(113), 105–110 (2019)
10. Phan, A.V., Chau, P.N., Nguyen, M.L., Bui, L.T.: Automatically classifying source code using tree-based approaches. Data Knowl. Eng. **114**, 12–25 (2018)
11. Nishioka, D., Kamiya, T.: Detecting semantic changes in source files with Doc2Vec and clustering. In: Proceedings of the JSSST Workshop on Foundation of Software Engineering, Foundation of Software Engineering 2021, vol. 2021, pp. 103–108
12. Onoue, Y., Baba, K., Koyamada, K.: Visualizing text structure of scientific articles using AI. J. Visualiz. Soc. Jpn. **38**(151), 23–37 (2018)
13. Yamanaka, K., Aman, H., Kawahara, M.: An automated evaluation of variable name using program slicing and Doc2Vec. Comput. Softw. **38**(4), 9–15 (2021)
14. Tanaka, K., Murakawa, T.: Constructing an online judging system to support programming learning. Joho Chishiki Gakkaishi. **32**(2), 307–312 (2022)
15. Yamamoto, T.: A study on source code search and classification using distributed representations. IEICE Tech. Rep. **116**(512), 67–72 (2017)
16. Zhang, J., Wang, X., Zhang, H., Sun, H., Wang, K., Liu, X.: A novel neural source code representation based on abstract syntax tree. In: ICSE 2019, pp. 783–794 (2019)
17. Le, Q., Mikolov, T., Com, T.G.: Distributed representations of sentences and documents. In: Proceedings of the 31st International Conference on Machine Learning, pp. 1188–1196 (2014)
18. Mou, L., Li, G., Zhang, L., Wang, T., Jin, Z.: Convolutional neural networks over tree structures for programming language processing. In: AAAI 2016, pp. 1287–1293 (2016)

Investing vs. Gambling: A Case Study on Retirement Planning and Wealth Management

Shanzhen Gao[1] and Weizheng Gao[2](✉)

[1] Reginald F Lewis College of Business, Virginia State University, Petersburg, VA 23806, USA
sgao@vsu.edu

[2] Department of Mathematics, Computer Science, and Engineering Technology, Elizabeth City State University, Elizabeth City, NC 27909, USA
wegao@ecsu.edu

Abstract. This case study examines the long-term financial consequences of two contrasting wealth management strategies pursued by twin professors over a 40-year academic career. Dr. Old exemplifies disciplined investing by contributing daily to a university retirement plan with employer matching, demonstrating how early planning and compound interest foster sustainable wealth. In contrast, Dr. Young spends an equivalent amount on lottery tickets, accumulates high-interest debt, and exhibits behavioral biases such as the gambler's fallacy, overconfidence, and present bias. The case challenges students to apply the time value of money (TVM) principles, compare the compounding effects of savings versus debt, and evaluate behavioral finance factors that shape financial decisions. Designed for courses in personal finance, investments, and financial literacy, the study underscores how consistent, informed behaviors promote long-term financial security.

Keywords: Behavioral Finance · Compound Interest · Financial Literacy · Gambler's Fallacy · Retirement Planning · Time Value of Money (TVM) · Wealth Management

1 Introduction and Literature Review

The pursuit of sustainable wealth management and long-term financial security rests upon the fundamental principles of time-value-of-money (TVM) and compound interest. These core financial concepts demonstrate that small, consistent contributions, when sustained over decades, yield exponential returns—making early and disciplined saving essential for effective retirement planning. Regular contributions, especially when paired with employer matching, significantly accelerate capital accumulation and reward patience far more than speculative, high-risk strategies. Prior work emphasizes that long-term financial success is driven primarily by disciplined saving behavior and "time in the market," rather than attempts to build wealth through rapid gains or chance-based

H. R. Arabnia et al. (Eds.): CSCI 2025, CCIS 2974, pp. 338–347, 2027.
https://doi.org/10.1007/978-3-032-28475-4_24

endeavors. Case-based and experiential financial instruction has been shown to help students transition from abstract TVM theory to practical decision-making and real-world application (Gao et al., 2025a). Furthermore, incorporating computational and algorithmic problem-solving into academic settings reinforces the quantitative rigor required for sound financial analysis and enhances students' analytical capabilities (Gao et al., 2025b).

Well-designed retirement systems and "choice architecture" play a pivotal role in shaping individual wealth accumulation outcomes. Automatic enrollment, default contribution rates, and employer-matching features significantly increase retirement plan participation, demonstrating how structural nudges can transform financial intentions into sustained saving habits (Madrian & Shea, 2001). Commitment-based saving programs, such as the Save More Tomorrow (SMarT) model, further leverage behavioral insights to help individuals gradually increase contribution rates without immediate financial strain, resulting in significantly higher long-term savings (Thaler & Benartzi, 2004). These principles align closely with the disciplined financial behavior exemplified by Dr. Old in this case study, illustrating how structured systems, combined with rational planning, can maximize the benefits of compounding.

Despite the compelling mathematics of wealth accumulation, rational financial decision-making is frequently undermined by cognitive and behavioral biases. Behavioral finance research reveals that individuals with typical impulsiveness present bias, limitations in self-control, and short-term impulsivity, prioritizing immediate consumption over long-term financial well-being. Hyperbolic-discounting theory formalizes this tendency, explaining why individuals may fail to save adequately for retirement despite long-run intentions to do so (Laibson, 1997). Prospect theory further demonstrates that individuals overweight small probabilities and underweight steady gains, making lottery participation psychologically attractive even when statistically irrational (Kahneman & Tversky, 1979). This pattern is evident in Dr. Young's repeated lottery purchases, despite the negative expected returns. Student learning research also suggests that cognitive reasoning and logical judgment can be influenced by human biases when solving financial or mathematical problems—highlighting the importance of integrating behavioral awareness into financial education (Gao et al., 2024).

Overconfidence further exacerbates bias-driven financial decisions. Empirical research has shown that overconfident individuals trade more frequently, take excessive risks, and ultimately earn lower net returns (Barber & Odean, 2001). In the context of this case, Dr. Young's belief that persistence and "luck" will eventually generate a major win reflects a combination of overconfidence and the gambler's fallacy—the mistaken expectation that a favorable outcome becomes more likely after repeated losses. Such behavior is consistent with findings in the human-factors literature, which identifies risk-taking tendencies and flawed judgment as common vulnerabilities in high-stakes decision environments (Malomo et al., 2025). Moreover, longstanding research on lottery behavior has shown that lotteries exploit psychological biases by offering high-payout, low-probability incentives that attract individuals despite deeply unfavorable financial outcomes (Clotfelter & Cook, 1989).

Financial literacy has been identified as a critical determinant of personal financial outcomes. Individuals with higher levels of financial literacy demonstrate stronger retirement planning behaviors, more effective investment decision-making, and lower susceptibility to high-interest debt traps (Lusardi & Mitchell, 2014). Conversely, individuals with limited financial knowledge are more vulnerable to predatory financial products, which can exacerbate debt cycles and lead to poor spending habits. Integrating TVM practice, fee awareness, and probability calibration into classroom instruction has been shown to improve financial decision-making competence. This teaching case addresses these needs by integrating TVM-based quantitative analysis with behavioral finance insights to strengthen students' ability to evaluate real-world financial choices through both mathematical reasoning and reflective judgment.

In contrast to Dr. Young, Dr. Old demonstrates disciplined financial planning through consistent retirement contributions, maximizing employer matching, early debt elimination, and adherence to a long-term investment strategy that fully leverages the power of compound interest. The stark divergence in the twins' financial outcomes over a 40-year academic career underscores the opportunity cost of speculation relative to consistent saving. To enhance student engagement and conceptual understanding, contemporary business and finance education increasingly incorporates computational tools, simulations, and generative AI to support data-driven financial learning, scenario modeling, and critical thinking development (Gao et al., 2025b). These tools enable students to visualize the mechanics of compounding, model debt amortization, and assess decision-making biases, thereby deepening their understanding of how daily choices influence long-term financial outcomes.

Overall, the literature demonstrates that four key drivers influence long-term financial outcomes: (i) structured saving systems and retirement plan design, (ii) behavioral and cognitive biases that distort rational judgment, (iii) financial literacy competency, and (iv) integration of computational and analytical tools to improve decision-making. This teaching case synthesizes these dimensions, providing a comprehensive instructional platform for analyzing both the mathematical foundations of wealth accumulation and the behavioral factors that contribute to financial success or failure—effectively setting the stage for the contrasting financial journeys of Dr. Old and Dr. Young.

2 Case Narrative

Background

Sixty-six years ago, twin boys with exceptional talent were born. Forty years ago, after earning Ph.D. degrees in Investment from a top university, both began their careers as assistant professors in the same Department of Investment. Their students affectionately called the elder twin Dr. Old and the younger twin Dr. Young. At the beginning of their careers, each earned an annual salary of $50,000.

Dr. Old's Path

Dr. Old enrolled in the university's retirement plan immediately. He contributed $60 per day, and the university matched his contribution with an additional $40 per day, for a total of $100 daily. He maintained this disciplined investment habit for 40 years (based on 365 days per year). Furthermore, he paid off all his student loans, mortgage, and car loans 30 years ago, achieving complete financial independence.

Through consistent saving and the power of compound interest, Dr. Old's wealth grew exponentially, demonstrating how discipline and time can transform modest contributions into extraordinary long-term success.

Dr. Young's Path

Dr. Young, in contrast, chose not to participate in the retirement plan. Instead, he spent $100 per day on the "You Will Be Very Rich Lottery," a routine he maintained for 40 years.

After four decades, he found himself with $2,000,000 in debt—from student loans, a mortgage, and other expenses—compounding at 15% annually. By chance, he won a $1,000,000 lottery prize today. However, instead of reducing his debts, he decided to reinvest the entire amount in lottery play and then use future winnings to reinvest, continuing the cycle indefinitely. He hopes that persistence might make him a billionaire.

His situation illustrates how cognitive biases, such as the gambler's fallacy, overconfidence, and present bias, can undermine rational financial behavior.

Financial Analysis Questions

1. If Dr. Old's retirement contributions earned an average daily return of 0.031%, what would his retirement account balance be today?
2. How much has Dr. Young spent on lottery tickets over the past 40 years?
3. If every $1 spent on the lottery yields an average return of $0.01, how realistic is Dr. Young's chance of becoming a billionaire?
4. If Dr. Young earns an annual salary of $150,000 for the next 10 years, can he pay off his debt if he allocates all remaining income (after $70,000 annual expenses) toward debt repayment?
5. What financial strategies ensure long-term security and sustainable wealth?

3 Teaching Objectives

This case enables students to:

Apply time-value-of-money (TVM) and compound-interest formulas to real-world investment and debt problems.
Compare the long-term outcomes of disciplined saving versus gambling behavior.
Evaluate the feasibility of repaying high-interest debt over time.
Identify behavioral finance biases that distort rational decision-making.
Recommend sustainable strategies for building and maintaining long-term financial security.

4 Key Formulas and Model Results

Dr. Old's Retirement Account Balance

This calculation utilizes the Future Value of an Ordinary Annuity formula, as contributions are made daily and the daily return is specified.

The formula for the future value of an ordinary annuity is:

$$FV = C \times [(1 + r)\hat{}N - 1]/r$$

Where:

C = \$100 (Daily contribution: \$60 personal + \$40 match)
r = 0.031% = 0.00031 (Daily interest rate)
N = 40 years × 365 days/year = 14,600 (Total number of periods)

Calculation:

FV = \$100 × [(1 + 0.00031)^14,600–1] / 0.00031 = 29,459,190.61.
Dr. Old's retirement account balance today would be \$29,459,190.61.

Dr. Young's Lottery Strategy
Given:

Daily spending = \$100
Average return = \$0.01 per \$1 (1%)
Time = 40 years × 365 days = 14,600 days

Formulas:

Total Spending = 100 × 365 × 40 = \$1,460,000
Expected Return = 1,460,000 × 0.01 = \$14,600

Iterated Lottery Reinvestment (Hypothetical Chain):

Iteration	Amount Invested	Expected Return (1%)
Initial win	\$1,000,000	\$10,000
2nd round	\$10,000	\$100
3rd round	\$100	\$1
4th round	\$1	\$0.01

Conclusion: Even with repeated reinvestments, the expected returns diminish exponentially. The probability of reaching \$1 billion through continual reinvestment is statistically negligible.

Dr. Young's Debt Repayment Analysis (Scenario Comparison)
Dr. Young's annual repayment capacity is \$150,000 (Income) – \$70,000 (Expenses) = \$80,000. The debt compounds at a high annual rate of 15%. The following analysis compares two scenarios: one without applying the lottery winnings, and one where the winnings are applied immediately to reduce debt.

Scenario 1: No Lottery Winnings Applied to Debt (Original Problem)
This calculation assumes Dr. Young ignores his \$1 million win and only makes the annual \$80,000 payment.

Year 1 Calculation:

Interest Accrued: $2,000,000 × 0.15 = $300,000
Balance Before Payment: $2,000,000 + $300,000 = $2,300,000
Ending Debt Balance: $2,300,000 − $80,000 = $2,220,000

Outcome After 10 Years:

Since the annual interest ($300,000) exceeds the annual payment ($80,000), the debt grows each year.

After 10 years:

Unpaid Debt: $6,466,818.01

Conclusion: Dr. Young is unable to pay off his debt. The balance is expected to grow to over $6 million.

Scenario 2: Lottery Winnings Applied to Debt
This scenario assumes Dr. Young makes the rational choice to immediately reduce his principal by $1,000,000, leaving a new starting principal of $1,000,000.

Year 1 Calculation:

Interest Accrued: $1,000,000 × 0.15 = $150,000
Balance Before Payment: $1,000,000 + $150,000 = $1,150,000
Ending Debt Balance: $1,150,000 − $80,000 = $1,070,000

Outcome after 10 years:

Unpaid Debt: $2,421,260.28
Required Annual Payment: To prevent the debt from growing, Dr. Young must make an annual payment of at least $150,000.

Conclusion: Even after receiving the $1 million lottery winnings, Dr. Young remains unable to pay off his debt. The balance continues to increase, though at a slower rate.

Overall Insight:
The contrast between Dr. Old and Dr. Young demonstrates the mathematical power of compound growth versus compound debt. Dr. Old's consistent savings create exponential wealth, while Dr. Young's speculative behavior and insufficient repayment plan result in accelerating debt accumulation. Small daily financial decisions, when compounded over time, lead to vastly different life outcomes.

5 Teaching Note

Case Synopsis
This teaching note accompanies a financial decision-making case that compares the choices of two twin professors, Dr. Old and Dr. Young, who make opposite decisions over 40 years. Both begin their academic careers with identical salaries and opportunities but develop contrasting financial outcomes.

Dr. Old invests $100 daily (including employer match) in his university retirement plan for 40 years, achieving multimillionaire status through the power of compound growth.

Dr. Young spends \$100 per day on the 'You Will Be Very Rich Lottery.' After 40 years, he has accumulated \$2,000,000 in debt at an annual interest rate of 15%. Even after winning a \$1,000,000 prize, he reinvests the winnings into gambling, hoping to become a billionaire.

The case enables students to apply time-value-of-money (TVM) principles, analyze compound interest and debt accumulation, and evaluate behavioral biases such as overconfidence, present bias, and gambler's fallacy.

Learning Objectives

Apply TVM and compound interest formulas to savings and debt.

Compare the outcomes of consistent investing with those of speculative gambling.

Evaluate debt repayment feasibility under high-interest conditions.

Analyze behavioral finance biases that distort rational decision-making.

Develop practical financial literacy strategies to achieve wealth and avoid debt.

Target Audience and Prerequisites

Target Courses: Personal Finance, Investments, Financial Planning, Behavioral Finance

Academic Level: Undergraduate or MBA

Prerequisites: Basic understanding of compound interest, annuities, and behavioral economics.

Teaching Plan

Section	Time (Minutes)	Activity	Purpose
1	10	Introduce background and story	Engage and contextualize
2	20	Quantitative analysis of investment vs. debt	Reinforce TVM and compounding
3	15	Compare outcomes and scenarios	Highlight opportunity cost
4	10	Discuss behavioral finance insights	Connect psychology to finance
5	5	Conclude and reflect	Reinforce financial literacy

Quantitative Model Solutions

Dr. Old's Retirement Account Balance:

$$FV = C \times [(1 + r)\hat{}N - 1]/r$$

Where:

$C = \$100$ (daily contribution)
$r = 0.00031$ (0.031% daily rate)
$N = 14{,}600$ days (40 years $\times$ 365)

$FV = 100 \times [(1.00031)^{14{,}600} - 1] / 0.00031 = \$29{,}459{,}190.61$

Dr. Old's Retirement Balance ≈ $29.46 million.

Dr. Young's Lottery Spending and Expected Return:

Total Spending = 100 × 365 × 40 = $1,460,000
Expected Return = 1,460,000 × 0.01 = $14,600
Expected Lifetime Gain ≈ $14,600.

Debt Repayment Scenarios:

Scenario 1 (No Lottery): After 10 years, Unpaid Debt ≈ $6,466,818.
Scenario 2 (With Lottery): After 10 years, Unpaid Debt ≈ $2,421,260.

Key Discussion Points

Compounding rewards consistency: small savings grow exponentially.
Identical cash flows yield opposite results due to behavioral choices.
High-interest debt compounds faster than most investments.
Behavioral biases—overconfidence, gambler's fallacy, present bias—explain irrational behavior.
Financial literacy and planning are crucial for achieving long-term financial independence.

Discussion Questions

Why does compounding favor time and consistency in wealth creation?
Which behavioral finance concepts explain Dr. Young's decisions?
What practical steps could Dr. Young take to improve his financial future?
What can institutions do to promote financial literacy and debt awareness?
How can individuals recognize and overcome the gambler's fallacy?

Assessment Criteria

Criterion	Weight	Description
Quantitative Accuracy	40%	Correct formulas and computations
Analytical Insight	30%	Depth of reasoning and understanding
Behavioral Finance Integration	20%	Recognition of psychological biases
Presentation Clarity	10%	Structure, flow, and professionalism

Teaching Takeaways

Compounding builds wealth: gambling destroys it.
Debt grows faster than income when interest rates are high.
Behavior and discipline, not luck, determine success.
Financial education reduces irrational decision-making.

6 Conclusion

This case study presents a compelling quantitative illustration of the power of compound interest, contrasting it with the destructive effects of compound debt and speculative gambling. The forty-year financial trajectories of twin professors, Dr. Old and Dr. Young, vividly demonstrate the superiority of disciplined saving and long-term planning over high-risk behavior. Dr. Old's consistent, matched contributions—amplified by time and the exponential force of compounding—produced a retirement balance exceeding $29.46 million. His success was achieved not through extraordinary returns but through sustained financial discipline and commitment to systematic investing.

Conversely, Dr. Young's experience highlights the steep costs of financial illiteracy and behavioral bias. His daily lottery spending, characterized by a negative expected return, combined with high-interest debt compounding 15% annually, generated liabilities that far outpaced his repayment capacity. The quantitative analysis confirms that even substantial windfalls are insufficient to reverse such debt growth, underscoring the dangers of speculative habits and cognitive distortions such as the gambler's fallacy and present bias.

Ultimately, this case serves as a powerful instructional resource for courses in personal finance, investment, and behavioral economics. It reinforces a fundamental truth of wealth management: long-term financial security rests not on luck, but on informed decision-making, consistent saving, and the disciplined application of time-value-of-money principles. The stark divergence in outcomes highlights how small, daily financial choices—when compounded over a lifetime—can significantly influence vastly different financial destinies.

Data Availability. The data used to support this study's findings are available from the corresponding author upon request.

Conflicts of Interest. The authors declare that they have no conflicts of interest.

Funding Statement. This study received no funding in any form.

References

Barber, B.M., Odean, T.: Boys will be boys: gender, overconfidence, and common stock investment. Q. J. Econ. **116**(1), 261–292 (2001). https://doi.org/10.1162/003355301556400

Clotfelter, C.T., Cook, P.J.: Selling Hope: State Lotteries in America. Harvard University Press (1989)

Gao, S., Gao, W., Allagan, J., Su, J.: Innovative teaching in business analytics: bridging theory, practice, and student engagement. J. Technol. Res. **12** (2025a). Retrieved from https://www.aabri.com/jtr.html

Gao, S., et al.: Problem-solving using logic and reasoning, mathematics, algorithms, python, and generative AI: part three. In: Arabnia, H.R., Deligiannidis, L., Shenavarmasouleh, F., Amirian, S., Ghareh Mohammadi, F. (eds.) Computational Science and Computational Intelligence: CSCI 2024 (Communications in Computer and Information Science), vol. 2504, pp. 165–181. Springer (2025b). https://doi.org/10.1007/978-3-031-94943-2_13

Gao, S., et al.: Exploring the interplay between AI and human logic in mathematical problem-solving. Online J. Appl. Knowl. Manag. **12**(1), 73–93 (2024). https://doi.org/10.36965/OJAKM.2024.12(1)73-93

Kahneman, D., Tversky, A.: Prospect theory: an analysis of decision under risk. Econometrica. **47**(2), 263–291 (1979). https://doi.org/10.2307/1914185

Laibson, D.: Golden eggs and hyperbolic discounting. Q. J. Econ. **112**(2), 443–477 (1997). https://doi.org/10.1162/003355397555253

Lusardi, A., Mitchell, O.S.: The economic importance of financial literacy: theory and evidence. J. Econ. Lit. **52**(1), 5–44 (2014). https://doi.org/10.1257/jel.52.1.5

Madrian, B.C., Shea, D.F.: The power of suggestion: inertia in 401(k) participation and savings behavior. Q. J. Econ. **116**(4), 1149–1187 (2001). https://doi.org/10.1162/003355301753265543

Malomo, O., et al. (2025). Managing human factors vulnerabilities in the risk assessment process. In Daimi, K., Arabnia, H.R., Deligiannidis, L. (Eds.) Security and Management and Wireless Networks. CSCE 2024. Communications in Computer and Information Science, vol 2254. Springer, Cham. doi:https://doi.org/10.1007/978-3-031-86637-1_11

Thaler, R.H., Benartzi, S.: Save more tomorrow™: using behavioral economics to increase employee saving. J. Polit. Econ. **112**(S1), S164–S187 (2004). https://doi.org/10.1086/380085

Leveraging AI-Generated Podcasts to Enhance Higher Education

Yunkai Liu(✉) and Mei-Huei Tang

Gannon University, Erie, PA 16506, USA
{Liu006,Tang002}@gannon.edu

Abstract. Podcasting has emerged as a significant medium in higher education, evolving from early use as lecture recordings to more sophisticated applications that supplement instruction and foster student creativity. This paper reviews literature and examines case studies across multiple disciplines to trace this progression, with particular attention to science and engineering contexts where podcasts support laboratory preparation, conceptual reinforcement, and science communication. A central focus of the paper is the rise of AI-generated podcasts, which introduce new possibilities for automation, personalization, and scalability in content delivery. Case studies from institutions such as Harvard, Yale, MIT, and Manchester illustrate established practices, while examples from Arizona State University and Georgia Tech highlight emerging models of AI-enhanced podcasting. The study contributes by (1) synthesizing podcast applications into a framework of substitutional, supplementary, and creative roles, (2) highlighting distinctive applications in STEM education, and (3) analyzing the opportunities and challenges of AI-driven podcast production. The findings suggest that podcasts—especially when augmented by AI—hold considerable potential for creating flexible, scalable, and pedagogically rich learning environments, provided that implementation is guided by careful attention to accuracy, ethics, and the indispensable role of human educators. This paper is expected to submit to the 2025 International Conference on Computational Science and Computational Intelligence – Research Track on Education (CSCI-RTED).

Keywords: Podcast-Based Education · Artificial Intelligence · Higher Education

1 Introduction

Podcasting has evolved from a niche medium into a widely adopted channel for communication, knowledge sharing, and education. In the United States, more than 160 million individuals were regular podcast listeners by 2023, with awareness exceeding 79% of the population [3]. This growth has been fueled by the ubiquity of smartphones and smart speakers, creating on-demand opportunities for audio-based learning [1, 2]. Beyond entertainment, podcasts now play meaningful roles in politics, professional development, and higher education.

Within education, podcasts have been embraced as versatile tools for supporting learning across disciplines. Research identifies three pedagogical roles: substitutional

H. R. Arabnia et al. (Eds.): CSCI 2025, CCIS 2974, pp. 348–358, 2027.
https://doi.org/10.1007/978-3-032-28475-4_25

(e.g., lecture recordings), supplementary (e.g., expert interviews, explanatory discussions), and creative (e.g., student-produced podcasts) [7, 8]. These roles demonstrate that podcasts are more than simple delivery mechanisms; they can foster active learning, critical thinking, and communication skills [9].

The rapid development of artificial intelligence (AI) extends these possibilities. AI tools such as Google's NotebookLM, Illuminate, and Microsoft Copilot automate the production of podcasts by summarizing complex texts, generating dialogues, and tailoring content to learners [5, 6, 30]. These technologies hold potential for improving comprehension, enabling large-scale personalization, and reducing the workload of instructors. Yet they also raise questions of accuracy, bias, and ethics that must be addressed.

This paper investigates the evolving role of podcasts in higher education, with a particular emphasis on AI-generated content. It addresses two guiding questions:

- What are the applications and benefits of podcasts within higher education, and how have these evolved over time?
- What distinctive impacts do AI-generated podcasts have on teaching and learning outcomes?

To answer these questions, the paper synthesizes literature on podcast use in higher education and examines case studies of both traditional and AI-driven applications. Together, these analyses illustrate the pedagogical strengths of podcasts, their growing relevance in STEM and professional education, and the opportunities and challenges introduced by AI-based content generation.

2 Methodology

A scoping review methodology was adopted to map the pedagogical applications of podcasts in higher education and to identify emerging practices involving AI-generated content. Following Arksey and O'Malley's framework, searches were conducted in Scopus, Web of Science, IEEE Xplore, and ERIC for the period 2007–2025. The search strategy employed keyword combinations including "podcast," "higher education," "STEM," "online learning," and "AI podcast."

The initial search returned approximately 150 articles. After screening abstracts and full texts for relevance to higher education pedagogy, 42 articles were retained for analysis. Studies were included if they (1) examined podcasts as a teaching or learning tool in higher education, or (2) addressed the use of AI for content delivery or instructional support. Articles focused exclusively on K-12, corporate training, or non-educational contexts were excluded.

Complementing the literature review, case studies were identified through institutional repositories and public-facing resources. Selection criteria emphasized disciplinary variety, innovation in podcast application, and relevance to STEM or AI integration. The final set includes Harvard Business Review's IdeaCast, Yale's The Happiness Lab, MIT's Curiosity Unbounded, the University of Manchester's Productivity Puzzles, and AI-enabled initiatives such as Georgia Tech's D2B2 and Arizona State University's AI for U.

This methodological design ensures both breadth, by surveying a wide body of literature, and depth, by analyzing representative case studies that illustrate contemporary practices and innovations in higher education podcasting.

3 Literature Survey on Podcasts in Education

Early research on podcasting in education focused on technical feasibility and student access. For example, Copley (2007) examined the challenges of recording and distributing lectures, emphasizing usability and reliability [7]. These early investigations established the foundation for integrating podcasts into higher education settings.

Building on this, McGarr (2009) proposed a tripartite framework that remains central today: substitutional, supplementary, and creative uses [8].

- Substitutional use refers to making lecture recordings available for students. This model supports asynchronous review, allowing learners to revisit explanations outside of class.
- Supplementary use expands learning by providing expert interviews, case discussions, or additional commentary that goes beyond core lecture material.
- Creative use involves students producing their own podcasts, which fosters communication skills, critical thinking, and ownership of learning.

Merhi (2015) highlighted further considerations, such as production quality, episode length, and integration with other technologies [9]. These factors shape the effectiveness of podcasts in practice, influencing how well they align with instructional goals.

Overall, the literature demonstrates a clear trajectory: from initial experiments with lecture capture to a broader recognition of podcasts as versatile, pedagogically significant tools. Their evolving role now encompasses not only content delivery but also the development of active learning, critical reflection, and communication competencies.

4 Applications for Podcasts in Higher Education

Podcasts have become powerful educational tools with diverse applications across disciplines. This section highlights several representative case studies, examines the specific uses of podcasts within various academic fields, and discusses their distinctive benefits for online and continuing education.

4.1 Podcast Case Studies

Institutional case studies illustrate how podcasts have been deployed for teaching, outreach, and innovation in higher education. Each example highlights a distinct dimension of academic podcasting and its impact.

Harvard Business Review's "IdeaCast" [12, 13] is one of the earliest and most established academic-related podcasts. Since 2006, it has featured interviews with business leaders and management experts, making scholarship relevant to a global professional audience. With millions of downloads annually, the series demonstrates how podcasts can extend institutional influence well beyond campus, positioning Harvard as a thought leader in management education.

Yale University's "The Happiness Lab" [14], launched in 2019 by Professor Laurie Santos, translates psychological research on well-being into accessible episodes for students and the public. The podcast has reached an audience of over 65 million listeners worldwide, indicating the potential of podcasts not only for course support but also for reshaping public engagement with social science research.

MIT's "Curiosity Unbounded" [15] showcases conversations with faculty engaged in cutting-edge research across science and technology. Rather than replicating lectures, the series emphasizes discovery and inquiry. Its uptake within MIT's alumni and public communities demonstrates how podcasts can strengthen institutional reputation as a hub for innovation while offering students and graduates an accessible way to remain connected to current research.

The University of Manchester's "Productivity Puzzles" [16], hosted by Professor Bart van Ark, situates academic research within policy and societal debates. The podcast has been cited in government and policy discussions, underscoring the capacity of academic podcasting to influence stakeholders beyond the university and contribute to national dialogue on economic issues.

Arizona State University's "AI for U" initiative [17] reflects a newer direction: using podcasts to disseminate institutional experimentation with artificial intelligence. By framing AI as both a topic and a tool, the initiative functions as professional development for staff and a public demonstration of innovation. Early evaluations suggest it has helped normalize the use of AI among faculty and administrators, offering a model for organizational learning in higher education.

Taken together, these case studies show that podcasts serve multiple roles: extending thought leadership, translating specialized research for wider audiences, strengthening institutional identity, influencing policy, and documenting organizational innovation. Their measurable reach and influence suggest that podcasts are not only teaching tools but also strategic assets in shaping how institutions engage with society.

4.2 Podcasts in Different Disciplines

Podcasts have been adopted across higher education, but their impact is particularly notable in science and engineering, where students face challenges in mastering abstract theories and technical procedures.

In laboratory-based courses, podcasts are used for pre-lab preparation. Short, targeted episodes can walk students through safety protocols, explain underlying principles, and outline experimental procedures [18]. This approach reduces cognitive overload during lab sessions and allows students to enter the lab better prepared.

In theory-intensive areas such as physics, chemistry, and computer science, podcasts provide conceptual reinforcement. Instructors can present complex topics in conversational form, complementing traditional lectures and giving students multiple pathways to comprehension [19].

Podcasts also support interdisciplinary integration in engineering and applied sciences. Episodes featuring industry professionals expose students to real-world applications of technical knowledge and help them understand how academic content translates into professional practice. This strengthens the relevance of coursework and broadens students' career awareness.

Another growing use is in student-created podcasts, where learners synthesize research findings and present them for a general audience. This exercise develops communication skills and reinforces the societal role of science by training students to explain specialized knowledge to non-experts. Such projects also encourage collaboration and active engagement with content.

Beyond STEM, podcasts are widely used in other disciplines. In the humanities and social sciences, they support language learning and public scholarship [21]. In business education, they serve as platforms for case studies and applied problem-solving [20]. Medical education uses clinical review podcasts to strengthen diagnostic reasoning [22], while law schools employ them to analyze precedent-setting cases and professional ethics [23].

Taken together, these examples show that podcasts serve both discipline-specific needs—such as lab readiness in STEM or case study discussion in business—and general pedagogical functions like reinforcing concepts, fostering active learning, and promoting communication skills.

4.3 Podcasts' Strengths in Online and Continuing Education

Podcasts have proven especially valuable in online education, where learners often face challenges of isolation and self-regulation. Unlike text-based materials, the human voice conveys tone, emphasis, and enthusiasm, which helps create a sense of instructor presence [25, 26]. This vocal immediacy reduces the impersonality of digital environments and can motivate students to remain engaged in courses that lack face-to-face interaction.

In addition, podcasts align naturally with flipped classroom models. Short, topic-specific episodes can prepare learners for synchronous activities by explaining foundational concepts in advance. This is particularly useful in STEM contexts, where students benefit from reviewing explanations multiple times before tackling complex problem-solving tasks [25].

For continuing education and professional learning, podcasts offer a flexible medium that integrates seamlessly into busy schedules. Healthcare professionals may use clinical podcasts to review recent case studies, while engineers and IT specialists can follow updates on emerging technologies. Because podcasts can be consumed during commuting, exercising, or routine activities, they allow adult learners to embed professional development into daily life [27].

Podcasts also align well with principles of adult learning (andragogy) by fostering self-directed and goal-oriented study. Learners can select episodes that address immediate workplace challenges or personal interests, which promotes intrinsic motivation. In this way, podcasts serve not only as instructional tools but also as resources for lifelong learning and career advancement.

Finally, from a cognitive perspective, podcasts can reduce extraneous load compared with multimedia resources such as video. Audio-only delivery allows learners to focus on explanations without the distraction of non-essential visual information, making them particularly effective for reinforcing conceptual understanding. When paired with transcripts or supplementary materials, podcasts provide multimodal support while maintaining efficiency.

Together, these strengths show that podcasts are particularly effective in online and continuing education because they provide emotional presence, flexible access, and cognitive efficiency, while also supporting the autonomy and motivation essential for lifelong learning.

4.4 Synthesis

Across case studies, disciplinary applications, and online or continuing education contexts, podcasts demonstrate a consistent pattern: they are most effective when aligned with specific pedagogical needs rather than treated as general-purpose supplements. Case studies highlight their role in extending institutional influence and public engagement; disciplinary analyses underscore their unique value in STEM, where they support laboratory preparation, conceptual reinforcement, and science communication; and online education examples illustrate their ability to create presence, accommodate adult learning, and reduce cognitive load. Together, these findings suggest that podcasts contribute not only to improved course outcomes but also to broader institutional strategies and lifelong learning agendas. This integrated perspective addresses the guiding research questions by showing how podcasts have evolved in scope and what new opportunities AI-generated formats may provide.

5 AI in Podcasts

The integration of artificial intelligence (AI) into podcasting represents a new stage in educational technology, extending the utility of traditional podcasts while introducing capabilities that were previously impractical. AI tools such as Google's NotebookLM, Illuminate, and Microsoft Copilot can summarize complex texts, generate dialogues, and tailor audio content to different learners [5, 6, 30]. These advances significantly reduce the time and expertise required to produce educational podcasts, enabling scalable and adaptive content delivery.

A notable example is Georgia Tech's "Deep Dive into Brain and Behavior (D2B2)" project, which uses NotebookLM to convert neuroscience papers into accessible podcast episodes [28]. The system generates conversational dialogues between AI hosts and produces explanatory segments that simplify technical findings. Importantly, human experts validate these outputs to ensure accuracy, highlighting the potential of human-in-the-loop models for AI-enhanced podcasting.

5.1 Pedagogical Benefits

AI-generated podcasts provide distinctive benefits beyond those of traditional formats:

- Automation streamlines content production, allowing instructors to focus more on teaching and mentoring.

- Personalization enables tailoring of content to learner needs; for instance, an engineering student struggling with thermodynamics might receive foundational explanations, while an advanced learner could access industry applications.
- Scalability supports large-enrollment courses by generating supplemental materials for thousands of students without proportionally increasing faculty workload.
- Rapid knowledge dissemination allows research findings to be transformed into podcasts for students, alumni, and the public, broadening institutional reach.

5.2 Challenges and Risks

Despite these advantages, AI-generated podcasts raise critical challenges. Maintaining content accuracy is essential, particularly in STEM fields where misrepresentation can have serious consequences. Bias in source material or algorithms may also distort narratives, requiring oversight to preserve academic integrity. Privacy concerns emerge when personalization relies on learner data, underscoring the need for clear ethical guidelines. Finally, there is a risk of over-automation: while AI can generate content efficiently, the mentoring and critical dialogue provided by educators remain irreplaceable.

5.3 Future Outlook

Looking ahead, AI-generated podcasts are likely to become increasingly interactive and multimodal. Emerging systems may allow students to query AI hosts in real time, creating dynamic, personalized learning experiences. Integration with learning management systems could connect podcast listening with assessments and analytics, offering richer feedback loops. As adoption grows, longitudinal studies will be needed to measure not only short-term comprehension but also long-term impacts on retention, critical thinking, and professional practice.

In sum, AI has the potential to transform educational podcasting by introducing automation, personalization, and scalability. However, realizing this promise requires a balanced model that combines AI efficiency with human oversight to safeguard quality, ethics, and the pedagogical role of the instructor.

6 Comparing Podcasts with Long or Short Videos

Podcasts share similarities with other educational media but differ in how they engage learners and what kinds of content they best support.

6.1 Podcasts and Long-Form Media (Books and Long Videos)

Like books, podcasts can explore topics in depth, but the spoken format introduces an interpersonal dimension: tone, pace, and emphasis create a sense of dialogue rather than solitary study [31]. Compared with long-form video lectures, podcasts impose lower cognitive demands since they rely solely on auditory processing. While long videos are valuable for demonstrating complex procedures—for example, step-by-step engineering calculations or lab techniques—they often require undivided attention and can overwhelm learners when too dense [32]. Podcasts, by contrast, reinforce conceptual understanding and allow students to revisit explanations without visual distraction.

6.2 Podcasts and Short Videos

Short-form videos have become popular in education because they combine visual and auditory elements in concise packages. They are particularly effective for showing procedural tasks, such as medical examinations, coding demonstrations, or laboratory methods [33, 34]. Podcasts, while lacking visuals, excel at delivering narrative explanations and context, making them more suitable for theoretical discussions or ethical debates where visuals add little value.

6.3 Complementarity in STEM Contexts

Rather than competing, podcasts and videos function best as complementary tools. In STEM education, for example, a short video may demonstrate how to set up an experiment, while a podcast provides the theoretical background, ethical considerations, or interviews with practitioners who apply these methods in industry. This pairing balances procedural learning with conceptual depth, offering a multimodal approach that accommodates diverse learning needs.

Podcasts contribute clarity, mobility, and narrative focus, while videos—especially short-form—provide visual precision for skill-based tasks. When integrated thoughtfully, these media create richer, more balanced learning environments than either could achieve alone.

7 Conclusion and Discussion

This paper has examined the evolution of podcasts in higher education, highlighting their pedagogical applications and the emerging role of AI-generated content. The review shows a clear trajectory: from early substitutional uses such as lecture recordings, to supplementary roles that extend learning through interviews and case studies, and finally to creative applications where students themselves produce podcasts. Case studies from institutions including Harvard, Yale, MIT, and Manchester demonstrate how podcasts have been used to extend research visibility, support disciplinary communication, and foster public engagement. More recent innovations at Arizona State University and Georgia Tech illustrate how AI tools are reshaping podcasting by enabling automated, adaptive, and interactive learning experiences.

7.1 Contributions

This study makes three contributions to the literature on educational technology:

- Framework of podcast roles – synthesizing their substitutional, supplementary, and creative uses across disciplines.
- STEM-focused applications – showing how podcasts address distinctive challenges such as laboratory preparation, conceptual reinforcement, and science communication.
- AI-driven innovation – analyzing how emerging tools like NotebookLM and Copilot expand podcasting through automation, personalization, and scalability while raising new ethical and pedagogical questions.

7.2 Implications for Educators and Institutions

For educators, podcasts should be treated not as supplemental add-ons but as integral components of multimodal course design. When paired strategically with videos or interactive platforms, they balance conceptual depth with procedural demonstration. For institutions, investing in podcasting infrastructure and training faculty in AI-enhanced tools can strengthen both instructional quality and institutional visibility, particularly in fields that require lifelong learning.

7.3 Future Directions

Future research should explore:

- Longitudinal outcomes, examining how podcasts affect retention, critical thinking, and professional practice over time.
- Disciplinary depth, especially in STEM fields where technical accuracy is critical.
- Interactive AI formats, allowing learners to query podcast hosts in real time.
- Ethical frameworks, ensuring that automation, personalization, and data use remain aligned with academic integrity and student privacy.

7.4 Closing Statement

Podcasts—especially when enhanced by AI—represent a powerful and adaptable educational medium. Their potential lies not only in extending learning beyond classrooms but also in reshaping how knowledge is produced, shared, and personalized. Realizing this promise will require careful balance: leveraging automation and scalability while preserving human mentorship and critical engagement. By doing so, higher education can position podcasts as both a contemporary learning tool and a foundation for the next generation of AI-driven pedagogy.

References

1. Duishekeeva, A., Perizat, D., Sharipa, N., Nurlangul, T., Aibek, T.: Transformation of traditional media in social media and the podcast phenomenon. Rotura Rev. Comun. Cult. Artes. **5**(1), 97–111 (2025)
2. DeMars, T.R.: Editor's remarks: historical and contemporary study in radio, audio and podcasting. J. Radio Audio Media. **31**(2), 287–293 (2024)
3. Berry, R.: Part of the establishment: reflecting on 10 years of podcasting as an audio medium. Convergence. **22**(6), 661–671 (2016)
4. Soto-Vásquez, A.D., Kim, F., Aram, S.: The turn to podcasts as a mass campaign medium. J. Radio Audio Media (special issue, forthcoming) (2025)
5. Kokala, A.: Revolutionizing content creation: leveraging AI-driven podcast generation with NotebookLM and personalized insights. Int. Res. J. Modernization Eng. Technol. Sci. **6**(12), 1798–1803 (2024)
6. Lee, Y.L., Nina, P.L.: Re-embracing orality in digital education: the pedagogical affordances of podcasting in the era of generative AI. Front. Educ. **9**, 1447015 (2024)
7. Copley, J.: Audio and video podcasts of lectures for campus-based students: production and evaluation of student use. Innov. Educ. Teach. Int. **44**(4), 387–399 (2007)

8. McGarr, O.: A review of podcasting in higher education: its influence on the traditional lecture. Australas. J. Educ. Technol. **25**(3) (2009)
9. Merhi, M.I.: Factors influencing higher education students to adopt podcast: an empirical study. Comput. Educ. **83**, 32–43 (2015)
10. Vajoczki, S., Susan, W., Nick, M., Katherine, H.: Podcasts: are they an effective tool to enhance student learning? A case study. J. Educ. Multimedia Hypermedia. **19**(3), 349–362 (2010)
11. Ng'ambi, D., Annette, L.: Using podcasting to facilitate student learning: a constructivist perspective. J. Educ. Technol. Soc. **15**(4), 181–192 (2012)
12. Waddingham, J.A., Miles, A.Z., David, J.K.: Insights on the go: leveraging business podcasts to enhance organizational performance. Bus. Horiz. **63**(3), 275–285 (2020)
13. Harvard Business Review: HBR IdeaCast. https://hbr.org/2018/01/podcast-ideacast (Accessed 2025)
14. Santos, L.: The happiness lab with Dr. Laurie Santos. https://www.pushkin.fm/podcasts/the-happiness-lab-with-dr-laurie-santos (Accessed 2025)
15. MIT News: Curious unbounded. https://news.mit.edu/podcasts/curiosity-unbounded (Accessed 2025)
16. Productivity Puzzles: Productivity puzzles podcast. https://pod.co/productivity-puzzles (Accessed 2025)
17. Enrollify: AI for U podcast. https://www.enrollify.org/podcasts/ai-for-u (Accessed 2025)
18. MacKenzie, L.E.: Science podcasts: analysis of global production and output from 2004 to 2018. R. Soc. Open Sci. **6**(1), 180932 (2019)
19. Alpay, E., Shelly, G.: Student-led podcasting for engineering education. Eur. J. Eng. Educ. **35**(4), 415–427 (2010)
20. Valladares, H.L., Guiselle, R.L.: Creating social value through problem-based learning: using podcasts in business education. In: The Palgrave Handbook of Social Sustainability in Business Education, pp. 227–243. Springer, Cham (2024)
21. Moore, T.: Pedagogy, podcasts, and politics: what role does podcasting have in planning education? J. Plan. Educ. Res. **44**(3), 1134–1147 (2024)
22. Malecki, S.L., et al.: Understanding the use and perceived impact of a medical podcast: qualitative study. JMIR Med. Educ. **5**(2), e12901 (2019)
23. Wall, I.: Podcast as assessment: entanglement and affect in the law school. Law Teach. **53**(3), 309–320 (2019)
24. Guertin, L.A.: Creating and using podcasts across the disciplines. Currents Teach. Learn. **2**(2), 4–12 (2010)
25. Bolliger, D.U., Supawan, S., Christine, B.: Impact of podcasting on student motivation in the online learning environment. Comput. Educ. **55**(2), 714–722 (2010)
26. Forbes, D., Elaine, K.: Voice over distance: a case of podcasting for learning in online teacher education. Distance Educ. **36**(3), 335–350 (2015)
27. McFadden, M., Crystal, S., Ashley, T., Shirley, V.: Effectiveness of podcasting in continuing education. PhD thesis, Stanbridge University (2020)
28. Georgia Tech Psychology: Psychology pioneers AI-generated podcast. https://psychology.gatech.edu/news/psychology-pioneers-ai-generated-podcast (Accessed 2025)
29. Kilz, L.: Student perspective on the use of AI in higher education: analysis of the students' podcast. In: IDIMT-2024: Changes to ICT, Management, and Business Processes through AI, pp. 325–334. Trauner Verlag, Linz (2024)
30. Schlicht, C.: AI in higher education: transforming education through innovative learning formats and enhanced student engagement. In: Tagungsband SPOTLIGHT FMplus 2025, pp. 210–218 (2025)
31. Zuraikat, L.: The podcast. In: Radio's Second Century: Past, Present, and Future Perspectives, pp. 39–52. Routledge, New York (2020)

32. Sullivan, J.L.: The platforms of podcasting: past and present. Soc. Media Soc. **5**(4), 2056305119880002 (2019)
33. Hsin, W.J., John, C.: Short videos improve student learning in online education. J. Comput. Sci. Coll. **28**(5), 253–259 (2013)
34. Zhang, T.: A brief study on short video platform and education. In: 2nd International Conference on Literature, Art and Human Development (ICLAHD 2020), pp. 543–547. Atlantis Press, Paris (2020)

Extracting the Differences Between a Presenter's Emphasis and Listeners' Attention Points

Misaki Abe[1(✉)], Sho Onami[1], Ryo Onuma[1], Hiroki Nakayama[1], Youzou Miyadera[2], and Shoichi Nakamura[1]

[1] Fukushima University, Kanayagawa 1, Fukushima 960-1296, Japan
{misaki,onami}@cs.sss.fukushima-u.ac.jp, {onuma,hnakayama, nakamura}@sss.fukusshima-u.ac.jp

[2] Tokyo Gakugei University, 4-1-1, Nukuikita, Koganei, Tokyo 184-8501, Japan
miyadera@u-gakugei.ac.jp

Abstract. Slide-based presentations are an important means of sharing the progress of intellectual activities such as research work, and training of the ability to refine presentations is required. To refine a presentation effectively, it is important to sufficiently grasp the areas to be improved, but this is a difficult task for the inexperienced presenters. In particular, although it is desirable to refine a presentation while understanding the listeners' responses—such as situations in which the presenter's emphasis points do not align with the listeners' points of attention understanding listeners' responses during presentation is difficult. This paper describes methods for estimating the differences between the presenter's emphasis points and the listeners' points of attention along with an overview of the support for presentation refinement. In addition, we describe an experiment conducted using a prototype system and discuss the effectiveness of the proposed methods on the basis of the results.

Keywords: Presentation Refinement Support · Presenter's Emphasis · Listeners' Attention Points · Gaze Analysis

1 Introduction

Presentations play a vital role as a means of communicating the progress of intellectual production tasks such as research and planning. In particular, slide-based presentations are widely used, and there is a growing need to develop the ability to revise them for clearer and more effective communication [1, 3, 11]. However, when revising slides, comments from listeners during rehearsals can be valuable references, but time constraints may sometimes prevent obtaining sufficient feedback. Furthermore, the unconscious reactions of listeners during rehearsals provide valuable hints about issues in the slides and should ideally be utilized, but practical constraints make this difficult. In response to these challenges, systems have been proposed that calculate listener attention rates based on gaze analysis [4, 5]. These approaches are interesting in that they attempt to evaluate presentations using audience attention data, but their analysis is limited to

H. R. Arabnia et al. (Eds.): CSCI 2025, CCIS 2974, pp. 359–364, 2027.
https://doi.org/10.1007/978-3-032-28475-4_26

the slide level and does not consider specific areas of attention within each slide. In addition, other studies have reported systems that detect objects attended to by viewers on web pages and highlight them on the original page [6, 10]. Although practical display methods are being considered, feedback regarding the relationship between the author's intent to emphasize and where the viewer's gaze falls has not been taken into account. Additionally, although previous studies have proposed methods for estimating the presenter's emphasis based on speech analysis [7, 8] and estimating the audience's attention [2, 9], no effective approach has yet been developed to support understanding the correspondence between the two.

Therefore, this study develops methods for estimating both the presenter's emphasized points and the audience's attention areas, and for visually feeding back the differences between them to the presenter. Ultimately, this approach aims to provide new support for slide refinement that considers the gap between the presenter's intended emphasis and the audience's actual perception.

2 Issues and Approach

2.1 Issues in Refinement of Presentations

We focus on the following three main issues in conducting the refinement of presentation through observation of actual presentation, slides used, and their refinement work.

I. It is difficult for the presenter to manage thoroughly which parts of the slides he/she intends to emphasize.
II. It is difficult for the presenter to grasp the audience's reactions during the presentation.
III. It is difficult to comprehend the differences between the presenter's emphasized areas and the audience's areas of attention.

2.2 Approach

To address the above issues, we initially develop a method to extract the presenter's points of emphasis on the basis of analyzing the structural features of slides and the oral explanations (i.e., audio data). Next, we develop a method to extract listeners' points of focus on the basis of analyzing their gaze patterns. Then, we develop a method to estimate the differences between the presenter's emphasis and listeners' attention points.

Based on these methods, we develop a system that support the refinement of presentations by visually feedbacking the presenter's emphasis, listeners' attention points, and their differences to the presenter. Figure 1 shows an overview of presentation refinement support.

3 Extraction of Differences Between a Presenter's Emphasis and Listeners' Attention Points

In this research, four types of emphasis expression are defined as follows.

A) Emphasis by font size: differentiation from other parts by using larger or smaller text.
B) Emphasis by color: differentiation from other parts by using text colors.
C) Emphasis by shapes: differentiation from other parts by using elements such as frames or callouts surrounding text.
D) Emphasis by text decoration: differentiation from other parts by using text styling such as bold, underlined, or italicized font.

We extracted the parts that contains these four types of emphasis as the presenters' emphasized area on the basis of analyzing presentation slides.

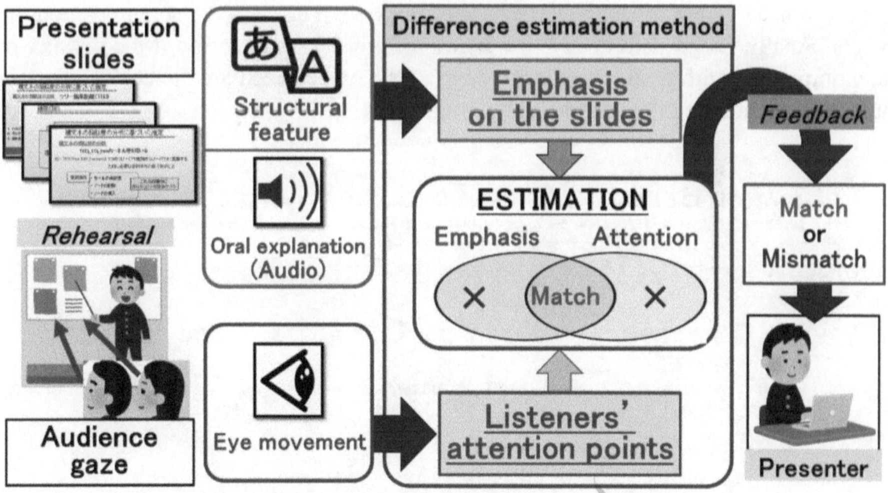

Fig. 1. Overview of Presentation Refinement Support Considering the Differences Between a Presenter's Emphasis and Listeners' Attention

In addition, the system extracts points of emphasis in the oral explanations from the audio data recorded during presentation rehearsals. Noting that pitch tends to rise in parts of presentations where presenters intend to emphasize, the system extracts the pitch from their oral explanations. Specifically, the system extract sections where the pitch increases and converts the oral explanations in those parts into text. Then, the system compares this text with the descriptions on the slides, and designate these segments as emphasized areas in the audio.

Ultimately, the system estimates the points of emphasis on the slides on the basis of comparing the candidates of emphasis extracted from structural features of slides and those of oral explanation.

Next, when people view screens or other displays, their gaze typically moves in a regular pattern following the steps: "upper left, upper right, lower left, lower right". In this research, this pattern has been considered the normal gaze behavior, we examine the following three types of gaze behavior during listening the presentation.

A. Fixation: a state in which the gaze remains on a specific point for a certain period of time or longer.

B. Revisit: a state in which the gaze returns to a location that has been viewed once before.
C. Exceptional initial gaze: a state in which the initial gaze position is outside the top-left region.

From these gaze behaviors, we estimate the fixation durations and gaze movement patterns while attending to the slide to determine the points of focus.

After these analyses, the system extracts the coordinate ranges of the presenter's emphasis areas and the listeners' attention points, and then identifies their differences based on the distances between these coordinates. Specifically, the system sets the coordinate range for the extracted highlighted sections on the slides. Next, the system compares the coordinates of the predefined emphasized areas with those of the acquired attention points. Finally, the system determines that a difference has occurred if the number of attention points within the coordinate range of the emphasized area is below a certain number, or if no attention points exist within that range (Fig. 2).

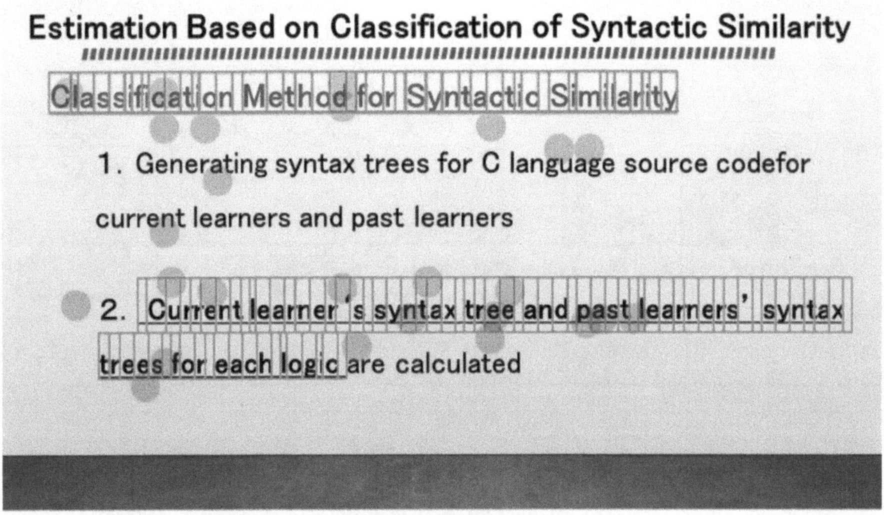

Fig. 2. Presentation of Differences Between Emphasis and Attention

4 Experiment and Discussion

4.1 Experiment Overview

First, the presenters delivered slide-based presentations and listeners' gaze was recorded during the presentations. After that, the system extracted the presenter's points of emphasis, the listeners' points of attention, and the differences between them on the basis of the proposed methods. On the other hand, we asked the presenter to identify the points they intentionally emphasized in advance, and the listeners to identify the points they focused on during the presentation after the experiment. Subsequently, the system's estimated results were compared with the actual responses to evaluate their validity.

4.2 Results and Discussion

Table 1 shows the results of the difference estimation. These are the results of two approaches: one is the estimation based solely on structural features and another is the estimation incorporating audio data. The precision values showed favorable results. On the other hand, the recall values were generally low. As we extracted all regions exhibiting structural features, the results may have been excessively influenced by it. Thus, further investigation is needed to narrow and refine the extraction targets. In addition, with respect to speech data, it is necessary to examine a method for setting the threshold of the pitch by each presenter to be regarded as emphasis.

Table 1. Results of estimating differences between emphasis and attentions

	Precision	Recall	F-score
Structure	0.67	0.40	0.50
Structure + audio	0.75	0.47	0.58

5 Conclusion

This paper described a method for extracting differences between the presenter's emphasis points and the listeners' points of attention. Specifically, we described methods for estimating the points of emphasis by the presenter on the basis of analyzing the structural features of slides and verbal explanations, and the points of attention by listeners on the basis of analyzing gaze patterns during presentations. Additionally, we explained methods for estimating the differences between these points. Furthermore, we described the experiment on extraction of these differences and discussed the basic effectiveness of proposed methods based on its results.

Future work includes further validation of the estimation methods by conducting additional experiments, as well as their improvements based on the results.

References

1. González-Mujico, F.d.L., Lasagabaster, D.: PowerPoint presentations in L1-medium and English-medium lectures: examining knowledge-building practices in multimodal slide content. Lang. Educ. **39**(2), 380–401 (2025)
2. Kwon, S.K., Yu, G.: The effect of viewing visual cues in a listening comprehension test on second language learners' test-taking process and performance: an eye-tracking study. Lang. Test. **41**(3), 649–680 (2024)
3. Yi, S., Matsugami, J., Yamasaki, T.: Assessment system of presentation slide design using visual and structural features. IEICE Trans. Inf. Syst. **105-D**(3), 587–596 (2022)
4. Takyo, Y., Kaji, K.: A system to directly feed back the audience's attention ratio to the presentation venue. In: Proceedings of International Workshop on Informatics, pp. 173–178 (2021)

5. Sharma, K., Giannakos, M., Dillenbourg, P.: Eye-tracking and artificial intelligence to enhance motivation and learning. Smart Learn. Environ. **7**(13), 19 (2020)
6. Nakamura, R., Akasaka, M., Yanagisawa, T., Ichimura, S.: Gaze visualizations for evaluating web pages based on eye-tracking and HTML document structure analysis. IPSJ J. **52**(12), 3868–3875 (2011)
7. Yamasumi, K., Kagomiya, T., Maki, Y., Maekawa, K.: The evaluation structure model of the relationship between impression ratings and speech properties in public speaking. Jpn. J. Sens. Eval. **11**(1–2), 30–36 (2007)
8. Ladd, D.R.: Intonational Phonology. Cambridge University Press (2008)
9. Han, S., Dong, Z., Yuan, R.: Can online meeting platforms be better utilized in teaching scenarios?—a study on the impact of online meeting layout on learners. In: Proceedings of 12th International Conference on Information and Education Technology, pp. 312–316 (2024)
10. Otomo, T., Mochiduki, S., Ishii, E., Hoshino, Y., Yamada, M.: Development of system that can obtain texts on web page based on gazing point. J. Jpn. Pers. Comput. Appl. Technol. Soc. **14**(1), 36–42 (2020)
11. Alley, M., Virginia, K.A.: Rethinking the design of presentation slides: a case for sentence headlines and visual evidence. Tech. Commun. **52**(4), 417–426 (2005)

Negative Result on Multiple Testing of Poisson's Distribution and Some Extensions for Students Providing Multiple Evaluations

Carmen Elena Stoenoiu, Oana Moldovan, and Lorentz Jäntschi(✉)

Technical University of Cluj-Napoca, 400641 Cluj-Napoca, CJ, Romania
carmen.stoenoiu@emd.utcluj.ro, moana@chimie.utcluj.ro,
lorentz.jantschi@chem.utcluj.ro

Abstract. Assessment in the educational process is essential for establishing the level of knowledge, increasing motivation and reducing dropout. The assessment system used allows for multiple assessments (students can self-assess several times). The sample of number of students by the number of assessments (having about 3000 students) was analyzed under 3 theoretical distribution hypotheses: classical Poisson, Zero Inflated Poisson and Exponentiated Poisson. The study was motivated by the need to identify the theoretical distribution, which helps to estimate the number of students who dropped out before the assessment. The study provided a negative result: all these distributions were rejected with a risk of being in error of 5%.

Keywords: Rare event modeling · Discrete count data · Student attrition · Goodness-of-fit testing

1 Introduction

We often operate with discrete count data (see [10,26]), non-negative integer values that represent the number of occurrences of an event within a fixed time period or space. Number of students having given a certain number of evaluations is a discrete count data.

Rare event modeling, like in [21,29] focuses on analyzing and predicting events that occur infrequently but often have significant consequences. For a student to give a large number of repeated evaluations is a rare event. Poisson process models are one of the most important alternatives in rare event modeling that are effective for events that occur randomly but at a constant average rate.

We considered here a detailed analysis of the distribution of students by the number of evaluations trying to fit our experimental data to three alternatives of Poisson processes: the classical alternative and two generalizations.

C. E. Stoenoiu and O. Moldovan—These authors equally contributed.

H. R. Arabnia et al. (Eds.): CSCI 2025, CCIS 2974, pp. 365–374, 2027.
https://doi.org/10.1007/978-3-032-28475-4_27

2 Background

The study is conducted in the context of seeking to estimate the number of students which abandoned the program of study before evaluation, being so called student attrition (see [2,8]). It is generally acknowledged that they have a significant impact on student success rates, institutional finances, and educational outcomes. Goodness-of-fit testing was involved into the distribution analysis. In the larger context, goodness-of-fit testing encompasses statistical methods used to determine how well observed data matches an expected probability distribution or model [11]. These tests help validating model assumptions and distribution choices. Includes:

- Pearson Chi-Squared Test [3];
- Kolmogorov-Smirnov Test [23];
- Anderson-Darling Test [17];
- Other tests (such as [14–16]), specifically devised for certain distributions, certain distribution types, or generally applicable;

In this study we used Pearson Chi-Squared test to measure the agreement between the observed distribution and certain theoretical alternatives.

We included into analysis (one may call it a search [28,30]) the classical Poisson distribution and two of its generalizations, Zero Inflated Poisson and Exponentiated Poisson.

3 Material and Method

When a student gave more than one evaluation, in establishing the evaluation mark, the most unfavorable one was eliminated, and so students were encouraged to give more than one evaluation. The rest of the evaluations were averaged, so in the same time giving evaluations without preparation was discouraged, and the students were motivate to keep their total number of evaluations to a reasonable small number. The system was designed to encourage self improvement.

The histogram of students by the number of evaluations was previously published in [12]. An update of it has been used here (Table 1).

Table 1. Students evaluation data

Number of evaluations	0	1	2	3	4	5	6	7	8	9	10
Observed number of students	s_0	s_1	s_2	s_3	s_4	s_5	s_6	s_7	s_8	s_9	s_{10}

$s_1 = 1369$, $s_2 = 1511$, $s_3 = 347$, $s_4 = 130$, $s_5 = 60$, $s_6 = 38$, $s_7 = 19$, $s_8 = 13$, $s_9 = 6$, $s_{10} = 2$.

Please note that the numbers given in Table 1 differs slightly from the numbers given in [12]. Explicitly, 5 students was evaluated meanwhile, from which one is new (and was evaluated twice). The other four come for one more evaluation, two of them having already one evaluation and other two having already four evaluations. Thus, s_1 is 1369 instead of 1371, s_2 is 1511 instead of 1508, s_4 is 130 instead of 132 and s_5 is 60 instead of 58. Finally, the total number of evaluated students ($\sum_{i=1}^{10} s_i$ in Table 1) has been increased with 1, from 3494 to 3495.

3.1 χ^2 Goodness-of-Fit Testing

Let be n_S the greatest number of evaluations. The agreement between the observed ($(s_i)_{1\leq i\leq n_S}$ in Eq. 1) and the expected ($(y_i)_{1\leq i\leq n_S}$ in Eq. 1) numbers are measured by χ^2 test.

X^2 value is calculated with Eq. 1:

$$X^2 \leftarrow \sum_{i=0}^{n_S} \frac{(s_i - y_i)^2}{y_i} \tag{1}$$

with $n_S = 10$ in Table 1.

We should consider all evaluations, but if the expected number of students for a certain number of evaluations (y_i in Eq. 1) then analysing the X^2 for the (s_i, y_i) pair should be avoided; furthermore, at least 80% of the values in the $(y_i)_{1\leq i\leq n}$ series should be greater than 5 ($y_i > 5$ in Eq. 1).

Having involved $n_S = 10$ observations constants (from s_1 to s_{10} in Table 1), the probability to observe the X^2 departure is drawn from χ^2 statistic with $n_S - 1 = 9$ degrees of freedom, as in Eq. 2:

$$p \leftarrow 1 - \mathrm{CDF}_{\chi^2}(X^2; n_S - 1) \tag{2}$$

where $\mathrm{CDF}_{\chi^2}(x;\nu) = \int_{t=0}^{x} \mathrm{PDF}_{\chi^2}(t;\nu)\mathrm{d}t$. (CDF is the cumulative distribution function and PDF is the probability density function).

We should recall that the χ^2 distribution (Eq. 3) is the distribution of a sum of the squares of independent standard normal random variables.

$$\mathrm{PDF}_{\chi^2}(x;\nu) = \frac{x^{\nu/2-1}\mathrm{e}^{-x/2}}{2^{\nu/2}\Gamma(\nu/2)} \tag{3}$$

where Γ is the Gamma function (Eq. 4, with $\Gamma(k) = (k-1)!$ for $k \in \mathbb{N}^*$).

$$\Gamma(z) = \int_0^{\infty} t^{z-1}\mathrm{e}^{-t}\mathrm{d}t \tag{4}$$

Three alternatives of theoretical distributions were considered here.

3.2 Poisson (PSS) Distribution

Poisson distribution [27] expresses the probability of a given number of events occurring in a fixed interval of time if these events occur with a known constant mean rate and independently of the time since the last event (PSS in Eq. 5).

$$\mathrm{PSS}(x;z) = \frac{z^x}{x!}\mathrm{e}^{-z} \tag{5}$$

Among first uses of Poisson distribution are [24] when Poisson distribution was fitted to the number of stars found in a unit of space and [4] where was shown that the frequency with which soldiers in the Prussian army were accidentally killed by horse kicks could be well modeled by a Poisson distribution.

3.3 Zero-Inflated Poisson (ZIP) Distribution

ZIP [1,5,6,18,20] is a model for count data with excess zeros; it assumes with a probability the only possible observation is 0, and the alternate event is a Poisson random variable - when manufacturing equipment is properly aligned, defects may be nearly impossible, but when it is misaligned, defects may occur according to a Poisson distribution (Eq. 6).

$$\mathrm{ZIP}(x;z,\nu) = \begin{cases} 1-\nu+\nu\mathrm{PSS}(0;z), & \text{for } x=0 \\ \nu\mathrm{PSS}(x;z), & \text{for } x>0 \end{cases} \tag{6}$$

3.4 Exponentiated Poisson (EPS) Distribution

Exponentiated Poisson (EPS) distribution [13,19,22] is defined by Eq. 7.

$$\mathrm{EPS}(x;z,\nu) = \begin{cases} (\mathrm{PSS}(0,z))^{\nu}, & \text{for } x=0 \\ \left(\sum\limits_{k=0}^{x}\mathrm{PSS}(k;z)\right)^{\nu} - \left(\sum\limits_{k=0}^{x-1}\mathrm{PSS}(k;z)\right)^{\nu}, & \text{for } x>0 \end{cases} \tag{7}$$

One should notice the inconvenient definition of the probability mass function from Eq. 7. Actually more convenient it is its definition via CDF. EPS is a typical example from a more general case. Imagine a distribution (continuous or discrete) having a certain CDF. Let be this one $f_{\mathrm{CDF}}(x;params)$, with $params$ its parameters. Then, with one more parameter (let be γ the new parameter) a new distribution can be imagined (let be this one $g_{\mathrm{CDF}}(x;\gamma,params)$), having as CDF the previous one raised to an arbitrary power (γ):

$$g_{\mathrm{CDF}}(x;\gamma,params) \leftarrow (f_{\mathrm{CDF}}(x;params))^{\gamma}$$

A such definition of a distribution is consistent in all instances when $\gamma > 0$, since at the limit of the domain: $f_{\mathrm{CDF}}(\sup\limits_{x\in X} x;params) = 1$, $1^{\gamma} = 1$, and thus $g_{\mathrm{CDF}}(\sup\limits_{x\in X} x;\gamma,params) = 1$.

4 Software Implementation

The calculation algorithm were implemented as follows:

- The observed values were given as a constant series ($(s_i)_{1\leq i\leq n_S}$ from Table 1);
- For the missing value (s_0 in Table 1) a variable was assigned (let be that variable a);
- X^2 function (Eq. 1) has been constructed using the observed frequencies (a and $(s_i)_{1\leq i\leq c_S}$) and calculating the expected frequencies as

$$(a+\sum_{i=1}^{n_S} s_i)\mathrm{PMF}(params) = (a+3495)\mathrm{PMF}(params), \tag{8}$$

where PMF($params$) is one of the alternatives given by Eq. 5, Eq. 6 and Eq. 7.

Explicitly, the three alternatives are:

- For Poisson alternative from Eq. 5:

$$X21(a,b) = \frac{(a-(a+\sum_{j=1}^{n_S} s_j)\mathrm{e}^{-b})^2}{(a+\sum_{j=1}^{n_S} s_j)\mathrm{e}^{-b}} + \sum_{i=1}^{n_S} \frac{(s_i-(a+\sum_{j=1}^{n_S} s_j)\mathrm{PSS}(i;b,c))^2}{(a+\sum_{j=1}^{n_S} s_j)\mathrm{PSS}(i;b,c)} \tag{9}$$

- For ZIP alternative from Eq. 6:

$$X22(a,b,c) = \frac{(a-(a+\sum_{j=1}^{n_S} s_j)(1-c+c\mathrm{e}^{-b}))^2}{(a+\sum_{j=1}^{n_S} s_j)(1-c+c\mathrm{e}^{-b})} + \sum_{i=1}^{n_S} \frac{(s_i-(a+\sum_{j=1}^{n_S} s_j)\mathrm{ZIP}(i;b,c))^2}{(a+\sum_{j=1}^{n_S} s_j)\mathrm{ZIP}(i;b,c)} \tag{10}$$

- For EPS alternative from Eq. 7:

$$X23(a,b,c) = \frac{(a-(a+\sum_{j=1}^{n_S} s_j)\mathrm{e}^{-bc})^2}{(a+\sum_{j=1}^{n_S} s_j)\mathrm{e}^{-bc}} + \sum_{i=1}^{n_S} \frac{(s_i-(a+\sum_{j=1}^{n_S} s_j)\mathrm{EPS}(i;b,c))^2}{(a+\sum_{j=1}^{n_S} s_j)\mathrm{EPS}(i;b,c)} \tag{11}$$

The parameters (a and b for PSS, a, b and c for ZIP and EP) were obtained by minimizing the objective functions ($X21(a,b)$, $X22(a,b,c)$, $X23(a,b,c)$), explaining, where necessary (in the Mathcad implementation), the constraints:

- For PSS alternative:

$$\frac{\partial}{\partial a}X21(a,b) = \frac{\partial}{\partial b}X21(a,b) = 0 \tag{12}$$

- For ZIP alternative:

$$\frac{\partial}{\partial a}X22(a,b,c) = \frac{\partial}{\partial b}X22(a,b,c) = \frac{\partial}{\partial c}X22(a,b,c) = 0 \tag{13}$$

- For EP alternative:

$$\frac{\partial}{\partial a}X23(a,b,c) = \frac{\partial}{\partial b}X23(a,b,c) = \frac{\partial}{\partial c}X23(a,b,c) = 0 \tag{14}$$

Mathcad, Mathematica and Matlab sheets and programs were created for the purpose of solving the problem. Solution has been obtained and verified transversely.

Table 2. Parameters for the selected distributions

Distribution	Initial parameters (guessed)	Final parameters (optimized)
PSS with $n_S = 10$	$a = 324$, $b = 2.0000$	$a = 732.3$, $b = 1.826$
PSS with $n_S = 9$	$a = 324$, $b = 2.0000$	$a = 764.9$, $b = 1.783$
ZIP with $n_S = 10$	$a = 7834$, $b = 1.8266$, $c = 0.7100$	$a = 6866$, $b = 1.8265$, $c = 0.5343$
ZIP with $n_S = 9$	$a = 7834$, $b = 1.8266$, $c = 0.7100$	$a = 6828$, $b = 1.7839$, $c = 0.5286$
EPS with $n_S = 10$	$a = 035$, $b = 2.000$, $c = 0.020$	$a = 23941$, $b = 1.332$, $c = 0.144$
EPS with $n_S = 7$	$a = 035$, $b = 2.000$, $c = 0.200$	$a = 4524$, $b = 1.051$, $c = 0.449$

Table 3. Observed vs. expected frequencies (rounded) in student's evaluation data

Evaluations	0	1	2	3	4	5	6	7	8	9	10
Hypothesis of PSS distribution for $n_S = 10$ (73% of expected frequencies greater than 5)											
Observed	732.3	1369.0	1511.0	347.0	130.0	60.0	38.0	19.0	13.0	6.0	2.0
Expected	680.8	1243.1	1135.0	690.8	315.4	115.2	35.0	9.1	2.1	0.4	0.1
X^2	3.9	12.7	124.6	171.1	108.9	26.4	0.2	10.6	57.1	73.5	48.0
p_{χ^2} (d.f. = 1)	4.8%	<1‰	<1‰	<1‰	<1‰	<1‰	62%	1‰	n.a.*	n.a.*	n.a.*
Hypothesis of PSS distribution for $n_S = 9$ (80% of expected frequencies greater than 5)											
Observed	764.9	1369.0	1511.0	347.0	130.0	60.0	38.0	19.0	13.0	6.0	2.0
Expected	716.1	1276.9	1138.5	676.7	301.7	107.6	32.0	8.1	1.8	0.4	0.1
X^2	3.4	6.7	122.3	160.4	97.6	21.0	1.1	14.5	68.9	88.5	58.5
p_{χ^2} (d.f. = 1)	6.6%	<1%	<1‰	<1‰	<1‰	<1‰	29%	<1‰	n.a.*	n.a.*	n.a.*
Hypothesis of ZIP distribution for $n_S = 10$ (73% of expected frequencies greater than 5)											
Observed	6866.2	1369.0	1511.0	347.0	130.0	60.0	38.0	19.0	13.0	6.0	2.0
Expected	6668.5	1294.1	1181.8	719.5	328.5	129.0	36.5	9.5	2.2	0.4	0.1
X^2	5.9	4.3	91.7	192.9	119.9	36.9	0.06	9.5	53.0	78.4	36.1
p_{χ^2} (d.f. = 1)	1.5%	3.7%	<1‰	<1‰	<1‰	<1‰	80%	2‰	n.a.*	n.a.*	n.a.*
Hypothesis of ZIP distribution for $n_S = 9$ (80% of expected frequencies greater than 5)											
Observed	6827.6	1369.0	1511.0	347.0	130.0	60.0	38.0	19.0	13.0	6.0	2.0
Expected	6645.4	1323.6	1180.6	702.0	313.1	111.7	33.2	8.5	1.9	0.4	0.1
X^2	5.0	1.6	92.4	179.6	107.1	23.9	0.7	13.1	65.4	84.6	56.0
p_{χ^2} (d.f. = 1)	2.5%	21.2%	<1‰	<1‰	<1‰	<1‰	40.5%	<1‰	n.a.*	n.a.*	n.a.*
Hypothesis of EPS distribution $n_S = 10$ (73% of expected frequencies greater than 5)											
Observed	2538.7	1369.0	1511.0	347.0	130.0	60.0	38.0	19.0	13.0	6.0	2.0
Expected	2430.9	1512.4	1033.1	592.2	287.1	118.5	42.0	12.9	3.5	0.8	0.2
X^2	4.8	13.6	221.1	101.5	86.0	28.9	0.38	2.9	25.8	33.8	16.2
p_{χ^2} (d.f. = 1)	2.9%	<1‰	<1‰	<1‰	<1‰	<1‰	53.7%	8.9%	n.a.*	n.a.*	n.a.*
Hypothesis of EPS distribution $n_S = 7$ (80% of expected frequencies greater than 5)											
Observed	2155.3	1369.0	1511.0	347.0	130.0	60.0	38.0	19.0	13.0	6.0	2.0
Expected	2061.5	1496.5	1046.2	595.9	282.2	112.6	38.4	11.3	2.9	0.7	0.1
X^2	4.3	10.9	206.5	104.0	82.1	24.6	0.004	5.2	35.2	40.1	36.1
p_{χ^2} (d.f. = 1)	3.9%	<1‰	<1‰	<1‰	<1‰	<1‰	94.9%	2.2%	n.a.*	n.a.*	n.a.*

*: probability associated with this frequency repartition is not reliable since $e_i < 5$ (see Table 1 and [3]).

5 Results and Discussion

The initial points and the solutions of the optimizations Eq. 12–Eq. 14 were obtained as in Table 2 while the expected frequencies are given in Table 3. In order to obey to the rule of 80% expected frequencies greater than 5, the calculation was restarted for all alternatives, considering less groups.

Analysing the individual probability values associated with frequency repartitions from Table 3:

- for PSS alternative the maximum number of evaluations was reduced from 10 to 9 to have at least 80% of the expected frequencies greater than 5; 6 probabilities (out of 10) reject the assumption of the distribution at 5% risk being in error; therefore we must reject this hypothesis;
- for ZIP alternative the maximum number of evaluations was reduced from 10 to 9 to have at least 80% of the expected frequencies greater than 5; 6 probabilities (out of 10) reject the assumption of the distribution at 5% risk being in error; therefore we must reject this hypothesis;
- for EPS alternative the maximum number of evaluations was reduced from 10 to 9 to have at least 80% of the expected frequencies greater than 5; 6 probabilities (out of 10) reject the assumption of the distribution at 5% risk being in error; therefore we must reject this hypothesis;

Since for three values of the observed variable - 8, 9, and 10 evaluations - all estimations provided expected frequencies smaller than 5, one can argue that we cannot get a definite answer from this study, and further investigation is required. Indeed, this problem with the Chi-Squared test is well known and documented [7,9,25]. However, the cause for rejection of the alternatives were identified in the larger part of the histogram, namely for number of evaluations from 2 to 5, where the expected frequencies exceeded 50 everywhere.

Comparing with the previously reported model - Lorentz Peak - see [12] the models are much less significant - in fact all of them were rejected as possible theoretical distributions.

6 Conclusions and Perspectives

A multiple-choice assessment system used from 2006 to 2024 to assess students in topics of chemistry. When a student gave more than one evaluation, the most unfavorable one was eliminated, and the rest were averaged and so students were encouraged to give more than one evaluation, but discouraged to give evaluations without preparation. One important statistic extracted is the number of students by the number of evaluations given. The supposition of PSS distribution for the number of students was tested, having three alternatives of distribution: Poisson—PSS, Zero-Inflated Poisson – ZIP, and Exponentiated Poisson – EPS. Chi-squared test has been used to assess the goodness-of-fit. The analysis revealed that all these three alternatives for the theoretical distribution of the number of students by the number of evaluations must be rejected. Further investigation is requiring, including other distributions. The statistical analysis revealed a systematic mismatch between the empirical data and all tested theoretical distributions PSS, ZIP, and EPS. Although parameter optimization improved the local fit (Table 2), the χ^2 goodness-of-fit test (Table 3) indicated significant deviations, with more than half of the frequency classes rejecting the null hypothesis at the 5% significance level. For all models, the requirement that at least 80% of the expected frequencies exceed 5 was met only after reducing the number of score groups (from $n_S = 10$ to 9 or 7), which limited the robustness of the inference. The deviations were most pronounced in the central range of the histogram (scores 2–5), where expected frequencies were high, suggesting a structural rather than sampling-related mismatch. The ZIP model, although accounting for zero inflation, did not significantly improve the fit, indicating that data asymmetry and multimodality—rather than an excess of zeros—explain the model's poor performance. Overall, none of the classical discrete distributions adequately represented the empirical pattern of student evaluations. The results confirm previous findings that alternative heavy-tailed or asymmetric models (e.g., Lorentz Peak) better capture the empirical heterogeneity and kurtosis inherent in such evaluation datasets. The systematic rejection of the PSS, ZIP, and EP distributions highlights that real-world behavioral or perception-based datasets rarely follow idealized theoretical patterns. Furthermore, the observed deviations concentrated in the modal range (scores 2–5) demonstrate that the lack of fit is not a consequence of random variation, but rather reflects structural heterogeneity within the data. For practitioners and researchers, this evidence supports the exploration of alternative models—such as asymmetric, mixed, or heavy-tailed distributions—that better capture the complexity of subjective evaluations. The significant deviations identified between the empirical data and the tested theoretical distributions (PSS, ZIP, and EPS) provide valuable insights into students' evaluative behavior and the broader educational process. First, the rejection of these models suggests that the distribution of scores is not random, but is influenced by cognitive, attitudinal, and contextual factors, such as students' subjective perceptions of course difficulty, teacher-student relationships, and motivation to engage. These elements introduce asymmetry and structural heterogeneity that classical probabilistic models fail to capture. From an educational perspective, these results indicate the diversity of learning experiences and expose the limitations of standardized assessment approaches. The lack of fit to theoretical distributions implies that student performance and perceptions are unevenly distributed, necessitating more flexible analytical and pedagogical tools. Therefore, the statistical results not only highlight the limits of theoretical modeling, but also encourage critical reflection on how assessment systems can be adapted to better represent the complexity of educational dynamics and motivational variability among learners. Such reflections are essential for developing more inclusive and context-sensitive assessment practices within higher education.

Disclosure of Interests. The authors declare that they have no competing financial interests or personal relationships that could have appeared to influence the work reported in this paper.

References

1. Arab, A., Holan, S.H., Wikle, C.K., Wildhaber, M.L.: Semiparametric bivariate zero-inflated Poisson models with application to studies of abundance for multiple species. Environmetrics **23**(2), 183–196 (2012). https://doi.org/10.1002/env.1142
2. Bean, J.P.: Dropouts and turnover: the synthesis and test of a causal model of student attrition. Res. High. Educ. **12**, 155–187 (1980). https://doi.org/10.1007/bf00976194
3. Bolboacă, S.D., Jäntschi, L., Sestras, A.F., Sestras, R.E., Pamfil, D.C.: Pearson-fisher chi-square statistic revisited. Information **2**(3), 528–545 (2011). https://doi.org/10.3390/info2030528
4. von Bortkiewitsch, L.: Das Gesetz der kleinen Zahlen. B.G. Teubner, Leipzig (1839). http://archive.org/details/dasgesetzderklei00bortrich
5. Boucher, J.P., Denuit, M., Guillen, M.: Number of accidents or number of claims? An approach with zero-inflated Poisson models for panel data. J. Risk Insur. **76**(4), 821–846 (2009). https://doi.org/10.1111/j.1539-6975.2009.01321.x
6. Costantino, F., Di Gravio, G., Patriarca, R., Petrella, L.: Spare parts management for irregular demand items. Omega **81**, 57–66 (2018). https://doi.org/10.1016/j.omega.2017.09.009
7. Fienberg, S.E.: The use of chi-squared statistics for categorical data problems. J. R. Stat. Soc., B: Stat. Methodol. **41**(1), 54–64 (1979). https://doi.org/10.1111/j.2517-6161.1979.tb01057.x
8. Golde, C.M.: The role of the department and discipline in doctoral student attrition: lessons from four departments. J. High. Educ. **76**(6), 669–700 (2005). https://doi.org/10.1080/00221546.2005.11772304
9. Haberman, S.J.: A warning on the use of chi-squared statistics with frequency tables with small expected cell counts. J. Am. Stat. Assoc. **83**(402), 555–560 (1988). https://doi.org/10.1080/01621459.1988.10478632
10. Hadi Safari-Katesari, S.Y.S., Zaroudi, S.: Modelling count data via copulas. Statistics **54**(6), 1329–1355 (2020). https://doi.org/10.1080/02331888.2020.1867140
11. Hein, J., Wiuf, C., Knudsen, B., Møller, M., Wibling, G.: Statistical alignment: computational properties, homology testing and goodness-of-fit. J. Mol. Biol. **302**(1), 265–279 (2000). https://doi.org/10.1006/jmbi.2000.4061
12. Jäntschi, L.: Online students evaluation systems: a time retrospective. In: ICERI2024 Proceedings, 17th Annual International Conference of Education, Research and Innovation, IATED, 11–13 November 2024, pp. 4719–4730 (2024). https://doi.org/10.21125/iceri.2024.1167
13. Jiang, R.: Discrete competing risk model with application to modeling bus-motor failure data. Reliab. Eng. Syst. Saf. **95**(9), 981–988 (2010). https://doi.org/10.1016/j.ress.2010.04.009
14. Jäntschi, L.: A test detecting the outliers for continuous distributions based on the cumulative distribution function of the data being tested. Symmetry **11**(6), 835 (2019). https://doi.org/10.3390/sym11060835
15. Jäntschi, L.: Detecting extreme values with order statistics in samples from continuous distributions. Mathematics **8**(2), 216 (2020). https://doi.org/10.3390/math8020216

16. Jäntschi, L., Bolboacă, S.D.: Performances of Shannon's entropy statistic in assessment of distribution of data. Ovidius University Ann. Chem. **28**(2), 30–42 (2017). https://doi.org/10.1515/auoc-2017-0006
17. Jäntschi, L., Bolboacă, S.D.: Computation of probability associated with Anderson-Darling statistic. Mathematics **6**(6), 88 (2018). https://doi.org/10.3390/math6060088
18. Kukar-Kinney, M., Scheinbaum, A.C., Orimoloye, L.O., Carlson, J.R., He, H.: A model of online shopping cart abandonment: evidence from e-tail clickstream data. J. Acad. Mark. Sci. **50**(5), 961–980 (2022). https://doi.org/10.1007/s11747-022-00857-8
19. Kuş, C.: A new lifetime distribution. Comput. Stat. Data Anal. **51**(9), 4497–4509 (2007). https://doi.org/10.1016/j.csda.2006.07.017
20. Lambert, D.: Zero-inflated Poisson regression, with an application to defects in manufacturing. Technometrics **34**(1), 1–14 (1992). https://doi.org/10.1080/00401706.1992.10485228
21. Leitgöb, H.: Analysis of Rare Events. SAGE Publications Ltd., London, March 2020. https://doi.org/10.4135/9781526421036863804
22. Louzada, F., Luiz Ramos, P., Henrique Ferreira, P.: Exponential-Poisson distribution: estimation and applications to rainfall and aircraft data with zero occurrence. Commun. Stat.-Simul. Comput. **49**(4), 1024–1043 (2020). https://doi.org/10.1080/03610918.2018.1491988
23. Massey, F.J., Jr.: The Kolmogorov-Smirnov test for goodness of fit. J. Am. Stat. Assoc. **46**(253), 68–78 (1951). https://doi.org/10.1080/01621459.1951.10500769
24. Newcomb, S.: Notes on the theory of probabilities. Math. Montly **2**(4), 134–140 (1860). http://jhanley.biostat.mcgill.ca/bios601/Intensity-Rate/Newcomb1860.pdf
25. Overall, J.E.: Sample size required to observe at least K rare events. Psychol. Rep. **21**(1), 70–72 (1967). https://doi.org/10.2466/pr0.1967.21.1.70
26. Plan, E.: Modeling and simulation of count data. CPT Pharmacometrics Syst. Pharmacol. **3**(8), 129 (2014). https://doi.org/10.1038/psp.2014.27
27. Poisson, S.D.: Probabilité des jugements en matière criminelle et en matière civile, précédées des règles générales du calcul des probabilités. Bachelier, Paris (1839). http://gallica.bnf.fr/ark:/12148/bpt6k110193z
28. Shen, M.L.: Search statistics of rare events. Nucl. Instrum. Methods **65**(2), 157–162 (1968). https://doi.org/10.1016/0029-554X(68)90556-9
29. Shyalika, C., Wickramarachchi, R., Sheth, A.P.: A comprehensive survey on rare event prediction. ACM Comput. Surv. **57**(3), 70 (2024). https://doi.org/10.1145/3699955
30. Tajer, A., Poor, H.V.: Quick search for rare events. IEEE Trans. Inform. Theory **59**(7), 4462–4481 (2013). https://doi.org/10.1109/TIT.2013.2253351

An Online PDF-Quiz Converter

Alen Curovic(✉) and Zizhong John Wang

Department of Computer Science Virginia Wesleyan University, Virginia Beach, USA
acurovic@vwu.edu, zwang@vwu.ed

Abstract. This paper presents a research project aimed at enhancing and simplifying students' learning by automating quiz creation from PDF versions of study materials. A secure web application was created for users to sign up and log in for continuous access. After uploading a PDF file, the system extracts and cleans the text, splits it into sentences, ranks them by importance, and clusters related context around those key sentences. It then uses natural language processing (NLP) models to generate natural-language questions and their answers. Finally, every quiz is stored privately and is accessible only to the user who created it.

Keywords: PDF · Quiz · NLP · PHP · MySQL

1 Introduction

In recent years, American students have faced increasing academic challenges. A 2024 article published by Harvard noted that "Declines in reading were also concerning, with 65 percent of third-graders performing on grade level, compared with 72 percent in 2019" [1]. This decline in academic performance reflects the harsh reality that American students are facing. However, strategies to combat this decline have been identified. A study from Iowa State found that "when students are quizzed over class material at least once a week, they tend to perform better on midterm and final exams compared to students who did not take quizzes." This emphasizes the effectiveness and importance of regularly using quizzes to reinforce students' knowledge and improve academic outcomes.

Motivated by the above facts, this project introduces a web-based application designed to support students' academic success. The goal with this work is to provide a free, accessible tool that turns study material into quizzes to help students learn.

The paper is organized as follows. In the next section, we introduce the system design, including the flowchart and the webpages that work with the database. Case studies are presented in Sect. 3, followed by a conclusion.

2 System Design and Implementation

This project combines web development with natural language processing (NLP) to create a web application that transforms PDFs into quizzes. The reasoning behind each technical choice will be further discussed.

H. R. Arabnia et al. (Eds.): CSCI 2025, CCIS 2974, pp. 375–380, 2027.
https://doi.org/10.1007/978-3-032-28475-4_28

The backend side of the online system is the quiz generator. The first part is text extraction. The process starts by extracting the raw text from user-supplied PDFs using PyPDF2, a pure-Python library for parsing PDF files ("PyPDF2"). PdfReader systematically reads each page's text stream, producing an initial block of unstructured text for further processing.

The next part of the quiz generator is text cleaning. Python's built-in re module is applied to remove spurious artifacts such as headers, footers, and embedded URLs. The re module removes multiple whitespace characters until a single space remains. It strips out page numbers and links and eliminates non-alphanumeric symbols, yielding a coherent and uniform text block [2].

The third part is the sentence segmentation. Cleaned text is then passed to spaCy, an industrial-strength natural language processing library for tokenization and sentence boundary detection (Explosion, "spaCy"). The en_core_web_md model is a statistical model that provides linguistically informed segmentation, ensuring that downstream modules operate on complete, human-readable sentences (Explosion, "*Models: en_core_web_md*") [3].

The fourth part is the relevance ranking. The TfidfVectorizer from scikit-learn is then used to convert each sentence into a vector of term-frequency inverse document frequency (TF–IDF) features. These features highlight frequent words within a sentence and across all the sentences [4]. NumPy is used to sum and sort these scores, allowing the computer to select the most informative sentences for quiz generation [5].

The fifth part is the question generation and refinement. Selected sentences are then fed into the Hugging Face Transformers library's pre-trained T5 models for text-to-text generation. A dedicated "question generation" checkpoint creates human-style questions from each sentence, while an answer-refinement model polishes the corresponding answers [6]. These steps produce high-quality Q&A pairs ready for user consumption.

The final process is serialization. The assembled list of questions and answers is serialized into JSON using Python's built-in json module [7]. This standard format ensures compatibility with web and mobile front ends. Utility modules sys, os, and logging manage command-line inputs, suppress extraneous warnings, and restrict console output to critical error messages [8–10]. The full processes are shown in Fig. 1.

The front platform is the online database. All user accounts, quizzes, and results reside in a MySQL database [11]. Its relational design and foreign-key constraints guarantee that each quiz result is reliably linked to its user.

The users table holds user account information in the database (see Fig. 2).

The details in the fields are as follows.

a) The id is a unique integer identifier for each user account.
b) The username is a varchar up to 50 characters long, and the name the user chooses to go by. It is not unique; multiple users could go by the same name.
c) The email is a varchar up to 255 characters long, and the email the user chooses to go by. It is a unique identifier, meaning no two accounts have the same email address.
d) The password_hash is a varchar up to 255 characters long, and the user's account password. Since the password is hashed within the database, it is invisible.
e) The is_admin is a boolean identifier that indicates whether the account signing in is an administrative account.

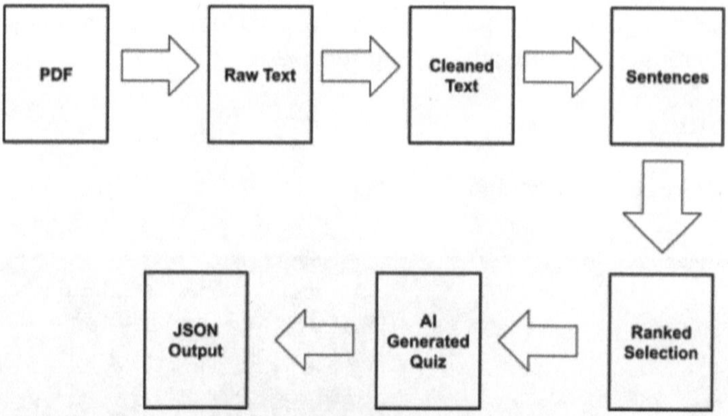

Fig. 1. The quiz generator

Field	Type	Null	Key	Default	Extra
id	int(11)	NO	PRI	NULL	auto_increment
username	varchar(50)	NO	UNI	NULL	
email	varchar(255)	NO	UNI	NULL	
password_hash	varchar(255)	NO		NULL	
is_admin	tinyint(1)	NO		0	
created_at	timestamp	NO		current_timestamp()	

Fig. 2. The User Table

f) The created_at records the time stamp of when the account was created.

The other table, quizzes, holds all the users' quizzes in the database, as shown in Fig. 3.

Field	Type	Null	Key	Default	Extra
id	int(11)	NO	PRI	NULL	auto_increment
user_id	int(11)	NO	MUL	NULL	
title	varchar(255)	NO		NULL	
questions	longtext	NO		NULL	
created_at	timestamp	NO		current_timestamp()	

Fig. 3. The Quizzes Table

The detailed information of the table fields is as follows.

a) The id is a unique integer identifier for each quiz.
b) The user_id is a unique integer identifier that indicates which quizzes belong to which users. It references the id from the users table.
c) The title is a varchar up to 255 characters long and stores the PDF's name in the database.

d) The questions is a longtext type. Longtext's are used to store a large amount of string data. Its purpose is to store quizzes in the database.

3 Case Study

The user's interface is shown in Fig. 4.

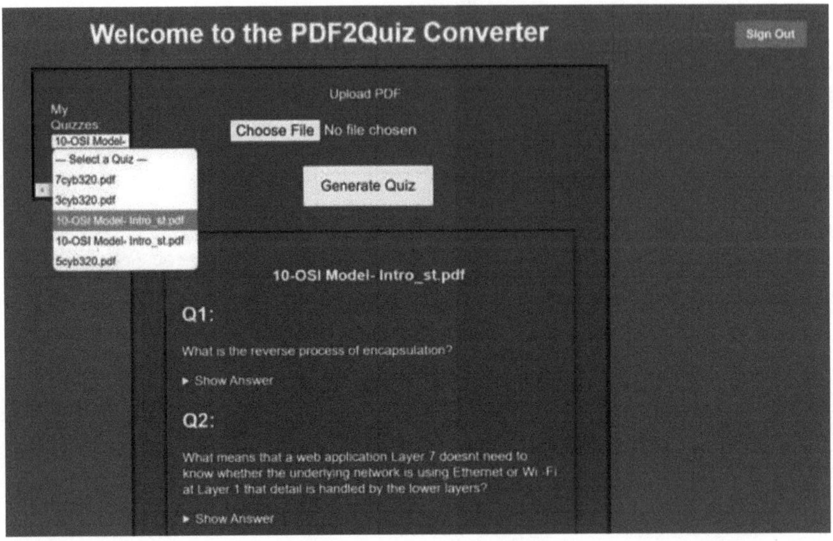

Fig. 4. The user's interface

When the user enters the home page, they can upload a PDF and then press "Generate Quiz." This will generate a quiz and place it in "My Quizzes," where the user can select it. All the quizzes are user-specific; even the administrator cannot view them. The "Sign Out" button will take the user to the registration page.

The screenshots of the signup and sign-in pages are shown in Fig. 5. When first entering the web application, the user must register. After successfully registering, the user will be sent to the Sign in page, or, if they have a previous account, they can choose to go directly to it.

To sign in, the user must input their email address and the password associated with that account. If they forget their password, they can always return to the registration page to create a new account.

The admin dashboard is shown in Fig. 6. The admin can delete any account. To make the process easier, the administrator can search for specific names or emails to delete.

The PDFs tested ranged from 3 to 17 pages; however, the limit set was 20 MB. After integrating a database, a set of two new documents, plus three from the initial batch, was tested for quality control. These files spanned diverse networking topics and lengths, ensuring breadth of coverage. The generator was iteratively refined until most quizzes produced questions and answers judged legitimate, indicating that the system met its core usability goals.

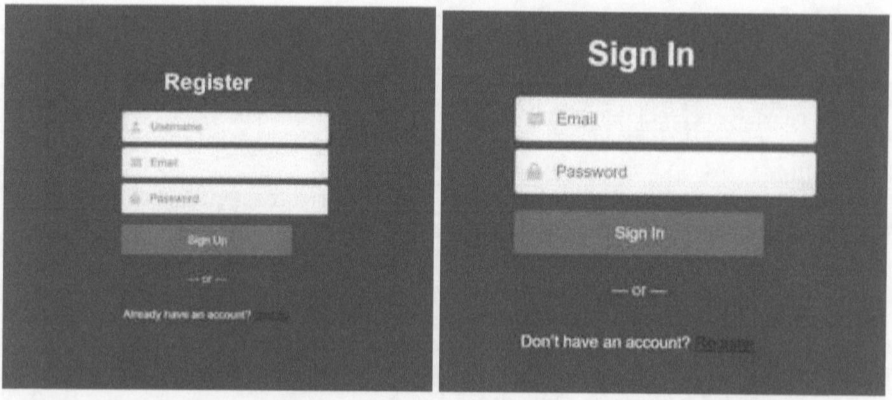

(a) (b)

Fig. 5. (a) The User's Signup Window; (b) The Sign-In Window

Fig. 6. The Admin Dashboard

4 Conclusion

This project aimed to create a web application that generates quizzes from study material, encouraging students to use it to improve their academics. The NLP pipeline was the foundation of the project's development. Building on that, a user-friendly web interface was then integrated to enhance the students' experience. However, certain limitations that present opportunities for enhancement were found. The NLP pipeline could be enhanced to provide quizzes faster and of higher quality, improve the user interface for usability, and implement more security features, such as denial-of-service protection and file scanning.

Disclosure of Interest. The authors have no competing interests to declare that are relevant to the content of this article.

***Author Information:.** Alen Curovic may be contacted at alen.curovic@outlook.com after graduation.

References

1. Center for Education Policy Research: The scary truth about how far behind American kids have fallen. Center for Education Policy Research at Harvard University (2024) https://cepr.harvard.edu/news/scary-truth-about-how-far-behind-american-kids-have-fallen. Accessed 4 May 2025
2. Python Software Foundation: re—regular expression operations. Python 3 Documentation. https://docs.python.org/3/library/re.html. Accessed 4 May 2025
3. Explosion: Models: en_core_web_md. spaCy Documentation. https://spacy.io/models/en#en_core_web_md. Accessed 4 May 2025
4. Pedregosa, F., et al.: Scikit-learn: machine learning in python. J. Mach. Learn. Res. **12**, 2825–2830 (2011). https://scikit-learn.org/stable/. Accessed 4 May 2025
5. Harris, C.R., et al.: Array programming with NumPy. Nature. **585**(7825), 357–362 (2020). https://numpy.org/. Accessed 4 May 2025
6. Wolf, T., et al.: Transformers: state-of-the-art natural language processing. In: Proceedings of the 2020 Conference on Empirical Methods in Natural Language Processing: System Demonstrations, pp. 38–45 (2020). https://huggingface.co/transformers/. Accessed 4 May 2025
7. Python Software Foundation: json—JSON encoder and decoder. Python 3 Documentation. https://docs.python.org/3/library/json.html. Accessed 4 May 2025
8. Python Software Foundation: sys—system-specific parameters and functions. Python 3 Documentation. https://docs.python.org/3/library/sys.html. Accessed 4 May 2025
9. Python Software Foundation: logging—logging facility for python. Python 3 Documentation. https://docs.python.org/3/library/logging.html. Accessed 4 May 2025
10. Python Software Foundation: os—miscellaneous operating system interfaces. Python 3 Documentation. https://docs.python.org/3/library/os.html. Accessed 4 May 2025
11. Wang, Z.: Web Programming with PHP/MySQL. VWU Bookstore (2023)

Designing an Integrated Undergraduate Course on Machine Learning and Parallel Computing

Xuguang Chen(✉)

Computer Science Department, Saint Martin's University, Lacey, WA 98503, USA
xchen@stmartin.edu

Abstract. This paper introduces the design of an undergraduate elective course titled Introduction to Parallel Computing and Machine Learning. Aimed at junior and senior students in computer science major, the designed course provides foundational knowledge of machine learning alongside hands on experience in parallel computing. It starts from core parallel programming techniques, followed by key machine learning methods. Emphasis is placed on exploring how to implement machine learning methods using the parallel computing techniques introduced in the course. The course will heavily use unplugged activities, simplified math, and coding exercises without external libraries to help deep comprehension of the machine learning methods, with assignments, group projects, and in-person exams as the coursework. It will first be offered as a pilot, with the goal of collecting feedback to refine and improve the course design before its official rollout.

Keywords: machine learning · supervised · unsupervised · enforcement learning · parallel computing · OpenMP · mpi

1 Introduction

Machine learning and parallel computing are playing an increasingly important role in a wide range of computing tasks, from real-time image recognition and recommendation systems to large-scale simulations and data analysis [1]. As the volume of data and model complexity continues to grow, so does the demand for high-performance computing techniques to make machine learning workflows efficient and scalable [2]. While machine learning helps extract insights from massive datasets, parallel computing provides the essential performance infrastructure to make such computations feasible in practice.

Today, the integration of these two areas is not only a technological necessity but also an important skill set for computer science students entering industry or pursuing research [3]. In response to these trends, many universities now offer courses in parallel computing or machine learning [4–6]. However, few curricula provide students with an opportunity to explore the intersection of the two – that is, how machine learning methods can be implemented and optimized using parallel programming techniques.

To address this gap, we designed a new upper-level undergraduate elective course titled Introduction to Parallel Computing and Machine Learning. It is intended for junior

H. R. Arabnia et al. (Eds.): CSCI 2025, CCIS 2974, pp. 381–390, 2027.
https://doi.org/10.1007/978-3-032-28475-4_29

and senior students who are in computer science major, interested in gaining practical experience in parallel programming, and eager to learn commonly used machine learning methods. The course is structured around two major components: parallel computing, especially OpenMP (Open Multi-Processing) [11], and MPI (Message Passing Interface) [12], and the machine learning methods from the areas of supervised, unsupervised, and reinforcement learning, emphasizing on how to parallelizing these selected methods.

The course is designed to be interactive and student-centered, with an emphasis on teamwork, hands-on coding, and unplugged activities to reinforce conceptual understanding. The course will be initially offered as a pilot in summer 2026, gathering feedback from students and refine the course structure and content. Various approaches, for example interviews and surveys, are planned to capture feedback for evaluating the effectiveness of the course design.

The remainder of this paper is organized as follows. The background and motivation for designing the course are in Sect. 2. Section 3 details the course design, including the selection of topics, instructional strategies, and major assignments such as team-based projects. Section 4 outlines the implementation plan and explains how the course's effectiveness will be evaluated through student feedback and project-based assessment. A summary and suggestions for future work are in Sect. 5.

2 Background and Motivation

2.1 Background

As of 2025, machine learning has become a foundational technology across nearly every major industry. For example, in healthcare, machine learning models can be used for cancer detection in medical imaging and outcome prediction using electronic health records [2, 10]. In finance, machine learning drives fraud detection, credit scoring, and high-frequency trading [9]. Autonomous systems such as self-driving vehicles and drones also heavily rely on deep learning and reinforcement learning for real-time perception and decision-making [7]. All in all, the proliferation of large-scale machine learning applications is expected to significantly accelerate in the coming years as models become more powerful and integrated into everyday life [1].

Fueled by advances in computer hardware and the parallel computing paradigms that harness it, parallel computing can allow large-scale computational tasks to be decomposed into smaller, independent subtasks that can be executed simultaneously. This dramatically improves execution time and resource utilization, particularly for data-intensive workloads [3]. With the rise of multi-core processors, general-purpose GPUs, and distributed computing platforms, parallelism has become central to scientific computing, engineering, and machine learning applications.

The connection between machine learning and parallel computing is becoming more and more strong today. Modern machine learning models - particularly deep neural networks and transformer-based architectures - require training on massive datasets with billions of parameters. Without parallel and distributed computing frameworks such as CUDA, OpenMP, MPI, and MapReduce, training these models would be computationally infeasible [2]. Tasks like matrix operations, gradient updates, and convolutional layers are highly parallelizable, and GPU clusters are now a standard component of machine learning pipelines [8].

As the reflection for this technological landscape, many universities have incorporated parallel computing into undergraduate and graduate computer science curricula. For instance, courses like *CS 267: Applications of Parallel Computing* at UC Berkeley [4] and *15-618: Parallel Computer Architecture and Programming* at Carnegie Mellon University [5] have blended theoretical instruction with practical applications, including parallel machine learning model training and distributed data processing. These courses acknowledge the growing importance of parallelism not only as a systems topic but as a critical enabler of real-world applications.

2.2 Motivation

Over the past few years, the author has taught parallel computing and artificial intelligence (AI) courses offered to undergraduate and graduate students. In the AI course, a range of machine learning techniques, such as decision trees and neural networks are covered. Throughout these courses, the author has noticed that many students taking the parallel computing course are highly interested in its real-world applications. On the other hand, although many students don't plan to choose machine learning as a focus for their careers or further studies, they are eager to get a solid overview of machine learning methods widely used in various areas.

Moreover, based on the teaching experience and student feedback from these courses, in my opinion, there are two limitations in these types of courses. In a typical machine learning course, students are usually introduced to a number of machine learning methods. Because these methods often involve heavy and time-consuming mathematical computations, they usually have to be handled using high-level libraries like TensorFlow, PyTorch, or scikit-learn rather than through manual calculations or the code written by the students themselves. While it can help students better learn how to use these industry-standard tools, it often limits their ability to deeply understand the computational processes underlying each model - especially how these processes scale or behave under different circumstances. In contrast, parallel computing courses primarily focus on the theories and techniques, however, without providing sustained exposure to practical domains where they can be fully applied. Examples like matrix multiplication or numerical simulations are commonly used in such a class, but they do not fully resonate with students interested.

As a way to address these limitations, a new course is designed, trying to integrate parallel computing and machine learning into a single upper-level elective for undergraduate students in computer science major. In author's opinion, machine learning algorithms are excellent candidates for parallelization. For example, in problems like linear regression and logistic regression, OpenMP can be used to parallelize loops for computing gradients, while MPI is often used to distribute training data across multiple nodes and aggregate partial gradients. When working on decision trees, MPI enables parallel construction by using distributed data to grow subtrees simultaneously. In convolutional neural networks (CNNs), OpenMP can help accelerate convolution and activation operations, whereas MPI can support either model parallelism or data parallelism by distributing images or feature maps across nodes.

In addition, teaching students how to parallelize machine learning algorithms will also equip them with both algorithmic intuition and practical system design skills. The computational demands of modern machine learning tasks provide a concrete motivation for performance optimization. As datasets and model sizes grow, parallel and distributed solutions become essential for feasible training times. In summary, machine learning offers a highly relevant, computationally demanding, and pedagogically engaging domain, through which to teach parallel computing concepts. Such integration should be able to better support deeper learning, fosters cross-disciplinary thinking, and better prepares students for modern computing challenges.

3 Course Design

3.1 Course Description

This course focuses on three areas: foundational concepts in parallel computing, machine learning methods that are commonly used, and the practical integration of parallel computing techniques into the implementation of those methods. The goal is to provide students with both a conceptual understanding and hands-on experience in applying parallel computing to optimize machine learning methods in real-world scenarios.

This course design mainly targets two types of junior and senior students majoring in computer science. The first type is the students who have not studied machine learning in depth and also do not plan to specialize in it. However, they are interested in learning the basics of widely used techniques such as regression, decision trees, neural networks, and k-means clustering. They also want to understand how these methods are applied in fields like healthcare, finance, and computer vision. The second type includes students who have a strong interest in parallel computing and want to learn how it is used in real-world applications. To succeed in the course, students should have completed a standard programming sequence (i.e., CS1 and CS2), be proficient in either C++ or Java, and have taken courses in calculus, linear algebra, and statistics.

The learning outcomes are that by the end of the course, students will be able to demonstrate a clear understanding of the core concepts, classifications, and applications of both parallel computing and machine learning. They will gain the skills to implement some machine learning methods and apply parallel computing techniques to improve their computational performance. Students will also learn to evaluate the efficiency of parallel programs and machine learning models using standard performance metrics, and to analyze the potential benefits and trade-offs involved in applying parallelism to machine learning workflows.

3.2 Topics and Course Schedule

As shown in Table 1, the topics to be covered are structured around two main components: parallel computing and machine learning, each making up roughly 50% of the lecture. The topics related to parallel computing are introduced first, followed by machine learning. The purpose is to better equip students with the parallel skills that are needed when working with machine learning methods.

Weeks 1–5: Parallel Computing The course begins by introducing students to the core principles of parallel computing. At first, students learn about the motivation behind parallelism, the basic types of parallel architectures (such as shared-memory and distributed-memory systems), and the different programming models that support parallelism, including their use in real-world machine learning applications. Then, it focuses on OpenMP, a standard API for programming shared-memory systems. Students explore fundamental constructs such as parallel loops, data scoping (shared vs. private variables), and synchronization mechanisms like critical sections, atomic operations, and barriers. They also learn how to identify and avoid common performance issues such as false sharing, poor cache locality, and load imbalance. Finally, the course transitions to MPI. Students practice initializing MPI programs, using point-to-point communication, understanding process ranks, and working with communicators. Key collective operations - such as MPI_Bcast, MPI_Reduce, MPI_Gather, and MPI_Scatter - are explored in depth. Advanced topics include defining custom MPI data types and optimizing communication to reduce overhead and latency. To wrap up this unit, students study performance metrics like speedup, efficiency, scalability, and important theoretical models like Amdahl's and Gustafson's Laws.

Weeks 7–11: Machine Learning Methods The second half of the course introduces students to foundational machine learning techniques and how they can be accelerated using parallel computing. It begins with an overview of machine learning concepts, including the major categories – for example, supervised, unsupervised, and reinforcement learning. After that, linear regression and logistic regression are introduced, focusing on model formulation, gradient-based optimization, and interpreting results. Then, it covers k-nearest neighbors (k-NN) and decision trees. Students learn how to build and evaluate models from scratch, and how distance computations in k-NN or tree construction can be parallelized for better performance.

In addition, more advanced models: neural networks and support vector machines (SVM) are taught, focusing on the architecture of multilayer perceptrons, the basics of backpropagation, and how neural nets are trained. The SVM portion covers the basic idea of SVM, margin maximization, and kernel functions. Finally, it explains k-means clustering - a classic unsupervised learning algorithm - and an introduction to Markov Decision Processes (MDPs), which are central to reinforcement learning.

Weeks 6 & 12: Project Presentations and Exams After the topics related to parallel computing are completed, students will present Project I and write their midterm exam. Then, when all topics of this course are introduced, the students deliver Project II and write their final exam. The project II will demonstrate the application of one or more machine learning methods, ideally enhanced through parallel computing techniques.

Table 1. Sample Topics and Schedule in Summer 2026

Week	Topics
1	• Overview of parallel computing • Parallel architectures • Parallel programming models • Applications in machine learning
2	• Introduction to OpenMP • Basic parallel constructs • Data sharing and privatisation
3	• Synchronization mechanisms in OpenMP, sucg as critical (critical section), atomic (atomic operations), and barrier • Performance considerations in OpenMP, such as false sharing, cache locality, thread load balancing.
4	• Introduction to MPI • Environment initialization and termination • Point-to-point communication • Communicators and process ranks • Collective communication in MPI, such as MPI_Bcast (broadcast), MPI_Reduce (reduction), MPI_Gather (gather), MPI_Scatter (scatter).
5	• MPI data types and derived data types. • MPI optimization: communication overhead, network latency, load balancing. • Performance evaluation concepts: speedup, efficiency, scalability, Amdahl's Law, Gustafson's Law.
6	• Project I Presentation and Midterm Exam
7	• Overview of machine learning • Types of machine learning • Classification details • Application scenarios • Potential for parallel computing in machine learning
8	• Linear regression and logistic regression,
9	• k-nearest neighbors (k-NN). • Decision tree
10	• Neural networks and support vector machine (SVM)
11	• k-means clustering and • Markov Decision Process
12	• Project II Presentation and Final Exam

3.3 Coursework

As shown in Table 2, the coursework of this class includes weekly assignments; term projects, midterm exam, and final exam, helping students acquire theoretical knowledge and practical skills to enhance the computational efficiency of machine learning applications. The details are below.

Weekly Assignments There will be ten assignments in the summer 2026. The problems in each assignment will primarily focus on the topics covered in the corresponding week, helping students practice and reinforce the material presented in class. The complexity of the assignment problems will be controlled such that students can complete them either through hand calculations or by writing their own code without relying on external software packages. Assignments will be provided on Mondays, and students will need to submit their solutions by the end of Sunday. Then, the course instructor will evaluate the answers of each student during the following week, providing both scores and written feedback. If necessary, the instructor will also review and explain common issues from the assignments in class.

Term Project This course will include two team-based projects. There will be 3 to 5 students in each group, with each student taking on specific responsibilities such as writing the project report, gathering data, writing code, or testing the software.

The topic for the first project will be the same for all groups that requires students to create a tutorial based on the parallel computing topics covered in the course. The goal is to help students thoroughly review the parallel computing concepts they've learned. The tutorial should not only explain the material clearly but also has to include simple, easy-to-understand examples. The second project asks students to select any two of the machine learning methods introduced in class and implement them using parallel programming techniques so as to help students acquire hands-on experience applying parallel computing to real-world problems. Other than the written report, each project will also have a presentation orally telling what each group has done in their project. In addition, after each group presents, all other groups will score the presentation, in order to encourage active participation.

Midterm Exam and Final Exam The course includes a midterm exam and a final exam. The midterm will evaluate students' overall understanding of the parallel computing topics, while the final will emphasize machine learning concepts and how parallel computing techniques are applied to those methods. Both exams will be closed-book and taken in person. Considering the fact that the material involves various formulas and theoretical concepts, students will be allowed to bring up to two pages of handwritten or typed notes (on standard letter-sized paper) for reference. However, the use of electronic devices such as cell phones or laptops will not be allowed during the exams.

Table 2. Sample Coursework in Summer 2026

Coursework	Grade
Assignments	10 assignments × 2%
Project I	15%
Project II	15%
Midterm Exam	20%
Final Exam	30%
Total	100%

4 Implementation Plan and Teaching Effectiveness Evaluation

This course will be offered on a trial basis in summer 2026. It will last twelve weeks; and there will be two meetings per week. The length of each meeting will be one hundred minutes.

Because this course involves a significant amount of parallel programming, the ideal teaching environment should resemble a computer lab. Each student should have access to a computer equipped with the necessary software for parallel programming, allowing students to write and test code while following the instructor's explanations in class. In addition, commonly used machine learning libraries should also be pre-installed so that students can compare the outputs of their own implementations or manual calculations with those from standard software tools. Beyond the technical setup, the classroom should be furnished with tables large enough for 4 to 6 students to sit around, enabling students to work together and collaborate more effectively during group activities.

The instructional approach will aim to be student-centered and example-driven, with an emphasis on teamwork, hands-on activities, and unplugged exercises. Many topics, especially those related to machine learning methods, often involve extensive mathematical derivations. In this course, such derivations will be simplified or reduced whenever possible, allowing students to focus more on the fundamental meaning of each method. When introducing each topic - especially machine learning methods - simple examples will always be provided by the instructor alongside the theoretical explanation to illustrate the corresponding applications. Whenever possible, these examples should be those suitable for hand calculation or unplugged implementation. After each topic is explained, either an unplugged activity or a programming exercise that can avoid using software packages will be provided for students to practice, helping students thoroughly understand each topic covered. The exercises in class will primarily be done as group exercises, that is, several students work together as a group to practice, for example practice communication patterns in parallel computing - such as broadcast, reduction, gather, and scatter - as well as machine learning methods like k-nearest neighbors (k-NN), decision trees, and neural networks.

The effectiveness of the course design will be assessed using multiple methods, as described below.

Standard Course Evaluations At the end of each semester, students in each class are required to complete a standard course evaluation form, which will be used to provide general feedback on the course and serve as a broad measure of the design's overall effectiveness.

Supplemental Feedback Survey In addition to the standard evaluation, students will also be asked to complete a separate questionnaire that aims at gathering more detailed feedback on specific elements of the course design. This survey will include questions such as which topics students find most or least useful, suggestions for additions or removals, opinions on instructional methods, and how well the coursework supported their understanding. The goal is to evaluate individual components of the course - specially the relevance of topics, the usefulness of assignments, and the clarity of teaching strategies.

Comparative Performance Review To better evaluate the impact of the course design, selected computing assignments previously used in other courses (e.g., those used in artificial intelligence, machine learning class, data analysis, or parallel computing classes) will be reused. The performance of students on the shared work will be compared with data from past cohorts, helping determine whether students in the new course show improved understanding or skills compared to those in earlier courses.

Select Student Interviews A subset of students from the course will be invited to participate in individual interviews. These interviews will focus on what students gained from the course, the challenges they faced, and any suggestions they have for improving the course design. The goal is to gather more in-depth feedback - especially on aspects that might not be fully captured by other evaluation methods - to help refine and improve the course going forward.

5 Summary and Future Work

This paper explains the design of integrating parallel computing and machine learning into a single course. It is designed to help students who are interested in parallel computing gain practical experience with real-world applications, while also filling a curricular gap for those who may not be specializing in machine learning but still need to understand how core machine learning methods work and how to optimize them for performance. As such, the first half of the course builds a strong foundation in parallel computing techniques, and the second half introduces fundamental machine learning techniques, showing how these methods can benefit from parallel execution in real-world contexts. A key pedagogical goal of this design is to make the material approachable while retaining rigor via simplifying mathematical treatments, unplugged activities, in-class examples, and exercises that avoid high-level libraries in favor of writing code from scratch.

As the future work, the design will be refined based on student feedback and teaching observations from its initial offering in summer 2026 so as to eventually develop a scalable and flexible curriculum that strengthens students' readiness for data-intensive computing tasks in research and industry.

References

1. Goodfellow, I., Bengio, Y., Courville, A.: Deep Learning. MIT Press, Cambridge (2016)
2. Dean, J., et al.: Large scale distributed deep networks. In: Advances in Neural Information Processing Systems, vol. 25, pp. 1223–1231 (2012)
3. Asanovic, K., et al.: The landscape of parallel computing research: a view from Berkeley. Technical Report EECS-2006-183, Electrical Engineering and Computer Sciences, University of California at Berkeley (2006). http://www2.eecs.berkeley.edu/Pubs/TechRpts/2006/EECS-2006-183.html
4. University of California, Berkeley: CS 267: Applications of parallel computing. https://www2.eecs.berkeley.edu/Courses/CSC267/
5. Carnegie Mellon University: 15-618: Parallel computer architecture and programming. https://www.cs.cmu.edu/~15418/

6. Massachusetts Institute of Technology: 6.172: Performance engineering of software systems. https://ocw.mit.edu/courses/6-172-performance-engineering-of-software-systems-fall-2018/
7. Bojarski, M., et al.: End to end learning for self-driving cars. arXiv preprint arXiv:1604.07316 (2016). https://doi.org/10.48550/arXiv.1604.07316
8. NVIDIA Corporation: CUDA toolkit documentation. Technical documentation (2023). https://docs.nvidia.com/cuda/
9. Heaton, J.B., Polson, N.G., Witte, J.H.: Deep learning for finance: deep portfolios. Appl. Stoch. Model. Bus. Ind. **33**(1), 3–12 (2017). https://doi.org/10.1002/asmb.2209
10. Esteva, A., et al.: Dermatologist-level classification of skin cancer with deep neural networks. Nature. **542**(7639), 115–118 (2017). https://doi.org/10.1038/nature21056
11. OpenMP Architecture Review Board: OpenMP application program interface, version 5.2. Specification document (2021). http://www.openmp.org/wpcontent/uploads/OpenMP-API-Specification-5-2.pdf
12. MPI Forum: MPI: A message-passing interface standard, version 4.0. Specification document (2021). http://mpi-forum.org/docs/mpi-4.0/mpi40-report.pdf
13. Rajkomar, A., et al.: Scalable and accurate deep learning with electronic health records. NPJ Digit. Med. **1**(1), 18 (2018). https://doi.org/10.1038/s41746-018-0029-1
14. Marques, L.S., von Wangenheim, C.G., Hauck, J.C.R.: Teaching machine learning in school: a systematic mapping of the state of the art. Inform. Educ. **19**(2), 283–321 (2020). https://doi.org/10.15388/infedu.2020.14

A Study on Commerce Channel Characteristics and Strategy: Focus on Domestic Department Stores in Korea

Chan Uk Jung[1], Hyeon Jo[2(✉)], and Won Jun Lee[1]

[1] Business School, Sung Kyun Kwan University,
104 Dong 703 Ho, Doosan WE'VE APT, Jeonjailro55, Bundang-gu, Seongnam-si, Kyeonggi-do, Republic of Korea
istjung@g.skku.edu

[2] Department of Strategic Planning, RealSecu,
IS Tower #809, 60 Centum buk-daero, Haeundae-gu, Busan 48059, Republic of Korea
sineoriz@gmail.com

Abstract. With the rapid advancement of technology, consumer shopping behavior has undergone substantial transformation. As online commerce platforms have diversified—from traditional internet-based channels to mobile environments—commerce channel strategy has emerged as a critical factor for retail success. Furthermore, the widespread adoption of mobile devices has led consumers to increasingly make purchases via mobile channels while simultaneously experiencing products in offline stores. This phenomenon underscores the need for traditional brick-and-mortar retailers to integrate online and mobile channels with physical stores as part of a comprehensive commerce channel strategy. Accordingly, the success of offline retailers today largely depends on their ability to effectively manage channel strategies and implement integrated channel services. This study analyzes the characteristics and strategies of commerce channels employed by leading domestic department stores in Korea, namely Lotte, Hyundai, and Shinsegae. Drawing on prior research, three types of commerce channels—single-channel, multi-channel, and cross-channel—are examined as the analytical framework. In addition, various documents and data related to omni-channel services offered by these department stores were collected to support the analysis. The study ultimately aims to provide insights that contribute to the development of effective commerce channel strategies in the Korean retail market, including large-scale marts, supermarkets, and convenience stores, as well as to inform future research in this area.

Keywords: Department store · commerce channel · Omni-channel · O4O · Retail

1 Introduction

1.1 Research Background

The global digital environment has rapidly evolved toward an integrated shopping ecosystem in which purchasing can be facilitated through multiple channels, including offline, online, and mobile platforms. In response, global retailers are adopting advanced

H. R. Arabnia et al. (Eds.): CSCI 2025, CCIS 2974, pp. 391–402, 2027.
https://doi.org/10.1007/978-3-032-28475-4_30

technologies and making substantial investments associated with the Fourth Industrial Revolution, thereby enabling personalized and intelligent shopping experiences. In the past, consumers primarily relied on visiting physical stores or contacting retailers directly to obtain product information. However, contemporary consumers access information through diverse channels, including online platforms, mobile applications, and social media, supported by increasingly varied digital devices. Accordingly, retail firms now operate multiple commerce channels, and the distribution paradigm is progressively shifting toward online and mobile environments. Recent statistics indicate that approximately 86% of global consumers utilize two or more purchasing channels, highlighting the growing importance of channel integration strategies that connect multiple touchpoints for sales and marketing purposes (Cosmetic Market, 2015).

Multi-channel commerce represents a next-generation distribution system that reflects evolving consumer behavior characterized by active information sharing and rational decision-making. Unlike traditional enterprise-centered distribution structures, which involved one-way communication with customers, next-generation commerce channels emphasize customer-centricity by providing consistent and seamless service across channels. For example, Macy's in the United States achieved an operating cash flow of $2.19 billion in 2014 following the adoption of an integrated channel strategy, significantly outperforming competitors that had not implemented similar approaches (Park, 2017). Such strategies—particularly multi-channel and cross-channel approaches—have been widely adopted by global department stores and large-scale retailers, and similar trends are increasingly observed in the Korean retail industry. Consumers can now conveniently purchase products via online and mobile platforms while also interacting with offline stores, and retailers actively utilize social media to disseminate product information and reviews. For instance, Lotte Group has emphasized the rapid development of IT technologies connecting department stores and has announced plans to expand digital services centered on multi- and cross-channel commerce (Kim, 2018b). Despite the growing importance of commerce channel strategies, prior research has predominantly focused on international contexts, such as the United States and Japan, with relatively limited attention given to the Korean retail industry.

Accordingly, this study aims to compare the current adoption and utilization of commerce channel strategies among leading domestic department stores in Korea and to provide insights that may contribute to the development of future channel strategies in the Korean retail sector.

1.2 Research Scope and Method

This is a strategic analysis study on the service implementation and promotion for revitalizing the domestic retail commerce channels based on the analysis of domestic and overseas cases of retail commerce channels. The scope of the study is limited mainly to the major domestic department stores (Lotte, Shinsegae, Hyundai Department Store), so large marts, convenience stores and home shopping companies in retail are excluded because they have not tried various approaches related to create and build new channel services yet, which are expanding into multi-channel and beyond channel strategies.

The study method is based on data collection from domestic and foreign papers, journals, and publications published to present theoretical considerations. Most of domestic department stores' commerce channel case analysis mainly used DB search through internet, and statistics for objective data are cited statistical data of related field.

2 Literature Reviews

2.1 Understand Retail Market in Korea

In the domestic retail market, various types of businesses such as large marts, chain supermarkets, convenience stores, internet shopping, and home shopping have entered the market since the first appearance of department store based on urbanization and economic growth in the 1990s (Kwak, 2018) (Table 1).

Table 1. The Types of Retail Operators

Category	Definition
Department Store	It refers to general retail store that has a sales facility divided into a number of stores and comprehensively retail various goods.
Large Mart	It means 'other large comprehensive retail business' according to the 9th amendment of Korea Standard Industry Classification. It is a general retail store that sells various types of products such as groceries and clothing, with large stores (excluding department stores), and includes large-sized marts, duty free shops and outlets.
Supermarket	It includes 'supermarket', other retailing business centered on other food and beverage products', other general retailing businesses. In Korea Standard Industry Classification, the 'supermarket', which refers to a retail store selling various miscellaneous goods mainly for foodstuffs, is represented as 'large supermarket' on a scale of 165 ~ 3,000 m^2. 'Other grocery retailers' with a size of less than 165m^2 are referred to as 'private supermarkets' and 'other general retailers' as 'other supermarkets'.
Convenience Store	It is a retail store that sells products 24 h a day through continuous supply of chain products and contracts.
Specialty Store	It refers to a retail store that specializes in the sales of certain products with a certain store and includes hardware, books, stationery office equipment, sporting goods, precious metals and arts retailing.
Non-store retail	It includes retailing businesses that sell Internet shopping, home shopping, and other visiting and delivering merchandise without opening direct-selling stores.

In particular, there was concern that domestic companies would go bankrupt due to the opening of the retail industry in 1996, and they would be transferred to foreign retailers. However, the domestic retail business continues to grow. As of 2018, the domestic retail market is expected to reach ₩330 trillion and is expected to grow about 7% from last year (Retail Industry White Paper, 2017) (see Fig. 1).

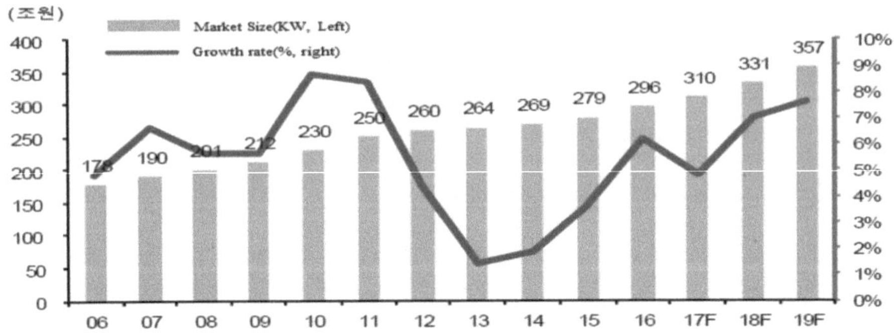

Fig. 1. Domestic retail market size trend & forecast

Traditional retailers such as department stores and large-scale marts are slowing down due to the regulation of large retailers and the increase in the number of single-person households, and supermarkets and convenience stores are replacing them. In addition, as the development of IT technology has led to the widespread use of the Internet and smartphones, the proportion of Internet shopping has also increased (Kwak, 2018) (see Fig. 2).

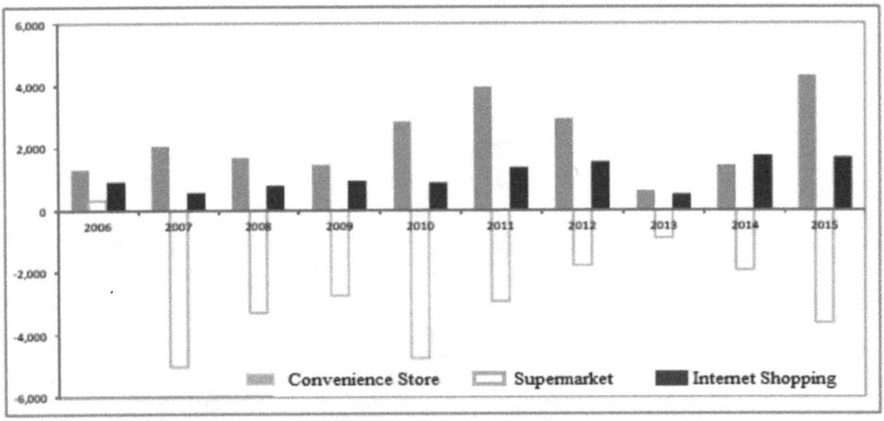

Fig. 2. Changes in the number of major retailers

The retail market is chaining rapidly with the development of IT technology. In line with this, each business is inevitably stolen without securing competitiveness and increasing productivity. Domestic retailers need to shift their strategies in order to increase productivity and enhance competitiveness so that they can invest in mobile channel technology. In particular, mobile channel is not merely a new form of commerce channel added to existing commerce channels, but it has a great impact on the retail industry, as mobile in combination with existing retail channels has caused structural changes in the retail industry and changes in consumer life.

2.2 Understand Various Types of Channel

Table 2 (Saeeun Kim, 2016) shows the terms of single channel, multi-channel and cross-channel, which mean commerce channels. From a strategic point of view, a single channel strategy is a single channel strategy that operates only one channel. It refers to a strategy that operate only in-store out of a large number of channels. Moreover, operating one of the offline stores, online malls, mobile apps and TV home shopping channels can be a single channel strategy as well.

Multi-channel strategy refers to multi-channel strategy that operates multiple channels by building one or more channels in online malls, mobile apps, offline stores and TV home shopping. Multi-channel strategy does not provide well connected among each other channels, but operates like a completely different store in the same company and competes between channels. With the term cross-channel appears here, the definition of cross-channel is interpreted in a slightly different direction, but it can be recognized as a transitional channel strategy from the multi-channel era to the omni-channel era. This is because there is a problem in that multiple contact points occur or are staggered in the course of merely channel expansion of multi-channel strategy and channel integration to solve it. That is, when the consumer tries to purchase the same product of the same brand, the multi-channel is referred to as a diversified distribution channel, and the cross-channel is a case where diversified channels are entangled with each other.

For example, when the same shoes of the same brand are sold in one shopping mall with different price policies according to the sellers, a seller who sells the same product in different policies according to the channel may be referred to as a cross-channel. From a customer's point of view, it is one brand, but the integration and connection between channels is not working smoothly.

The concept that emerged as a solution to this is the omni-channel strategy. The omni-channel strategy is to provide a consistent customer experience so that each channel can receive the same services and benefits regardless of the customer's access to that store. Omni-channel is not just shopping on one channel when shopping, but it is able to do real-time shopping anytime and anywhere by crossing various channels from shopping to comparison, search, purchase, alternative evaluation. It means that shopping in an online mall, a mobile app, a TV home shopping store and an in-store connected to a store of each channel is an omni-channel shopping environment.

As for the distribution process until the omni-channel shopping environment appears in Korea, it can be seen from the single channel era that operates offline store only, through both the multi-channel era that operates stores in multiple channels and the cross-channel era in which stores are operated from multiple channels, and from the year 2015, companies are trying to build a shopping environment for omni-channel. From that time on, omni-channel era can be said.

Table 2. The Term of various Types of channel

Channel type	Definition
Single Channel	– Run one distribution channel
Multi-Channel	– The brand manages the several sale distribution channels
	– The store which is independent and separated according to the distribution channel
Cross-Channel	– The brand manages the several sale distribution channels
	– The distribution channel of the part has ties and the distribution channel of the part is not combined
Omni-Channel	– Provide the shopping circumstance which is identical in all distribution channels

2.3 Understand Omni-Channel Service

The term omni-channel was first introduced in November 2011 by the US Retail Consortium when it announced its "Mobile Retailing Blueprint 2.0" and in January 2016 at the US Retail Consortium Convention in New York, USA, it is expressed as "Flexible shopping, Flexible shipping" (Kim & Kim, 2017). In other words, omni-channel shopping is a service that enables consumers to browse and purchase products while crossing various routes including online and mobile.

Omni-channel features a unified brand message centered on the customer by integrating all front and back channels related from planning stage of product to purchasing such as product, service, organization, system, process, and logistics without difference of online and offline stores. In the omni-channel environment, all channels provide customers with a consistent message and experience to prevent brand image diversification and unnecessary competition between channels.

Looking at the changes in domestic commerce channels, in the past, the sales and purchasing of products were conducted only in offline channels such ss shops and visiting sales. With TV home shopping in 1995 and Internet shopping mall in 1996, TV home shopping and online shopping market expanded and commerce channel became multi-channel era of 'TV, Online and Offline' channels. By the end of 2009, since mobile shopping became possible with the launch of smartphones, it became the era of shopping on multi-channels from TV, online PC, offline and mobile so that companies have been expanding their channels indiscriminately. Therefore, the phenomenon of management of the same customer, which is a problem that occurs in multiple channels, has to be separately managed for each channel, and sales competition for each channel has occurred despite the same product and the same store. To illustrate the phenomenon of multiple channel and redundant channels, the concept of "cross-channel" appeared, and shoppers who crossed multiple channels were called "cross-shoppers". The concept of' 'omni-channel' is a solution to the problem of redundant channels and multiple channels, and it aims t combine and integrate each channel.

Previous studies related to omni-channel have mainly focused on research on conceptual research and introduction of omni-channel (Ailawardi & Farris, 2017; Piotrwicz & Cuthbertson, 2014), research on economic effect of adopting omni-channel services (Bell et al., 2015), comparison research of omni-channel with multi-channel and cross-channel (Beck & Rygl, 2015).

These studies mainly focus on the conceptual studies and consumption behavior of introducing new channel services. Therefore, the results of presenting the specific characteristics of consumers with differentiated characteristics and the convenience of attracting consumers have been rarely done.

3 Domestic Traditional Offline Retailer' Channel Strategy

3.1 Commerce Channel Positioning

In order to respond to changes in consumption patterns, which are breaking down the boundaries between offline and online, large retailers are strengthening various channel services. In this situation, major department stores are analyzed in terms of current channel and competency positioning, and it is shown in Table 3.

Lotte Department Store has made it possible to shop conveniently using various commerce channels and has been actively responding from cross-channel by utilizing online and offline. Shinsegae Department Store has a system that integrates independently operated channels such as internet, mobile, and offline store so that customers can enjoy shopping any-time and -where. Lastly, Hyundai Department Store's channel positioning is still on the offline centering on purchasing, not in the multi-channel and online competence.

Table 3. Channel Positioning of Department Stores

Corporate	Positioning	Contents
Lotte Department	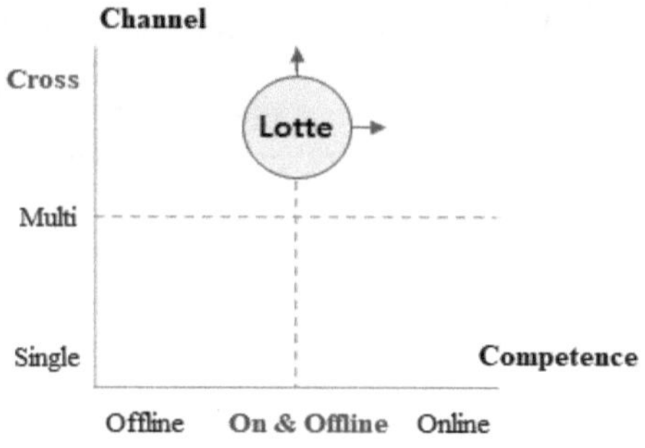	– Expand online shopping channel based on strong offline channels – Absorb customers by expanding online channel Ex) Providing personalized information and services

(*continued*)

Table 3. (*continued*)

Corporate	Positioning	Contents
Sinsegae Department	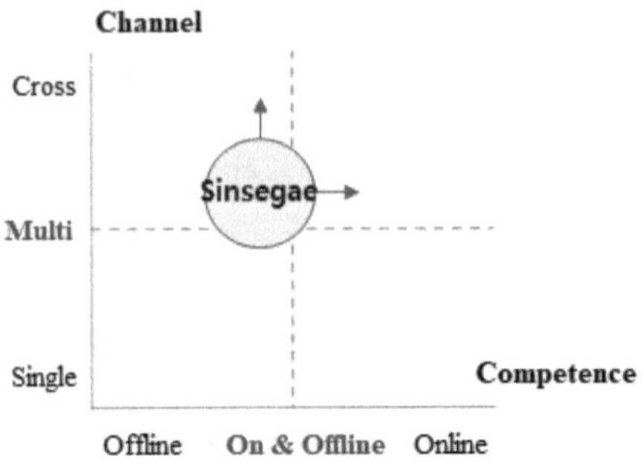	– Expand the services that were available online only at the offline store, making the offline store like online Ex) Provide various digital experiences at the store
Hyundai Department	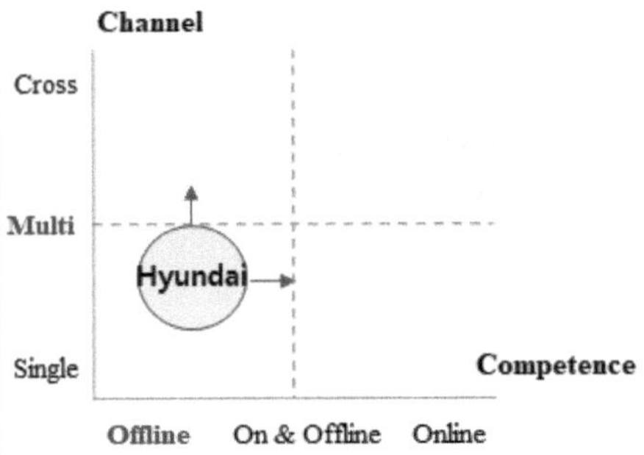	– Focus on the advantages of offline channel Ex) Provide mobile DM to offline customers – Expand online sales by continuously entering other online malls

3.2 Omni-Channel Service

Table 4 shows the major omni-channel services and contents of department stores that are pursuing omni-channel marketing strategies that respond to changing retail environments.

Lotte Department Store, which is the leading retail omni-channel services player in Korea, is actively promoting 'smart peek' service through 'living with omni' advertisement. It is a service that can pick up, return and exchange in offline store if you choose the product online. Hyundai Department Store has opened 'The Hyundai.com', an online store that sells department store products exclusively, and offers 'store pick' services that can be ordered online and visited offline store. Some brands include VR(Virtual Reality) Store service is provided so that customers can view offline stores as 3D video online. Shinsegae has an SSG site by integrating online sites such as department stores, e-marts, and traders, and provides a 'magic pickup' service that allows offline purchases of products purchased from SSG.com.

Table 4. Omni-Channel Services of Department Stores

Corporate	Services	Contents
Lotte Department Store	Smart Pick	– Smart Desk: order in online mall and receive in the offline mall – Smart locker: the manless pickup locker at seven-eleven convenience stores and at MVG(Most Valuable Guest) parking lot
	Mobile store	– Using QR code to compare the price of products at offline mall
	Smart beacon service	– Providing personalized and benefit information through mobile coupon
	Location based service	– Using location-based information, nearby stores and product information
Hyundai Department Store	VR store	– Providing Virtual reality store for customer to have digital experience
	Store pick	– Placing an order in online mall and picking up in offline mall
	Mobile App	– Personalized direct mail service based on purchasing history – Providing mobile parking ticket, membership card etc.
	Beacon Service	– Providing benefit information by store location
Shinsegae Department Store	Magic-pickup	– Placing an order in online mall and picking up products in offline mall
	SSG	– Online shopping mall integrated with Shinsegae mall, E-mart mall and Traders mall
	Electronic wallet S wallet	– Purchasing products, checking membership points and receipt, exchanging gift certification through smart phone
	Smart order	– Ordering and settling menu with smart phone, shorten order waiting time

(*continued*)

Table 4. (*continued*)

Corporate	Services	Contents
	Shinsegae app	– Providing special events, sales information, coupon and store information etc.

3.3 Common Characters of Commerce Channel Implementation

In this section, commerce channel strategy derived from the analysis of channel positioning and omni-channel service can be classified into types with common characteristics (Table 5).

First, in the upper analysis, the classified commerce channel strategy can be named 'Channel-connected System' which connects all channels together. Customers can buy or experience all the branded products of company through a single website or app by a way of integrating the online, offline, and mobile shopping mall of the company's brand. It is also a multi-channel delivery service that is formed by passing through multiple channels. This is a product delivery service that combines the advantages of online and offline, such as purchasing online and picking up offline, or picking up at other stores that have stock after purchase.

Second characteristic is named 'Digitalized IT service' as a channel strategy that incorporate IT technology. This is a type of strategy that utilizes a variety of smart devices, including smart phone, digital screen and so on. It provides online store experience that can directly search various information by integrating smart devices into offline stores. In addition, the service providing the virtual experience opportunity of the product to be purchased by utilizing the augmented reality technology is an example in which the customer can conveniently experience through the mobile application.

The channel strategy to create the simplest purchasing environment is named 'Customer engagement environment' as the third characteristic. The typical example of this strategy is location-based services with beacon service provide shopping information such as discounts and events at appropriate location according to the customer's location. In addition, based on customer information formed by data analysis, personalized marketing strategy will lead to more effective purchase. Finally, App card service, a simple payment system, can make it possible to purchase products easier and more convenience by integrating and simplifying payment methods.

Table 5. Channel Characteristics of Department Stores

Commerce channel characteristics		Lotte Department Store	Hyundai Department Store	Shinsegae Department Store
Channel-connected System	Integrated shopping mall	○	○	●

(*continued*)

Table 5. (*continued*)

Commerce channel characteristics		Lotte Department Store	Hyundai Department Store	Shinsegae Department Store
	Multi-channel product delivery	●	●	●
Digitalized IT service	Smart devices	●	●	●
	Augmented reality	○	●	○
Customer engagement environment	Customer management	●	●	●
	Location-based service	●	●	○
	Simplified payment	○	○	●

●: Available, ○: Not Available

4 Conclusion

This study defined the various types of channel terms related to 'channel' and investigated cases with the status of 'various channel position' and 'omni channel service' from domestic department store in order to derive the direction of 'Channel Strategy' as a countermeasure strategy against the changing retail environment. As s result of analyzing the channel strategies of domestic department stores through mutual linkages between channels, domestic department stores are expected to have some degree of business model and idea about commerce channel strategy and start implementing from 2015 and 2016, but it seems to be the situation that only channel extension is done. Consumers still perceive each channel as a different store, and department store vendors are also unable to integrate policies across channels. Offline stores, online malls, and mobile apps have different product configurations, discount policies and ongoing events are not integrated. In addition, most department stores sell multiple brands on a single site by accepting the concept of 'omni-channel' as integration. This is not the proper strategy both because it provides the unnecessary information to the customers when products that are not suitable for age, price, and the use purpose are searched and because it is not possible to imprint the unique identity of the brand to the consumers and the opportunity to experience the brand through the characteristics of online mall and the mobile app is insufficient. In order to overcome these problems, this study can present the proper direction of commerce channel strategy for domestic department store companies. It is 'O4O(Online for Offline)' business model. This business model is to connect online and offline by using the strengths of operating the offline stores so that it is possible for customers to be connected to offline stores by finding connection points between online and offline such as push alarm for promotional coupons, earning points or pick up service and so on. The purpose of this study is to present the characteristics and direction of the channel strategy for domestic department stores by exploring the various channels positioning and omni-channel services. In future, it is expected that it will become the basic data of research for the other retail markets for large-sized mart, convenience stores, and supermarket etc.

References

Kwak, E.-K.: Retail Market Status and Implications. The Center for Free Enterprise (2018)

Han, H.: An analysis on the omni-channel strategy of domestic and international cosmetics companies. Graduate School of Art & Design, Konkuk University (2017)

Park, H.G.: Change of ICT retail channel—omni-channel, IT trend & marketing issue, p 44 (2015)

Yang, J.-H.: Cosmetic and Retail Industry White Paper. Meritz Research (2017)

Kim, J.: Connected markets: digital retailing in the offline world. School of Business, Yonsei University (2018a)

Son, J.Y., Kang, I.W.: A study on the applicability of omni-channel service. e-Bus. Stud. **18**(5), 19–35 (2017)

Goo, J.: A study on the changes in the ecology of the retailing sector according to emerging new distribution channels. Korea Institute for Industrial Economics & Trade, Research report 2015-762 (2015)

Ha, K.-O., Lee, J.-H., Hwang, S.-H.: A study on consumer's shopping channel choice according to consumer purchase characteristics. J. Channel Retail. **20**(4) (2015)

Park, M.Y: Beauty, omni channel speed up AMOREPACIFIC (2017). http://www.dt.co.kr/contents.html?article_no=2017091102100976817001&ref=naver/

Kim, N.K.: Online operators who watch offline. LGERI (2018b)

Kim, S., Kim, M.: A study on omni-channel strategy in fashion industry. J. Korean Soc. Costume. **67**(1), 40–55 (2017)

Cha, S.S.: A comparative study on purchasing behavior of on/offline multi-channel and single-channel customers. Department of Digital Management, The Graduate School of Korea University (2016)

Lee, S.Y.: A study on omni channel case and strategy in domestic large-scale distributors. J. Korean Soc. Des. Cult. **21**(4), 477–488 (2015)

Strojkovic, D., Lovreta, S., Bogetic, Z.: Multichannel strategy—the dominant approach in modern retailing. Econ. Ann. **LXI**(209) (2016)

Deng, W.Q., Seo, Y.W.: A study on online channel integration in offline shops. Korean Manag. Sci. Rev. **33**(4), 77–89 (2016)

Internet Links

http://news.bizwatch.co.kr/article/consumer/2014/12/02/0035 (2014. 12. 03)

https://news.joins.com/article/16823469 (2014. 12. 31)

www. Etnews.com/20150114000112 (2015. 01. 14)

http://www2.deloitte.com/content/dam/.../kr_cbnt_issue-highlights_20150727.pdf (2015. 07. 27)

www.digitalretail.co.kr (2016. 03. 02)

http://news.mt.co.kr/mtview/php?no=2016022615043261299 (2016. 03. 02)

http://clomag.co.kr/article/1620 (2016. 06. 04)

www.ezyeconomy.com/news/articleView.html?idxno=62464 (2016. 09. 26)

http://www.ebn.co.kr/news/view/869898 (2017. 01. 02)

www.polinews.co.kr/news/article.html?no=306280 (2017. 03. 08)

http://news.mk.co.kr/newsRead.php?year=2017&no=232528 (2017. 04. 06)

http://www.morningeconomics.com/news/articleView.html?idxo=4140 (2018. 01. 29)

http://www.wikileaks-kr.org/news/articleView.html?idxno=22301 (2018. 05. 15)

Smart Glasses in Education

Mehdi Zargham[1(✉)] and Jie Cheng[2]

[1] University of Dayton, Dayton, OH, USA
mzargham1@udayton.edu
[2] Miami University, Hamilton, OH, USA

Abstract. Smart glasses have the potential to revolutionize education by enhancing student engagement, accessibility, and personalized learning. Although they are not perfect yet, like cell phones, they are becoming more refined to the point where it is easy to wear them everywhere without attracting the attention of those around you. Smart glasses incorporated with AI agents could guide us throughout our daily activities. We can learn about subjects of interest, ranging from math to historical events, as we walk, wait, work, or visit a site. The glasses can constantly correct or advise us during learning or performing a task. This paper examines current smart glasses technologies, including Meta's Ray-Ban and Apple's Vision Pro, as well as emerging AR glasses, analyzing their educational applications, from hands-free learning to real-time collaboration. While challenges such as privacy concerns, battery limitations, and social acceptance remain, the research demonstrates that smart glasses offer transformative opportunities for immersive learning experiences that surpass traditional classroom boundaries. The findings suggest that successful implementation requires collaborative efforts between educators, policymakers, and technology developers to create frameworks for context-aware AI assistants and address current limitations. As these technologies mature, smart glasses are self-assured to enable continuous, personalized learning 24/7 with our daily activities, potentially reshaping the future of college education.

Keywords: AR glasses · smart glasses · AI · education · future of education

1 Introduction

This paper focuses on smart glasses and their influence on education. These glasses enable students to overlay the physical world with the virtual (digital) world, providing a better understanding of the subject at hand. By being able to overlay digital information onto the real world, students are able to observe abstract concepts as tangible and interactive. It becomes clearer that smart glasses will eventually replace our cell phones. They are becoming lighter and are increasingly indistinguishable from regular glasses. It will become harder to recognize whether someone is wearing them. The first generation of such glasses is on the market and becoming popular; some have high-quality speakers and video displays, while others have simple text displays. In the near future, they will have holographic displays allowing us to see objects in 3D that we are interacting with.

H. R. Arabnia et al. (Eds.): CSCI 2025, CCIS 2974, pp. 403–411, 2027.
https://doi.org/10.1007/978-3-032-28475-4_31

In this way, smart glasses are able to deliver a sense of social presence while providing real-time personalized AI. As we do our daily activities, AI will see and hear those activities and will learn our personalities. Google is working on AI apps that can analyze anything we see. If we are looking at a monitor with someone typing code, we can ask about the function of that code, and the AI will be able to answer. In fact, we can ask about things we have seen in the past, and it can retrieve or answer a particular question for us.

Imagine that you are traveling around the world, passing through many amazing landscapes. You don't have any coding experience, but you decide to make a website capturing your experiences at different locations. You could simply tell AI to code for you and develop the first page of a website as you tell it what to do. Perhaps you say, "I want the first page to include my name, email, permanent address, and a bio about me. I want the background to be a faded image of the landscape that I am seeing now (of course, through AR glasses), and the arrangement of the page to be something similar to John Smith's and very professional. Let me see it through the Google Chrome browser. It looks good, but can you make it more beautiful and eye-catching? Also, insert a small image of me in the top left corner."

Or imagine when a surgeon operates on complex surgery—the expertise of the whole world will be at his/her ears and eyes. The glasses can constantly correct or advise the surgeon during the operation. Of course, it is possible to achieve the same outcome by using cameras and monitors, but the glasses simplify everything and follow the surgeon at all times, even when he/she is looking at X-rays. Or, imagine how much you can learn from the environment by just walking around. Therefore, given the current technology and the prospect of advances that we will have in just a few years, educational reform is essential. This has already happened, but not fast enough to catch up with state-of-the-art technology. Many students are becoming experts in utilizing AI systems, and we should encourage more rather than prevent it. If there is a writing course, how can we make assignments more exciting and collaborative, even if the student uses AI? The instructor should ask himself/herself, "How can I give a writing assignment that requires AI to enhance the final product?" Perhaps to be more self-reflective, collaborative writing, or personal views on current events that happened in the classroom, and so on. As a Japanese philosopher stated about ethics, the same concepts can be applied between teacher and student. He says ethics is between people and not individuals. For example, if a teacher weighs the final exam at about 80% of the grade, then the student has a tendency to cheat more. While if the majority of the class grade was based on students' creativity on certain projects, the philosopher makes a point that all the guilt should not be loaded on the student.

As the government is increasing the number of apprenticeship opportunities, more students may take advantage of them. This implies that some high school students, and even college students, are better off if they obtain a diploma/through work-based learning. This model, which follows apprenticeship success in Switzerland, has been propagated through many states, particularly Indiana. The idea behind this concept is that many well-paying jobs do not require a college degree, and students will also be debt-free and not worry about high college tuition. However, some parents are still resistant to their teenager working during high school and much prefer they go to a four-year college, and perhaps at that time they will participate in work-based learning. The nature of the workforce is changing rapidly, as companies need to be more innovative, more responsive, more connected, and more technology-driven. To this effect, some companies are moving toward hiring employees with certificates in specific fields rather than requiring a college degree. In addition, they will provide opportunities for employees to gain new knowledge/education as needed. The learning becomes more personalized toward achieving different certificates rather than a specific degree. A certificate implies a fixed set of knowledge and experiences and is hardly dynamic. There may be an AI entity that is a student's lifetime teacher. It knows how much the student knows about what they want to do, what else they need to know, and how and at what speed and method to teach and train them. This is more like a coordinated method, but it is personal and dynamic. The AI entity from an employer will teach the student's AI entity what is needed, and the student's personal AI will train the student!

One may say that there is a good chance that AI companies, which are limited to four or five, will be able to develop super-intelligent AI in a few years from now. Such an achievement could reform society to one where many jobs are automated. If we just look at computer science skills, we may just need a few employees who are very strong in the sciences to develop more efficient techniques for training large language/agent models. There may not be a need for basic programmers, and many careers in computer science can be done by AI or just a few people. This will reduce the number of colleges drastically. The colleges that survive are the ones that can educate top students in their fields. Many entry-level jobs will be automated, and if they are not, they will just require students to get training as they work.

The rapid advancement of artificial intelligence suggests that the handful of major AI development companies, currently dominated by four or five key players, may achieve super-intelligent AI within the next few years. This technological breakthrough could fundamentally transform society by automating a vast array of jobs across multiple sectors. For example in the computer science sector, in a future dominated by AI capabilities, organizations may only require a small number of highly skilled specialists, those with deep expertise in advanced sciences and mathematics, to develop increasingly sophisticated techniques for training and optimizing large language agent models. The traditional computer science careers could downfall, with entry-level and intermediate programming positions becoming obsolete as AI systems become capable of handling routine coding tasks, software maintenance, and even complex algorithm development. This transformation will likely trigger a dramatic changes in higher education. Many colleges and universities may struggle to justify their existence when their graduates can no longer secure employment in fields that have been automated. Only institutions capable of producing top-tier talent, those that can nurture the exceptional minds needed to advance AI research, manage complex systems, or work in fields that remain resistant to automation, will likely survive this transition. The employment landscape will be divided into two categories: jobs that can be fully automated by AI systems, and those that cannot. For positions that resist automation, traditional four-year degree programs may become unnecessarily lengthy and expensive. Instead, organizations will likely shift toward apprenticeship models and on-the-job training programs, where workers develop specific skills as they work, adapting quickly to evolving technological requirements.

2 Available Smart Glasses

At present, several smart glasses are on the market, and promising technologies are emerging for future glasses. One well-known example is the Ray-Ban | Meta Smart Glasses, which support hands-free voice commands through integrated AI while preserving the classic Ray-Ban design, such as the Wayfarer and Headliner styles. These glasses have high-quality built-in audio, and in the new models, they can project text/video. One of the main concerns about these glasses is their acceptance by society, as they record all the time, raising privacy concerns [10–14].

The RayNeo Air 3 s [10] focuses on providing users with a 201-in. virtual display experience, creating the impression of viewing content on a large screen while wearing lightweight glasses. The device offers quality built-in speakers for clear audio at an affordable price. However, the glasses require a connection to external devices such as smartphones or computers to function, making them dependent on additional hardware. The major issue is that the glasses function primarily as a large virtual screen, and the XR/spatial-computing features are very limited or nonexistent—they work more like a large virtual screen than a full mixed reality experience [15–19].

Viture Pro XR glasses, marketed as "multi-platform XR glasses" and similar to RayNeo Air, offer virtual display features, such as a Full HD (1920 × 1080) panel, and support a simulated ~135-in. virtual screen at up to 120 Hz, allowing users to experience immersive visual content through their lenses. Their compatibility with a wide range of devices, e.g., PCs, gaming consoles, and smartphones, makes them more appealing in many areas. However, the main limitation is their reliance on external power sources and computing devices, which restricts mobility and makes them less practical for everyday use than standalone smart glasses [20–22].

Solos AirGo 3, unlike Meta Ray-Ban, does not include cameras in order to address privacy concerns. It similarly supports AI connectivity, including ChatGPT-powered "Solos Chat," via voice commands and speakers. The audio performance is decent, including built-in speakers and approximately 7–10 h of battery life. It is compatible with prescription lenses. The main drawback is that it is more like "smart audio glasses," since it lacks the augmented reality capabilities enabled by cameras, limiting its functionality compared to glasses with cameras [23–28].

XREAL Air glasses deliver high-resolution display capabilities directly on the lenses, combined with augmented reality features and motion tracking. They are very light; for example, the Air model weighs approximately 79 g and includes high-quality built-in audio. A significant drawback is that many models lack internal batteries and depend on external devices such as smartphones or computers for processing power. In addition, several reviewers have noted the device's limited "true AR" functionality. Many features merely rely on screen mirroring rather than comprehensive spatial computing (e.g., static screens lacking world tracking), despite promotional claims of full AR capabilities [29–31].

Even Realities G1 glasses offer enhanced functionality with features such as voice commands, a microphone, a camera, and a text display on the lenses, while remaining lightweight and socially acceptable. However, the text display is rather basic compared to full-color video displays. Additionally, feature responsiveness is a concern. For example, reviewers have pointed out delays and slow AI responses. As TechRadar comments, "the novelty will quickly wear off ... you'll reach for your phone for anything more involved." This indicates that current performance issues may limit the device's usefulness and negatively affect user engagement [36–39].

Xreal One Pro is an AR glasses model that improves on previous versions, offering a more comfortable design for longer wear, comprehensive augmented reality experiences with enhanced motion tracking, high-resolution video displays, and improved audio. However, they suffer from shorter battery life due to their advanced processing and display requirements; the devices have to rely on external power sources for specific AR functions [32–35].

Rokid Max Series glasses support dual full-HD micro-OLED displays operating at 120 Hz, with a 50° field of view and a virtual screen equivalent to 215″ at a 6-m viewing distance. The device offers improved visual clarity and comfort over earlier Rokid models, weighing approximately 75 g and providing built-in myopia adjustment along with certified eye-comfort features. These characteristics make the Rokid Max suitable for entertainment and productivity applications. However, the glasses rely on an external host device for computation and do not provide full spatial tracking, thereby positioning the product as a high-quality wearable display rather than an autonomous AR system [40–42].

The Apple Vision Pro represents a new phase and benchmark in wearable spatial computing by combining ultrahigh-resolution micro-OLED displays, advanced eye, hand, and voice-based interaction, and a video-see-through AR/VR hybrid architecture within a single headset running visionOS. The platform enables both immersive virtual environments and contextually anchored digital content in real physical spaces, thereby offering dramatically expanded workspace and interaction paradigms beyond traditional 2D monitors or smartphone screens. Nevertheless, its reliance on tethered external power, use case maturity, ecosystem support, and cost remain significant barriers to widespread adoption in educational and enterprise settings [43–45].

3 Use of Smart Glasses

Although AI support can be utilized through smartphones and/or mobile devices, AR glasses could become the most effective tool in comparison. Perhaps their most advantageous feature would be hands-free learning and realistic remote collaboration. This allows anyone to have access to learning materials without interrupting daily activities, including workflow in some cases. We are able to learn while walking or lying down. AR glasses could facilitate the learning process by allowing the user to obtain remote help from a teacher/mentor/coach in real time in such a way that it appears the help provider is standing next to the user. For example, the user can do a science/engineering/art experience while being monitored by a teacher during the process, while the user receives AR clarifications from the instructor. The instructor would be able to utilize 3D visualization/modeling to present guidance to the user. This could be a very effective learning process in many situations, such as a medical student performing complex surgery on a corpse while the teacher directs him/her remotely. Other effective uses of AR glasses could be in learning historical events by imposing them on an actual location where the user is standing or walking, or taking your friends along to any place you visit through a real-time video feed.

The potential and challenges of smart glasses use in education and the commercial industry are well analyzed and documented [1–6]. Regarding their use in education, they emphasize that smart glasses can help in capturing class lectures, important information such as videos or images, or discussions. They can also support active group learning, direct feedback in the classroom between student and teacher, fast retrieval of information through stored data/lectures/internet, and real-time learning with the augmented reality concepts of smart glasses. Reference [6] presents an approach to use smart glasses in teaching experimental physics. They developed an app for high school students to study the relationship between the frequencies of generated sound when they hit a set of glasses with different levels of water in them. The smart glass can show them a graph representing water level against sound frequency. Reference [7] proposes a new learning environment for programming using AR glasses. Reference [8] indicates that students in STEM education who have experienced smart glasses and AR educational apps have shown positive learning outcomes. They conclude that smart glasses with AR functionality have the potential to facilitate the learning process of scientific concepts.

To utilize the smart glasses' potential fully, there is still much work to be done. We need to develop frameworks for creating context-aware AI assistants and collaborative human-AI learning [9]. Smart glasses should be accepted by society regarding privacy and become lightweight and independent from other computing devices. Also, for smart glasses to function in real time with high resolution, better communication networks and high-speed, stable Wi-Fi should become the norm.

In general, smart glasses will have many uses, a few could be: breaking language barriers for students with different native languages—listening to a talk in real time in a different language; assisting students with communication, hearing impairments, visual impairments, and learning disabilities; facilitating students to explore different locations such as museums, historical scenes, underwater ecosystems, scientific experiences, and nature without leaving their physical location; learning through games by getting engaged and excited to overcome a task at hand; accessibility to information instantly to resolve a problem at hand through sound and video; tailoring lessons to individual learning styles; and providing a 24/7 tutor/companion for students. Furthermore, as smart glasses become more and more like regular glasses and incorporate intelligent agents, they will offer big transformative opportunities in education—from immersive AR lessons to AI-powered personalization. With current technology, the opportunities for informal learning increase significantly. Students could learn about subjects they are interested in during their daily activities, such as walking, shopping, or waiting for an event. A personalized AI agent will take them through their study in the most efficient and effective way. The formal way of learning, going to a classroom, may become less and less attractive and demanding.

4 Conclusion

Smart glasses and AI represent a paradigm shift in educational technology, offering unprecedented opportunities for immersive, personalized, and accessible learning experiences. While some current smart glasses face limitations such as privacy concerns, battery life, and dependency on external devices, the rapid advancement in this technology promises solutions to these challenges. The integration of AI agents with smart glasses creates a powerful platform for continuous learning that exceeds traditional classroom boundaries, enabling students to engage with educational content seamlessly throughout their daily lives.

The potential applications in education are vast, from providing real-time assistance to students with disabilities to enabling remote collaboration and immersive historical experiences. However, the successful implementation of smart glasses in education will require careful consideration of privacy issues, development of robust frameworks for human-AI collaboration, and widespread acceptance by educational institutions and society at large.

As we move forward, it is crucial for educators, policymakers, and technology developers to work collaboratively to exploit the full potential of smart glasses while addressing their current limitations. The future of education may well depend on our ability to integrate these emerging technologies thoughtfully and effectively into our learning environments, ultimately creating more engaging, accessible, and personalized educational experiences for all students.

References

1. Kumar, N.M., Krishna, P.R., Pagadala, P.K., Kumar, N.M.S.: Use of smart glasses in education-a study. In: Proceedings of the Second International Conference on I-SMAC (IoT in Social, Mobile, Analytics and Cloud) (I-SMAC 2018) (2018)
2. Kumar, N.M., Singh, N.K., Peddiny, V.K.: Wearable smart glass: features, applications, current progress and challenges. In: Second IEEE International Conference on Green Computing and Internet of Things (ICGCIoT 2018), Bangalore (2018)
3. Lee, L.H., Hui, P.: Interaction methods for smart glasses: a survey. IEEE Access. **6**, 28712–28732 (2018)
4. Syberfeldt, A., Danielsson, O., Gustavsson, P.: Augmented reality smart glasses in the smart factory: product evaluation guidelines and review of available products. IEEE Access. **5**, 9118–9130 (2017)
5. Molnár, G., Sik, D.: Smart devices, smart environments, smart students—a review on educational opportunities in virtual and augmented reality learning environments. In: 10th IEEE International Conference on Cognitive Infocommunications—CogInfoCom 2019, Naples (2019)
6. Kuhn, J., Lukowicz, P., Hirth, M., Poxrucker, A., Weppner, J., Younas, J.: gPhysics—using smart glasses for head-centered, context-aware learning in physics experiments. IEEE Trans. Learn. Technol. **9**(4) (2016)
7. Hsu, S.-Y., Chin, K.-Y., Chang, H.-L.: Enhancing programming learning environment with augmented reality smart glasses: a blended learning approach for IoT. In: 2024 IEEE International Symposium on Consumer Technology (ISCT) (2024). https://doi.org/10.1109/ISCT62336.2024.10791175

8. Santana, R., Rodríguez, A., Rybarczyk, Y., Vera, F., Méndez, G.G., Rossi, G.: A study on user experience of smart glasses for higher education students. In: 2022 17th Iberian Conference on Information Systems and Technologies (CISTI), Madrid (2022)
9. Janaka, N.: Towards Intelligent Wearable Assistants. In: UbiComp Companion '24, Melbourne (2024)
10. Meta/EssilorLuxottica: Introducing the New Ray-Ban | Meta Smart Glasses. Meta Newsroom (2023). https://about.fb.com/news/2023/09/new-ray-ban-meta-smart-glasses/. Last accessed 20 Nov 2025
11. EssilorLuxottica: Introducing Ray-Ban | Meta—the next generation of smart glasses. EssilorLuxottica Newsroom (2023). https://www.essilorluxottica.com/en/newsroom/stories/introducing-ray-ban-meta/. Last accessed 20 Nov 2025
12. Meta: New Ray-Ban | Meta Smart Glasses styles and Meta AI updates. Meta Blog (2024). https://about.fb.com/news/2024/04/new-ray-ban-meta-smart-glasses-styles-and-meta-ai-updates/. Last accessed 20 Nov 2025
13. WIRED: Ray-Ban Meta smart glasses review: fine audio and video, privacy issues. WIRED (2024). https://www.wired.com/review/review-ray-ban-meta-smart-glasses/. Last accessed 20 Nov 2025
14. Ray-Ban: Ray-Ban smart glasses—features & prices. Ray-Ban Brand (2025). https://raybanbrand.com/ray-ban-smart-glasses/. Last accessed 20 Nov 2025
15. Tom's Guide: TCL RayNeo Air 3s review. Tom's Guide (2024). https://www.tomsguide.com/wearables/ar-glasses/tcl-rayneo-air-3s-review. Last accessed 20 Nov 2025
16. Jesse Hollington: RayNeo Air 3S review: these OLED AR glasses really pop. Digital Trends (2024). https://www.digitaltrends.com/mobile/rayneo-air-3s-review/. Last accessed 20 Nov 2025
17. L. Smith: TCL RayNeo Air 3s review: I brought these wearable displays to my extended work-cation. Android Police (2024). https://www.androidpolice.com/tcl-rayneo-air-3s-review/. Last accessed 20 Nov 2025
18. Best Buy: RayNeo Air 3S smart glasses USB-C—Space Gray. Best Buy Product Page (2025). https://www.bestbuy.com/site/rayneo-air-3s-smart-glasses-usb-c-space-gray/6598389.p. Last accessed 20 Nov 2025
19. Hector, H.: RayNeo Air 3s review. TechRadar (2025). https://www.techradar.com/wearables/ar-glasses/rayneo-air-3s-review. Last accessed 20 Nov 2025
20. E. Nield: VITURE Pro XR glasses review. How-To Geek (2024). https://www.howtogeek.com/viture-pro-xr-glasses-review/. Last accessed 20 Nov 2025
21. The Wearable Guru: VITURE Pro XR glasses review—the ultimate XR gaming and productivity solution. Wearable Guru (2024). https://thewearableguru.com/viture-pro-xr-glasses-review-the-ultimate-xr-gaming-and-productivity-solution/. Last accessed 20 Nov 2025
22. VITURE: VITURE Pro XR glasses—product specifications. VITURE Official (2024). https://www.viture.com/. Last accessed 20 Nov 2025
23. TechRadar: Solos AirGo 3 review: these smart glasses are more than ChatGPT speakers for your face. TechRadar (2024). https://www.techradar.com/wearables/smart-glasses/solos-airgo-3-review. Last accessed 20 Nov 2025
24. S. Tuohy: Solos AirGo 3 review. Wall Street Journal (2024). https://www.wsj.com/style/fashion/solos-airgo-3-review. Last accessed 20 Nov 2025
25. The Verge: Solos AirGo 3 smart glasses review. The Verge (2024). https://www.theverge.com/2024/solos-airgo-3-review. Last accessed 20 Nov 2025
26. Engadget: Solos AirGo 3 review. Engadget (2024). https://www.engadget.com/solos-airgo-3-review/. Last accessed 20 Nov 2025
27. Wired: Solos AirGo 3 review. WIRED (2024). https://www.wired.com/review/solos-airgo-3/. Last accessed 20 Nov 2025

28. Solos: AirGo 3—product details. Solos Official (2025). https://www.solosglasses.com/. Last accessed 20 Nov 2025
29. TechRadar: XREAL Air glasses review. TechRadar (2023). https://www.techradar.com/wearables/ar-glasses/xreal-air-review. Last accessed 20 Nov 2025
30. Digital Trends: XREAL Air review. Digital Trends (2024). https://www.digitaltrends.com/mobile/xreal-air-review/. Last accessed 20 Nov 2025
31. XREAL: XREAL Air—product specifications. XREAL Official (2025). https://www.xreal.com/. Last accessed 20 Nov 2025
32. The Verge: XREAL One Pro hands-on. The Verge (2024). https://www.theverge.com/xreal-one-pro-hands-on. Last accessed 20 Nov 2025
33. Road to VR: XREAL One Pro preview. Road to VR (2024). https://www.roadtovr.com/xreal-one-pro/. Last accessed 20 Nov 2025
34. Rokid: Rokid Max AR glasses—product details. Rokid Official (2024). https://www.rokid.com/max. Last accessed 20 Nov 2025
35. Apple: Apple Vision Pro—overview. Apple Official (2024). https://www.apple.com/apple-vision-pro/. Last accessed 20 Nov 2025
36. Even Realities: G1 AR glasses—product specifications. Even Realities Official (2024). https://www.evenrealities.com/. Last accessed 20 Nov 2025
37. XR Today: Even Realities G1 review. XR Today (2024). https://www.xrtoday.com/ar/even-realities-g1-review/. Last accessed 20 Nov 2025
38. AR Insider: Even Realities G1 hands-on. AR Insider (2024). https://arinsider.co/2024/even-realities-g1/. Last accessed 20 Nov 2025
39. TechXplore: Even Realities G1 analysis. TechXplore (2024). https://techxplore.com/news/2024-even-realities-g1.html. Last accessed 20 Nov 2025
40. Rokid: Rokid Max Product Specifications. Rokid Official (2024). https://global.rokid.com/products/rokid-max
41. TechRadar: Rokid Max AR Glasses Review: Another Passable Pair of Smart Specs. TechRadar (2023). https://www.techradar.com/computing/virtual-reality-augmented-reality/rokid-max-ar-glasses-review-another-passable-pair-of-smart-specs
42. PhoneArena: Rokid Station Review: Full Android TV on Your Face, Supercharged Rokid Max. PhoneArena (2024). https://www.phonearena.com/reviews/rokid-station-review_id5890
43. Cheng, R., Wu, N., Varvello, M., Chai, E., Han, B.: A first look at immersive telepresence on Apple Vision Pro. In: Proceedings of the ACM Internet Measurement Conference (IMC '24) (2024). https://dl.acm.org/doi/10.1145/3646547.3689006
44. Zhu, X., Yang, L., Duan, H., Min, X., Zhai, G., Le Callet, P.: ESIQA: Perceptual Quality Assessment of Vision-Pro-based Egocentric Spatial Images. arXiv preprint (2024). https://arxiv.org/abs/2407.21363
45. Blackman, T., Harley, D.: Interpreting Apple's visions: examining the spatiality of the Apple Vision Pro. Platforms Soc. **2**(1) (2024). https://journals.sagepub.com/doi/10.1177/29768624241283913

Determinants of Behavioral Intention to E-Learning During the COVID-19: System Factors and Risk Perception

Hyeon Jo(✉)

Department of Strategic Planning, RealSecu, Busan, Republic of Korea
sineoriz@gmail.com

Abstract. Since the outbreak of COVID-19, the use of e-learning has increased significantly. E-learning now serves not only an educational function but also a public health function by reducing social contact. This study aims to identify the factors influencing behavioral intention to use e-learning during the COVID-19 pandemic. For the empirical analysis, a total of 360 responses were collected from university students in South Korea and Vietnam. The proposed model was validated using partial least squares structural equation modeling (PLS-SEM). The results indicate that behavioral intention is significantly influenced by system interactivity, system responsiveness, and affective risk perception. The findings provide meaningful implications for sustaining education during pandemic conditions.

Keywords: e-learning · behavioral intention · system factors · risk perception · PLS

1 Introduction

Global society has undergone substantial changes since the outbreak of COVID-19. Social distancing measures have been widely implemented [1, 2]. Restaurants and bars have faced restrictions on operating hours, large-scale gatherings have been prohibited, and crowded events have been limited. In daily life, office workers increasingly rely on remote communication platforms, and educational institutions have adopted non-face-to-face teaching methods [3, 4]. Consequently, e-learning has become an essential tool for delivering education in this context [5]. Greater adoption of e-learning can help ensure the continuity of education during global crises. In this context, the present study seeks to identify the determinants of students' behavioral intention to use e-learning.

E-learning provides a convenient environment for electronic learning and plays a critical role in connecting instructors and students during the pandemic. As an information-processing system, e-learning relies heavily on stable network performance to ensure effective educational delivery [6]. In particular, interactivity and responsiveness are crucial in systems where two-way communication is central [7]. Accordingly, this study focuses on system interactivity and system responsiveness as key technical factors of e-learning.

H. R. Arabnia et al. (Eds.): CSCI 2025, CCIS 2974, pp. 412–420, 2027.
https://doi.org/10.1007/978-3-032-28475-4_32

E-learning, as a form of distance education [8], also contributes to infection prevention by minimizing physical contact. Students who perceive a higher risk of COVID-19 may demonstrate stronger intentions to use e-learning. Prior research has conceptualized risk perception as comprising both affective and cognitive dimensions [9, 10]. Therefore, this study proposes that both affective risk perception and cognitive risk perception influence behavioral intention.

The contributions of this study are as follows. First, it examines e-learning within the unique context of the COVID-19 pandemic, offering insights that may differ from those observed under normal conditions. Second, the study integrates both functional system factors and risk perception, providing a more comprehensive and balanced understanding of behavioral intention in e-learning environments.

2 Theoretical Background and Hypotheses

Figure 1 presents the theoretical framework for examining the key determinants of behavioral intention in the e-learning context. The conceptual model incorporates system-related factors and risk perception to explain behavioral intention to use e-learning. Specifically, this study posits that behavioral intention is influenced by system interactivity, system responsiveness, affective risk perception, and cognitive risk perception.

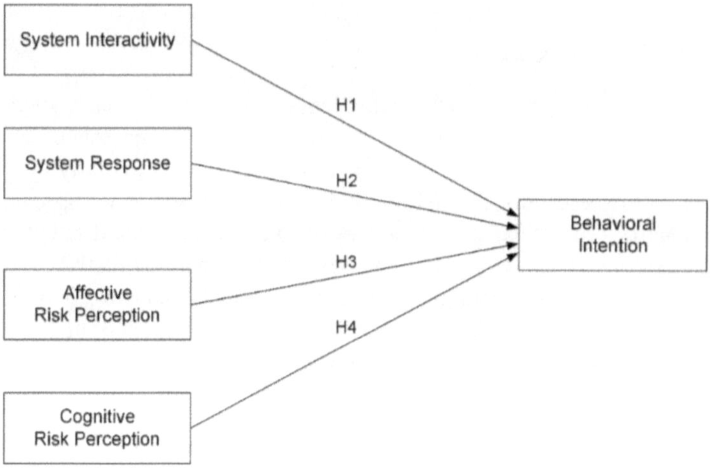

Fig. 1. Research model

2.1 System Interactivity

System interactivity refers to the extent of interactive communication between instructors and students, among students, and through communication tools embedded in the system

[7]. Interaction has been identified as a critical factor for the success of e-learning environments [11]. Prior research indicates that system interactivity significantly influences the use of distance education systems [7]. In the context of the COVID-19 pandemic, students may require a higher level of interactivity than in face-to-face settings, as e-learning platforms function primarily as communication-based systems. Higher levels of system interactivity are therefore expected to enhance students' willingness to engage with e-learning. Accordingly, this study posits that system interactivity positively influences behavioral intention.

Hypothesis H1. System interactivity is positively associated with behavioral intention.

2.2 System Responsiveness

System responsiveness reflects the extent to which users perceive the e-learning system as fast, stable, and consistent in its responses [12]. During the COVID-19 pandemic, many countries have faced challenges in implementing e-learning due to constraints such as inadequate infrastructure, unstable networks, and outdated devices [13, 14]. These factors directly affect system responsiveness. When students perceive higher levels of responsiveness, they are more likely to develop favorable intentions toward using e-learning. Therefore, this study proposes that system responsiveness positively influences behavioral intention.

Hypothesis H2. System responsiveness is positively associated with behavioral intention.

2.3 Affective Risk Perception

Affective risk perception refers to an individual's emotional response, such as worry or fear, regarding exposure to risk [15]. Prior studies have demonstrated that affective risk perception significantly influences behavioral intention, particularly in the context of social distancing behaviors, through mechanisms such as social norms and perceived behavioral control [16]. It has also been shown to affect behavioral intention in contactless service contexts, including tourism, both directly and indirectly [9]. Given that e-learning reduces physical interaction and supports social distancing, it can be viewed as a preventive alternative during the COVID-19 pandemic. Accordingly, this study posits that higher levels of affective risk perception increase students' intention to use e-learning.

Hypothesis H3. Affective risk perception is positively associated with behavioral intention.

2.4 Cognitive Risk Perception

Cognitive risk perception refers to an individual's assessment of the severity of and susceptibility to a given risk [15]. It has been shown to significantly influence both subjective norms and behavioral intention in contexts such as non-contact tourism [9]. In the context of COVID-19, higher cognitive risk perception may motivate individuals to adopt preventive behaviors. E-learning can be considered one such preventive measure, as it minimizes physical contact. Therefore, students with higher levels of cognitive risk perception are more likely to use e-learning. This study thus proposes that cognitive risk perception positively influences behavioral intention.

Hypothesis H4. Cognitive risk perception is positively associated with behavioral intention.

3 Research Methodology

3.1 Instrument Development

This study employed a cross-sectional research design to examine the determinants of university students' intention to use e-learning. All measurement items were adopted from prior studies to ensure construct validity, with minor modifications to reflect the e-learning context. Each item was measured using a 7-point Likert scale ranging from 1 to 7. A pilot study was conducted to assess the reliability and validity of the measurement instruments. Detailed information on the measurement items and their sources is provided in Table A1.

3.2 Data Collection

Data were collected from university students in two Asian countries, South Korea and Vietnam. Survey responses were obtained with the assistance of instructors who distributed the questionnaire during their classes. After data screening and cleaning, a total of 360 valid responses were retained for analysis. Among the respondents, 283 (78.6%) were from Vietnam, and 77 (21.4%) were from South Korea. In terms of gender distribution, 77 respondents (21.4%) were male. The majority of participants (294, 81.7%) were between 20 and 23 years old. Table 1 presents the detailed descriptive statistics of the sample.

Table 1. Demographic information of a sample

Demographics	Item	Subjects (N = 360)	
		Frequency	Percentage
Nationality	Korea	77	21.4%
	Vietnam	283	78.6%
Gender	Male	128	35.6%
	Female	232	64.4%
Age	19 or younger	48	13.3%
	20–23	294	81.7%
	24 or older	18	5.0%

4 Analysis Results

This study employed structural equation modeling (SEM) to evaluate both the measurement and structural models. The proposed research model was analyzed using partial least squares (PLS-SEM). The PLS approach is advantageous as it does not require strict normality assumptions and is relatively robust to sample size limitations [17]. SmartPLS 3.0 was utilized to perform the analysis [18].

4.1 Measurement Model

The measurement model was assessed in terms of convergent validity, discriminant validity, and reliability. The results of the measurement model evaluation are presented in Tables 2 and 3.

Reliability was assessed using composite reliability (CR) and Cronbach's alpha. Both indices exceeded the recommended threshold of 0.70, indicating a high level of internal consistency among the measurement items. Convergent validity was examined to determine whether the observed indicators adequately represent their respective latent constructs. This was assessed using average variance extracted (AVE) and factor loadings. AVE values exceeding 0.50 indicate that the construct explains more than half of the variance of its indicators [19, 20], while factor loadings should be greater than 0.70 [19]. The results show that all AVE values and factor loadings exceed the recommended thresholds, thereby confirming adequate convergent validity.

Table 2. Measurement model evaluation results

Construct	Items	Mean	St. Dev.	Factor Loading	Cronbach's Alpha	CR	AVE
Behavioral Intention	BIT1	5.378	1.528	0.935	0.942	0.963	0.897
	BIT2	4.978	1.725	0.947			
	BIT3	5.228	1.591	0.959			
System Interactivity	SIT1	4.644	1.652	0.906	0.849	0.909	0.770
	SIT2	4.222	1.772	0.907			
	SIT3	4.767	1.578	0.816			
System Response	SRP1	5.228	1.456	0.897	0.913	0.945	0.852
	SRP2	5.100	1.446	0.920			
	SRP3	5.189	1.398	0.952			
Affective Risk Perception	ARP1	4.844	1.867	0.921	0.900	0.937	0.832
	ARP2	5.533	1.610	0.914			
	ARP3	4.961	1.762	0.901			
Cognitive Risk Perception	CRP1	4.561	1.697	0.880	0.798	0.906	0.828
	CRP2	4.444	1.777	0.939			

Discriminant validity was evaluated using the Fornell–Larcker criterion, which compares the square root of AVE for each construct with its correlations with other constructs. The square root of AVE should be greater than the corresponding inter-construct correlations [21]. The results indicate that the square roots of AVE exceed the off-diagonal correlation coefficients, demonstrating satisfactory discriminant validity.

Table 3. Discriminant validity tests results

Constructs	1	2	3	4	5
1. Behavioral Intention	0.947				
2. System Interactivity	0.609	0.877			
3. System Response	0.634	0.551	0.923		
4. Affective Risk Perception	0.384	0.397	0.292	0.912	
5. Cognitive Risk Perception	0.221	0.359	0.167	0.642	0.910

Note: Diagonal values are the square root of AVE

4.2 Structural Model

The structural model was evaluated by examining the significance of the hypothesized relationships. Path coefficients and their significance levels were estimated using a bootstrapping procedure with 500 resamples [19]. The results of the structural model analysis are presented in Table 4.

Table 4. Summary of the SEM results

H	Predictor	Outcome	β	t	Hypothesis
H1	System Interactivity	Behavioral Intention	0.344	4.757	Supported
H2	System Response	Behavioral Intention	0.405	5.432	Supported
H3	Affective Risk Perception	Behavioral Intention	0.187	2.706	Supported
H4	Cognitive Risk Perception	Behavioral Intention	−0.090	1.481	Not Supported

5 Conclusion

5.1 Summary

T This study examined the factors influencing users' behavioral intention to adopt e-learning. Specifically, it proposed that system interactivity, system responsiveness, affective risk perception, and cognitive risk perception are key determinants of behavioral intention. Data were collected from university students in South Korea and Vietnam, and the proposed model was empirically validated using PLS-SEM. The findings reveal that system interactivity, system responsiveness, and affective risk perception significantly influence behavioral intention, whereas cognitive risk perception does not exhibit a significant effect.

5.2 Implications for Theory and Practice

This study offers several theoretical and practical implications. First, it contributes to the e-learning literature by integrating system-related factors with COVID-19-related contextual factors. Prior research has predominantly examined e-learning from the perspectives of information system success and technology acceptance [22–24]. By incorporating individual risk perception into the model, this study enhances its explanatory power and provides a more comprehensive understanding of user behavior. Future research is encouraged to adopt integrative frameworks that jointly consider technological and contextual influences. From a practical standpoint, service providers and marketers should emphasize not only system functionality but also the health-related benefits of e-learning to potential users.

Second, this study contributes to the literature by distinguishing between affective and cognitive dimensions of risk perception. The results demonstrate that these two dimensions exert different effects on behavioral intention, highlighting the importance of treating risk perception as a multidimensional construct. Future studies should further explore these distinctions to deepen theoretical insights. Practically, e-learning providers may benefit from segmenting users based on their levels of risk perception and tailoring their strategies accordingly.

5.3 Limitations and Future Research

Despite its contributions, this study has several limitations that suggest directions for future research. First, the model includes a limited set of exogenous variables. Since e-learning adoption is influenced by a wide range of individual and social factors, future studies should incorporate additional variables to improve explanatory power. Second, this study focuses solely on behavioral intention as the outcome variable. Future research could extend the model by including constructs such as actual usage, continuance intention, and user satisfaction to provide a more comprehensive understanding of e-learning behavior.

Appendix A

Table A1. Measurement Items

Construct	Item	Description
Behavioral Intention [25]	BIT1	I intend to use e-learning for learning in the near future.
	BIT2	I plan to use e-learning for learning in the near future.
	BIT3	I am willing to invest time and money to use e-learning for learning in the near future.
System Interactivity [7]	SIT1	The e-learning system enables interactive communication between instructors and students.
	SIT2	The e-learning system enables interactive communication among students.
	SIT3	The communication tools provided by the e-learning system (e.g., email, bulletin boards, chat rooms) are effective.
System Response [7]	SRP1	When using the e-learning system, the system responds quickly.
	SRP2	In general, the response time of the e-learning system is consistent.
	SRP3	In general, the response time of the e-learning system is reasonable.
Affective Risk Perception [26]	ARP1	I am worried that I may contract COVID-19.
	ARP2	I am concerned that my family members may contract COVID-19.
	ARP3	I am worried about the spread of COVID-19 in my region.
Cognitive Risk Perception [9, 26]	CRP1	There is a high likelihood of contracting COVID-19 in general.
	CRP2	The likelihood of contracting COVID-19 is higher than that of other diseases.

References

1. Wilder-Smith, A., Freedman, D.O.: Isolation, quarantine, social distancing and community containment: pivotal role for old-style public health measures in the novel coronavirus (2019-nCoV) outbreak. J. Travel Med. **27**(2) (2020)
2. Haque, A., et al.: Understanding social distancing intention among university students during COVID-19 outbreak: an application of protection motivation theory. Test Eng. Manag. **83**, 16360–16377 (2020)
3. Hoq, M.Z.: E-learning during the period of pandemic (COVID-19) in the kingdom of Saudi Arabia: an empirical study. Am. J. Educ. Res. **8**(7), 457–464 (2020)

4. Shahzad, A., et al.: Effects of COVID-19 in E-learning on higher education institution students: the group comparison between male and female. Qual. Quant. **55**(3), 805–826 (2021)
5. Mseleku, Z.: A literature review of E-learning and E-teaching in the era of COVID-19 pandemic. SAGE. **57**(52), 6 (2020)
6. Sá, M.J., Serpa, S.: The COVID-19 pandemic as an opportunity to foster the sustainable development of teaching in higher education. Sustainability. **12**(20), 8525 (2020)
7. Pituch, K.A., Lee, Y.-k.: The influence of system characteristics on E-learning use. Comput. Educ. **47**(2), 222–244 (2006)
8. Suzianti, A., Paramadini, S.A.: Continuance intention of E-learning: the condition and its connection with open innovation. J. Open Innov.: Technol. Mark. Complex. **7**(1), 97 (2021)
9. Bae, S.Y., Chang, P.-J.: The effect of coronavirus disease-19 (COVID-19) risk perception on behavioural intention towards 'untact' tourism in South Korea during the first wave of the pandemic (March 2020). Curr. Issue Tour. **24**(7), 1017–1035 (2021)
10. Savadori, L., Lauriola, M.: Risk perception and protective behaviors during the rise of the COVID-19 outbreak in Italy. Front. Psychol. **11**, 3822 (2021)
11. Palloff, R.M., Pratt, K.: Building Learning Communities in Cyberspace: Effective Strategies for the Online Classroom. Jossey-Bass Publishers, San Francisco (1999)
12. Bailey, J.E., Pearson, S.W.: Development of a tool for measuring and analyzing computer user satisfaction. Manag. Sci. **29**(5), 530–545 (1983)
13. Aristovnik, A., et al.: Impacts of the COVID-19 pandemic on life of higher education students: a global perspective. Sustainability. **12**(20), 8438 (2020)
14. Ebner, M., et al.: COVID-19 epidemic as E-learning boost? Chronological development and effects at an Austrian university against the background of the concept of "E-Learning Readiness". Future Internet. **12**(6), 94 (2020)
15. Sjöberg, L.: Worry and risk perception. Risk Anal. **18**(1), 85–93 (1998)
16. Adiyoso, W., Wilopo, W.: Social distancing intentions to reduce the spread of COVID-19: the extended theory of planned behavior. BMC Public Health. **21**, 1836 (2020)
17. Budiardjo, E.K., et al.: The impact of knowledge management system quality on the usage continuity and recommendation intention. Knowl. Manag. E-Learn. Int. J. **9**(2), 200–224 (2017)
18. Ringle, C.M., Wende, S., Becker, J.-M.: Smartpls 3. Hamburg: SmartPLS. https://www.smartpls.com (2014). Accessed 4 Aug 2021
19. Hair Jr., J.F., et al.: A Primer on Partial Least Squares Structural Equation Modeling (PLS-SEM). Sage Publications (2021)
20. Henseler, J., Ringle, C.M., Sinkovics, R.R.: The use of partial least squares path modeling in international marketing. In: New Challenges to International Marketing, pp. 277–319. Emerald Group Publishing Limited, Bingley (2009)
21. Fornell, C., Larcker, D.F.: Evaluating structural equation models with unobservable variables and measurement error. J. Mark. Res. **18**(1), 39–50 (1981)
22. Aparicio, M., Bacao, F., Oliveira, T.: Grit in the path to E-learning success. Comput. Hum. Behav. **66**, 388–399 (2017)
23. Al-Emran, M., Arpaci, I., Salloum, S.A.: An empirical examination of continuous intention to use m-learning: an integrated model. Educ. Inf. Technol. **25**(4), 2899–2918 (2020)
24. Mohammadi, H.: Social and individual antecedents of m-learning adoption in Iran. Comput. Hum. Behav. **49**, 191–207 (2015)
25. Ajzen, I.: The theory of planned behavior. Organ. Behav. Hum. Decis. Process. **50**(2), 179–211 (1991)
26. Brug, J., et al.: SARS risk perception, knowledge, precautions, and information sources, the Netherlands. Emerg. Infect. Dis. **10**(8), 1486–1489 (2004)

Poster Papers

Silent Sensors: Exploring Cemeteries as IoT Nodes in Smart Urban Networks

Alexis Brown and James Braman(✉)

Community College of Baltimore County, Baltimore, MD 21237, USA
{abrown9,jbraman}@ccbcmd.edu

Abstract. Cemeteries represent an overlooked yet significant space within the design of cities, intersecting environmental, cultural, and operational concerns. This paper introduces our project called *Silent Sensors*, an experimental Internet of Things (IoT) project designed to monitor various environmental aspects of cemetery environments. While IoT applications are commonly deployed for traffic, infrastructure, and environmental monitoring, few studies have considered their use in cemeteries. By embedding low-cost sensors within culturally sensitive enclosure, the Silent Sensors project demonstrates the feasibility of discreet, sustainable monitoring in cemeteries. A conceptual framework is presented to highlight the intersections of environmental stewardship, cultural awareness, and cemetery management. This work broadens the discourse on thanatechnology and smart cities, underscoring how integrating IoT into memorial landscapes can advance both sustainability and cultural preservation.

Keywords: IoT · smart sensors · smart cities · thanatechnology

1 Introduction

An often-overlooked dimension of the data-enhanced decision-making processes of smart city design is the human reality of death and loss. As urban populations grow, the number of deaths will inevitably increase, making cemetery environments an important yet neglected part of the smart city framework. While IoT technologies are widely applied to monitor infrastructure, safety, and environmental factors such as air and water quality, traffic, and climate, little attention has been given to their use in cultural and memorial spaces. Building on our previous work [1], we are developing an IoT-based system, "Silent Sensors," to track and potentially improve key aspects of cemetery environments. This paper presents the early stages of this project, with a focus on the rationale for deploying IoT-based sensors in cemeteries as a component of the broader vision for smart cities.

Smart cities will increasingly rely on the Internet of Things (IoT) to monitor various aspects of daily life, including infrastructure, environmental conditions, pollution, public safety, and more, through networks of connected sensors. With the numerous possibilities of these IoT sensors embedded in urban spaces, cemeteries in particular remain an overlooked space that can be effectively utilized. Despite their importance as green

H. R. Arabnia et al. (Eds.): CSCI 2025, CCIS 2974, pp. 423–427, 2027.
https://doi.org/10.1007/978-3-032-28475-4_33

zones, cemeteries are cultural heritage sites that contain vulnerable monuments, various memorials, and often their own distinct microclimates. To date, there are few, if any, IoT case studies situated in cemeteries, leaving a significant gap in both research and practice. Some projects have aimed to use IoT for mapping graves [2], while others have combined UAV imagery with images and other data to establish databases of gravestone locations [3]. Other researchers [4] point out initiatives that focus on digital memorialization artifacts, such as QR codes and other technologies embedded into headstones. Other efforts have been focused on creating digital memorials for preservation. There is a wide range of digital memorialization possibilities and considerations related to cemeteries [5].

This paper addresses that gap by presenting an experimental IoT project designed to monitor variables such as temperature, humidity and vibration at a cemetery location. The idea of this Silent Sensor network has various applications. As part of this research, we propose that (1) cemeteries can be used as novel testbeds for IoT research; (2) we demonstrate the feasibility of deploying a low-power IoT prototype in this context; and (3) we discuss how similar projects can be integrated into broader smart city infrastructures, including heritage preservation and environmental monitoring.

2 Related Work

The use of IoT technologies in smart cities has been widely studied, with applications ranging from traffic optimization to waste management and environmental monitoring [6]. Sensor networks are increasingly deployed to track variables such as air quality, water pollution, and noise levels, providing data-driven insights for urban planning and sustainability initiatives [7, 8]. Over the years, numerous innovations have emerged to address various problems. Beyond infrastructure and environmental monitoring, IoT has also been integrated into the preservation of cultural heritage. Museums and historical monuments have benefited from sensor-based systems that monitor structural stability, humidity, temperature, and visitor impact. These applications highlight IoT's potential not only to improve efficiency but also to safeguard spaces of cultural significance [9]. This type of research can also be observed in the monitoring of other green spaces, such as parks and gardens, where it is essential to track soil conditions, pollution, temperature, and other environmental factors. Components of a city's infrastructure can be designed with data collection in mind to improve traffic, wastewater management, power usage, and more [10].

3 Silent Sensor Design

The experimental device being developed for the *Silent Sensors* project is designed to monitor basic environmental conditions in an outdoor cemetery setting based on a Raspberry Pi (Model 5) with various accessories. The prototype integrates a suite of low-cost, low-power sensors, including a soil moisture sensor to track ground saturation levels, a temperature and humidity sensor and a tilt/vibration sensor to detect subtle shifts in headstones. The device is powered by rechargeable batteries supplemented with small solar panels, ensuring sustainable long-term deployment with minimal maintenance during the initial testing phase. Currently, we are housing the main components inside a traditional solid bronze memorial vase that would typically be found in a lawn cemetery or memorial park, having flat headstones with optional use built in vases that can be secured. The top of the vase will be sealed, with some wired components that can be attached from the outside to be concealed. This placement strategy allows the monitoring of both environmental conditions (e.g., soil moisture in grassy areas), temperature, humidity, and tilt/vibration while maintaining a secure location that would not be easily disturbed. To balance data richness with power efficiency, the system is configured to collect readings at 15-min intervals, though frequency can be adjusted depending on site-specific needs. The data recorded is stored on the device on local storage until collected. The initial pilot deployment (pending final approval) will be at a small local cemetery near the main college campus. This location provides direct access and presents unique conditions, including diverse vegetation, localized microclimates, and local heritage structures for Essex, Maryland, making it an ideal setting to test the robustness of IoT monitoring technologies.

Several challenges have emerged in the design and early deployment phases. Weatherproofing is crucial for protecting sensor components from rain, snow, and temperature fluctuations. Based on several smaller tests, we have selected the bronze metal enclosure with a 3D-printed top and internally sealed components as both the primary and secondary solutions. Artificial flowers will be placed in the vase to conceal the device even further, blending it into the cemetery environment. Additional custom 3D printed enclosures will ensure the safety of the temperature and humidity sensor from moisture damage. Another major issue is that of internet connectivity. Since the device will not fully be able to access the campus network, data collected will be stored locally until the device is collected. In the future, we will assess the feasibility of additional power sources and the possibility of an independent cellular-based internet connection. Equally important are cultural sensitivities. Cemeteries are spaces of mourning and heritage, and deployment in any location will require careful attention to placement, concealment of devices, and communication with stakeholders to ensure that monitoring does not appear intrusive or disrespectful.

4 Broader Impacts

This study introduces a novel perspective by situating cemeteries within the broader scope of smart city IoT deployments. Positioning cemeteries as smart city IoT sites underscores the potential to broaden the scope of data collection beyond efficiency and infrastructure to include spaces of memory and meaning. The diagram below (Fig. 1) illustrates three main components of the multidimensional impacts of the overall framework for consideration (Environment, Culture and Cemetery Maintenance). The central overlap highlights the integration of monitoring, respectful cultural preservation, and efficient operational management based on collected data.

Environmental: The environmental component is a key aspect of the project, as this is based on the collection of data through IoT. This can include readings for Soil moisture, air quality, humidity, biodiversity, and acoustic ecology based on the sensors used and

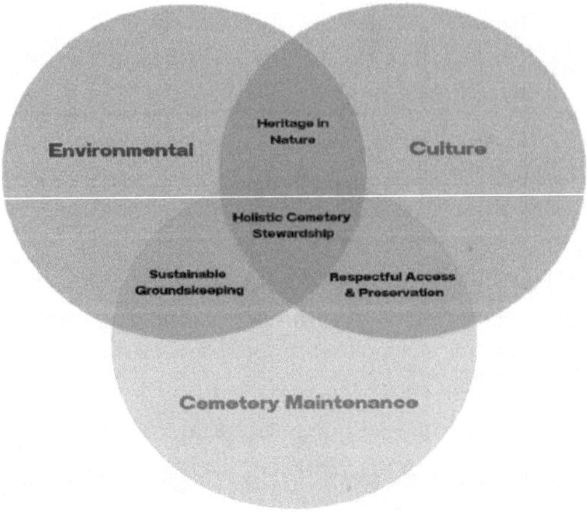

Fig. 1. Conceptual Framework: Overlap of Environmental, Cultural, and Maintenance Aspects

connectivity. Within the smart city framework, this can help preserve green spaces, reduce water waste, and contribute to a broader urban climate monitoring initiative.

Culture: It is also important to preserve and respect sacred spaces such as cemeteries and other cultural markers. By monitoring these special spaces, improved protections can be implemented. Combined with other cultural changes related to death and dying, the integration of IoT can improve privacy for visitors and provide long-term protection through the integration of these spaces [1], as they are important spiritual and cultural artifacts.

Cemetery Maintenance: Through monitoring and data collection, cemeteries can more effectively predict changes, soil erosion, vandalism, or other forms of damage. With often limited staff, this data can provide groundskeepers with better information to protect these sacred spaces. In addition, landscaping costs can be minimized, and the potential for using fewer chemicals on surfaces is reduced.

The societal value of this approach is multifaceted. From a heritage perspective, sensors can assist in the preservation of vulnerable monuments by detecting subtle shifts, vibrations, or environmental stressors before significant damage occurs. Additionally, IoT-enabled monitoring may support public safety by identifying hazardous conditions such as unstable headstones or pathways, thereby informing proactive urban management strategies.

5 Concluding Remarks

This project aims to explore the feasibility and value of integrating IoT technologies into cemetery environments as part of the broader smart city framework. By deploying sensors to monitor soil moisture, structural stability, and microclimatic conditions, we highlight how cemeteries can serve as novel testbeds that bridge cultural heritage preservation with environmental monitoring and public safety. The Silent Sensors project illustrates that these traditionally overlooked spaces can provide actionable insights for both heritage management and urban sustainability. Future work will expand this prototype through larger-scale deployments with smaller IoT sensors. In addition, we plan to add several additional sensing modalities, including acoustics and visitor traffic monitoring. Recognizing their role as cultural and environmental assets opens new opportunities for smart city research and practice.

References

1. Braman, J., Brown, A.: Death analytics and the potential role in smart cities. In: Wang, J. (ed.) Encyclopedia of Data Science and Machine Learning, pp. 2906–2916. IGI Global (2023). https://doi.org/10.4018/978-1-7998-9220-5.ch173
2. Ye, X.W., Pan, Y.Y., Peng, Y.Y., Tan, G.: Big data grave register information management system outside cemeteries under internet of things. In: Lin, Y.B., Deng, D.J., Yang, C.T. (eds.) Smart Grid and Internet of Things. SGIoT 2021. Lecture Notes of the Institute for Computer Sciences, Social Informatics and Telecommunications Engineering, vol. 447, pp. 142–153. Springer, Cham (2022)
3. Pál, M., Hajdú, E.: Methodological innovations for establishing cemetery spatial databases a UAV-based workflow helping small communities. ISPRS Int. J. Geo Inf. **13**(2), 57 (2024). https://doi.org/10.3390/ijgi13020057
4. Nansen, B., Arnold, M., Gibbs, M., Kohn, T.: The restless dead in the digital cemetery. In: Digital Death: Mortality and Beyond in the Online Age, pp. 111–124. Palgrave Macmillan, London (2014)
5. Braman, J., Dudley, A., Vincenti, G.: Memorializing the deceased using virtual worlds: a preliminary study. In: Stephanidis, C. (ed.) Online Communities and Social Computing. HCII 2017. Lecture Notes in Computer Science, vol. 10262, pp. 321–330. Springer, Cham (2017). https://doi.org/10.1007/978-3-319-58562-8_25
6. Boulos, M., Al-Shorbaji, N.: On the internet of things, smart cities and the WHO healthy cities. Int. J. Health Geogr. **13**(1), 10 (2014). https://doi.org/10.1186/1476-072X-13-10
7. Zanella, A., Bui, N., Castellani, A., Vangelista, L., Zorzi, M.: Internet of things for smart cities. IEEE Internet Things J. **1**(1), 22–32 (2014)
8. Perera, C., Zaslavsky, A., Christen, P., Georgakopoulos, D.: Context aware computing for the Internet of Things: a survey. IEEE Commun. Surv. Tutor. **16**(1), 414–454 (2014)
9. Piccialli, F.: The Internet of Things supporting the cultural heritage domain: analysis, design and implementation of a smart framework enhancing the smartness of cultural spaces. PhD Thesis, Università degli Studi di Napoli Federico II (2016)
10. Bawany, N.Z., Shamsi, J.A.: Smart city architecture: vision and challenges. Int. J. Adv. Comput. Sci. Appl. **6**(11), 1–5 (2015)

Author Index

H. R. Arabnia et al. (Eds.): CSCI 2025, CCIS 2974, pp. 429–430, 2027.
https://doi.org/10.1007/978-3-032-28475-4

FSC
www.fsc.org
MIX
Papier | Fördert gute Waldnutzung
FSC® C083411

Kolibri 360 GmbH
Ferdinand-Jühlke-Straße 7
99095 Erfurt, Deutschland
produktsicherheit@kolibri360.de